Low Temperature and General Plasma

SCIENTIFIC COMMITTEE OF 16th SPIG

Z.Petrović	Chairman, Institute of Physics, Belgrade
M.Milosavljević	Vice-Chairman, NI "Vinča", Belgrade
V.Čadež	Institute of Physics, Belgrade
M.Dimitrijević	Astronomical Observatory, Belgrade
T.Grozdanov	Institute of Physics, Belgrade
N.Konjević	Faculty of Physics, Belgrade
B.Marinković	Institute of Physics, Belgrade
A.Mihajlov	Institute of Physics, Belgrade
J.Purić	Faculty of Physics, Belgrade
M.Škorić	NI "Vinča", Belgrade
I.Vince	Astronomical Observatory, Belgrade
B.Vujičić	Faculty of Physics, Novi Sad

ORGANIZING COMMITTEE

M.Milosavljević	Chairman, NI "Vinča", Belgrade
M.Stojanović	Vice-Chairman, NI "Vinča", Belgrade
V.Parežanin	BK Company, Belgrade
B.Gaković	NI "Vinča", Belgrade
Lj.Milinković	NI "Vinča", Belgrade
D.Peruško	NI "Vinča", Belgrade
Z.Rakočević	NI "Vinča", Belgrade
S.Krsmanović	BK Faculty for Management, Belgrade

ACKNOWLEDGEMENTS

THE XVI SUMMER SCHOOL AND INTERNATIONAL SYMPOSIUM ON THE PHYSICS OF IONIZED GASES (16th SPIG) *was organized by*

The Institute of Nuclear Sciences "Vinča", Belgrade, Yugoslavia
Laboratory of Atomic Physics and Physics Laboratory
and co-organized by
The Institute "Braća Karić", Belgrade, Yugoslavia

under the auspices of

The Union of Mathematicians, Physicists and Astronomers of Yugoslavia
The Ministry of Science and Technology of the Republic of Serbia
The Ministry of Science, Technology and Development of FR Yugoslavia

and sponsored by

BK Company, Belgrade, Yugoslavia - General Sponsor
Credibel Bank, Belgrade, Yugoslavia
Progres, Belgrade, Yugoslavia
H.C.Yugoimport SDPR, Belgrade, Yugoslavia

Low Temperature and General Plasma

M. Milosavljevic and Z. Petrovic

(Editors)

Nova Science Publishers, Inc.

Art Director: Maria Ester Hawrys
Assistant Director: Elenor Kallberg
Graphics: Kerri Pfister and Susan A. Boriotti
Manuscript Coordinator: Phylis Gaynor
Book Production: Tammy Sauter, Gavin Aghamore,
Joanne Bennette and Christine Mathosian
Circulation: Irene Kwartiroff and Annette Hellinger

Library of Congress Cataloging–in–Publication Data
available upon request

ISBN 1-56072-399-8

6080 Jericho Turnpike, Suite 207
Commack, New York 11725
Tele. 516-499-3103 Fax 516-499-3146
E Mail Novasci1@aol.com

Printed in the United States of America

CONTENTS

LOW TEMPERATURE PLASMAS

GENERAL PLASMA

PREFACE

This book contains articles of invited lectures and progress reports from the XVI Summer School and International Symposium on the Physics of Ionized Gases (16th SPIG), held from 25-28 September 1993, in Belgrade, Yugoslavia.

The programme of 16th SPIG covered the following topics: Atomic Collision Processes, Particle and Laser Beam Interaction With Solids, Low Temperature Plasmas and General Plasma. The speakers (invited lectures and progress reports) were proposed to and selected by the Scientific Committee. This volume includes the papers dealing with low temperature and general plasmas.

We are indebted to the speakers for participating at the Conference and for preparing their manuscripts for this book. We hope that the book will be a valuable source of information in the considered topics.

On behalf of the Scientific and the Organizing Committees we would like to express our gratitude to all individuals, bodies and institutions who have helped and supported the organization of 16th SPIG and preparation of this book.

Editors

Low Temperature and
General Plasmas

ISBN 1-56072-399-8

CHAOTIC INSTABILITIES IN SINGLE-AND MULTI-MODE ELECTRONIC DEVICES

V.B. Ryabov, and D.M. Vavriv
Institute of Radio Astronomy, 4 Krasnoznamennaya St., 310002 Kharkov, Ukraine.

1. INTRODUCTION

It has become clear now, that the time behavior and spatial patterns of electromagnetic field in microwave devices, such as, for example, single and multi-mode generators, parametric amplifiers and oscillators, etc., can be rather complicated due to arising of turbulent states. We use this descriptive term in order to stress the analogies between such states and turbulent motion of liquid and gas being well known in hydrodynamics. Unlike the latter case, the studies on the complicated behavior of electromagnetic field has begun intensively only about two decades ago, when a new hypothesis was put forward that chaotic instabilities can give rise to the irregular field oscillations.

The study of dynamical chaos was carried out for a long time without separate consideration of the weakly nonlinear systems. It was apparently so because of an opinion that the chaotic behavior was principally nonlinear phenomenon demanding strong nonlinearity to be present. For the most practical microwave devices and circuits this condition fails to occur and, therefore, the chaos onset was supposed to be not typical for this case. However, the recent results have shown that, in fact,, chaotic oscillations can arise under quasilinear excitation conditions as well and, thus, can define the stability, output noise level, and sensitivity of the devices to a great extent.

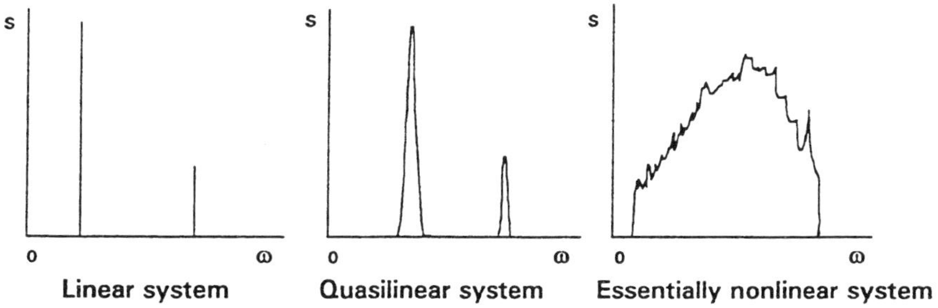

Fig.1. Distinction between power spectra of chaotic states in weakly- and strongly-nonlinear systems.

The major contribution of our study is the establishment of the existence of numerous chaotic states in weakly nonlinear single- and multi-mode systems under quasilinear excitation conditions which are dominant for many practical situations. We have revised the main classical results concerning field behavior in single- and two-mode oscillators, amplifiers, and parametric devices of similar nature. We have found out that most of these devices exhibit chaotic oscillations in some regions of control parameters. The theoretical studies were carried out, together with experiments on investigation of the above mentioned devices in different wavelength bands (from meter to millimeter). The qualitative and quantitative agreement between experimental and analytic results has been revealed.

The theoretical investigations are based on the application of the mode expansion that, as is well known, allows to reduce the initial Maxwell's equations for fields. to the study of the equations of motion of oscillators. We consider here the case of not higher than third order nonlinearity of the medium and take into account both linear and nonlinear dissipation terms, as well as nonisochronism of oscillations. Various analytical approaches such as averaging technique, Melnikov's method, and current Lyapunov techniques approach are applied in order to find the analytical conditions of the chaotic states arising.

We have found out that weakly nonlinear physical systems are characterized by specific conditions and scenarios of the chaotic motion onset. As it turned out, the chaotic instabilities can appear in quasilinear limit if, first, an n-dimensional torus with $n >= 2$ exists in phase space of the system, and, second, if the nonisochronism of oscillations goes over some critical value. Only under these conditions chaotic oscillations can arise in the weak nonlinearity limit.

2. CAVITY WITH NONLINEAR MEDIUM

The basic element of many of the devices considered is a cavity with some kind of nonlinear medium inside of it (Fig.2).

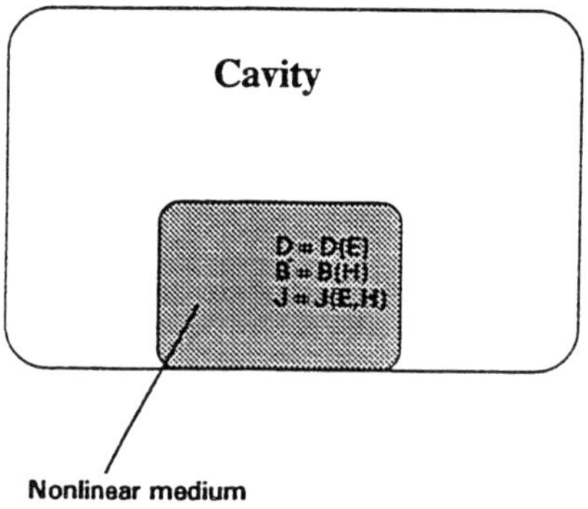

Fig.2. Schematic of a cavity.

The sought for solution of the electromagnetic field in the cavity can be expanded by the set of natural modes of a resonator with linear medium [2, 3]

$$\mathbf{X}(\bar{r},t) = \sum_n E_n(t)\, \mathbf{x}_n^e(\bar{r}) + \sum_n H_n(t)\, \mathbf{x}_n^h(\bar{r}) \qquad (1)$$

where $\mathbf{x}_n^e$, $\mathbf{x}_n^h$ constitute the mode basis representing spatial distribution of the field $\mathbf{X}(\bar{r},t) = \begin{pmatrix}\mathbf{D}\\ \mathbf{B}\end{pmatrix}$, **D** and **B** being electric and magnetic fields correspondingly, whereas the functions $E_n(t)$, $H_n(t)$ are mode amplitudes.

In general, the determination of the field evolution requires the infinite-dimensional system of coupled nonlinear differential equations to be solved.

$$\frac{d^2C_n}{dt^2} + \omega_n^2 C_n = \varepsilon\left[N_n\left(\bar{C}\right) + f_n(t)\right] \qquad (2)$$

C_n are either E_n or H_n, $N_n\left(\bar{C}\right)$, functions describing nonlinearity and dissipation, f_n, the external force, ε - nonlinearity parameter.

The averaging technique permits to make use of the weak nonlinearity property and pass from governing equations (2) for oscillators to so-called averaged equations which in the simplest case look like

$$\frac{da}{d\tau} = \left(-\alpha_0 - \alpha_1 a^2\right)a + P(\tau)\sin(\varphi)$$
$$\frac{d\varphi}{d\tau} = -\Delta + \beta a^2 + (P(\tau)/a)\cos(\varphi) \qquad (3)$$

where α_0 and α_1 are linear and nonlinear damping, β, parameter of nonisochronism, $P(\tau)$, external force, $\tau = \omega t$, slow time, a and φ, amplitude and phase of n-th mode.

In most cases of practical importance we can restrict our consideration to finite number of modes, neglecting their coupling to other ones. Such a possibility is conditioned by the fact that frequency band within which the effective medium-field interaction is possible is as a rule rather narrow, and the small number of modes natural frequencies fall into it. The number of essential modes depends also on the cavity shape and dissipation properties of its surface. In result, the behavior of only few modes has to be analyzed.

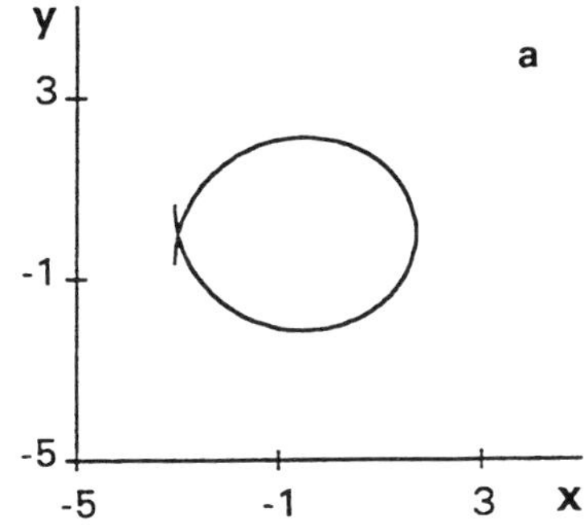

Fig. 3a. Homoclinic trajectory.

3. MELNIKOV'S CRITERION

The basic analytical tool for studying chaos arising conditions in equations of type (3) was chosen to be Melnikov's method. It is applicable to the dynamical systems having the form

$$\dot{\bar{x}} = \bar{F}(\bar{x}) + \varepsilon\bar{G}(\bar{x},t), \quad \varepsilon \ll 1 \qquad (4)$$

where point means time derivative.

At $\varepsilon = 0$ the system (4) should posses a homoclinic loop, i.e. a trajectory that comes to and from the saddle point as $t \to \pm\infty$. Such a loop is shown in Fig.3a.

When ε becomes non vanishing and the perturbation $\bar{G}(\bar{x},t)$ is small, the situation depicted in fig 3b takes place, when the lines, corresponding to the loop in Fig.3a do not intersect. With increase in perturbation the

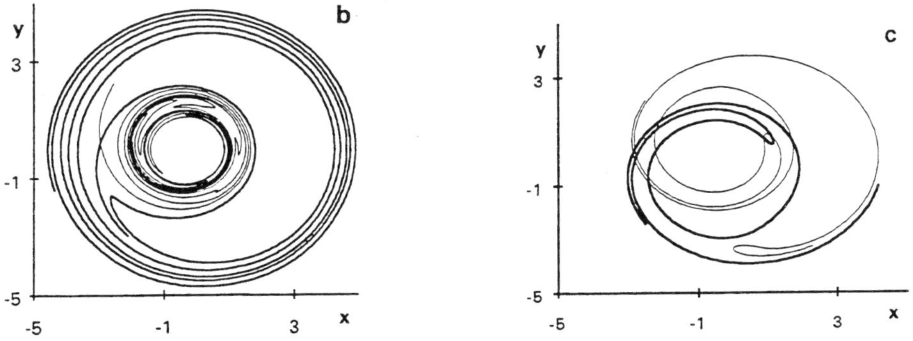

Fig.3b,c. Cross-section of manifolds before (b) and after (c) intersection.

intersection occurs (Fig.3c) giving rise to highly complicated behavior of the dynamical system at hand. These self-intersecting manifolds constitute a homoclinic structure that is considered at the present time as a precursor for chaotic attractors to exist in the phase space. The Melnikov's criterion permits to calculate the control parameters values where the intersection takes place. However, it should be stressed that this criterion is neither necessary nor sufficient for chaos arising. The only way to verify the predictions made on its basis is to perform detailed numerical experiments.

4. SINGLE-MODE SYSTEM

The following equation describes the dynamics of a passive mode. It has been proven to be the adequate mathematical model of many physical systems

$$\frac{d^2x}{dt^2} + \omega_0^2 x = \varepsilon\left(F_1 \cos(\omega_1 t) + F_2 \cos(\omega_2 t) - 2\delta \frac{dx}{dt} - \gamma x^3 \right) \quad (5)$$

The behavior of its solutions under single frequency excitation can be chaotic at rather large values of external force magnitude. However, it has been shown recently that two-frequency excitation of such system may result in considerable lowering of chaos onset. Following the well-known scheme of averaging procedure we put $x = U\cos(\omega_1 t) + V\sin(\omega_1 t)$ and obtain the system of averaged equations that permits to investigate in detail the solutions in the vicinity of different resonances. The case of principal resonance is investigated in the present report. Here we have in the limit $\varepsilon << 1$

$$\frac{dU}{d\tau} = -\delta U + \left[\Delta + \beta\left(U^2 + V^2\right)\right]V + P_2 \sin(\Omega\tau)$$

$$(6)$$

$$\frac{dV}{d\tau} = -\delta V - \left[\Delta + \beta\left(U^2 + V^2\right)\right]U + P_1 + P_2 \cos(\Omega\tau)$$

where $\Omega = (\omega_2 - \omega_1)/\varepsilon$; $\Delta = (\omega_0 - \omega_1)/\varepsilon$; $\tau = \varepsilon t$; $\beta = 3\gamma/(8\omega_1)$; $P_1 = F_1/(2\omega_1)$; $P_2 = F_2/(2\omega_1)$.

At $P_2 = 0$ (one frequency excitation) in the certain region of parameters Δ and P_1 in the phase space three equilibrium positions exist - two foci and a saddle (herefrom we shall call them primary). When P_2 becomes non vanishing the situation depicted in Fig.4 takes place. Here, in the area restricted by Curve 1 two attractors appear in the vicinity of the focus corresponding to the lower branch of response curve. Alongside with them the attractor corresponding to the upper branch of the response curve exists. Chaotic attractors arise independently in the vicinity of the foci through period doubling bifurcations (Curves 4,3 and 2,5) and may coexist. Such situation occurs, for example, at $\delta = 1$, $\beta = -1.54$, $\Delta = 15.05$, $\mu = -2.8$, $P_1 = 3.75$, $\Omega = 8.4$, and the cross-section of corresponding attractors (two chaotic - A2, A3, and periodic - A1) is shown in Fig.5.

It was recently shown that chaos onset in this system can be derived analytically by means of Melnikov's method through detecting the parameter values at which the manifolds of the perturbed primary saddle point intersect transversely (curve 7 in Fig.4).

$$\frac{P_2}{P_1\delta} > \left| \frac{1}{\pi}\left[3\Omega_0 + 2\Delta\theta^{\pm}\right] \mathrm{sh}\left(\pi\frac{\Omega}{\Omega_0}\right) \exp\left(\theta^{\pm}\frac{\Omega}{\Omega_0}\right) \right| \quad (7)$$

where $\Omega_0 = \sqrt{\left[\Delta + \beta U_s^2\right]\left[\Delta + 3\beta U_s^2\right]}$, $\theta^{\pm} = \pm \arccos\left(\mp\sqrt{-\left(\frac{\Delta + \beta U_s^2}{2\beta U_s^2}\right)}\right)$, and U_s is the coordinate of the saddle point defined from the solution of the following cubic equation

$$\beta U_s^3 + \Delta U_s - P_1 = 0$$

However, it should be noted that chaos occurs at much less values of P_2 than can be expected from the Melnikov criterion. The corresponding saddle orbit intersects the plane of Fig.5 at the point S3, and it is clear that the homoclinic structure associated with it is not attracting. The saddle points S1 and S2 correspond to the period one orbits that lost their stability at Curves 2 and 4 of Fig.4, i.e., at period doublings. So, these new period-one saddle points play the principal role in dynamics of the system.

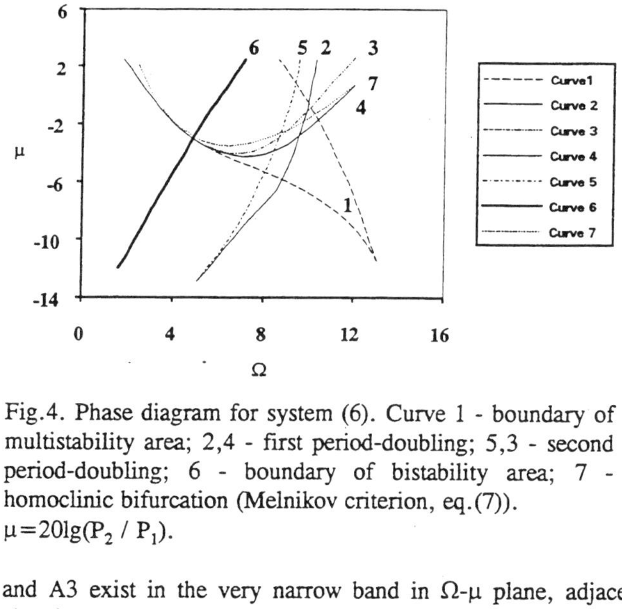

Fig.4. Phase diagram for system (6). Curve 1 - boundary of multistability area; 2,4 - first period-doubling; 5,3 - second period-doubling; 6 - boundary of bistability area; 7 - homoclinic bifurcation (Melnikov criterion, eq.(7)). $\mu = 20\lg(P_2 / P_1)$.

Moreover, the strange attractors A2 and A3 exist in the very narrow band in Ω-μ plane, adjacent to period-doublings. Both chaotic attractors undergo the crisis, caused by the extremely complicated structure of the phase space. The numerical experiments indicate that the unstable manifolds of periodic orbits S1 and S2 coinciding with the two chaotic attractors shown in Fig.5 demonstrate the sudden transition at Ω=8.25 and look like those depicted in Fig.6. In order to illustrate the structure of the phase space, the stable and unstable manifolds of saddle orbits S3 and S4 are also given in Fig.7. The fine fractal structure of basin boundaries of different attractors makes the system in hand extremely sensitive to the

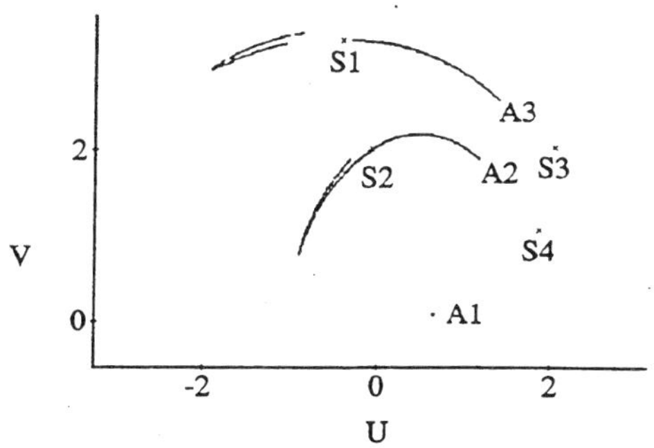

Fig.5. Poincare cross-section for system (6). A1-A3 - coexisting attractors, S1-S4 - period-one saddle points.

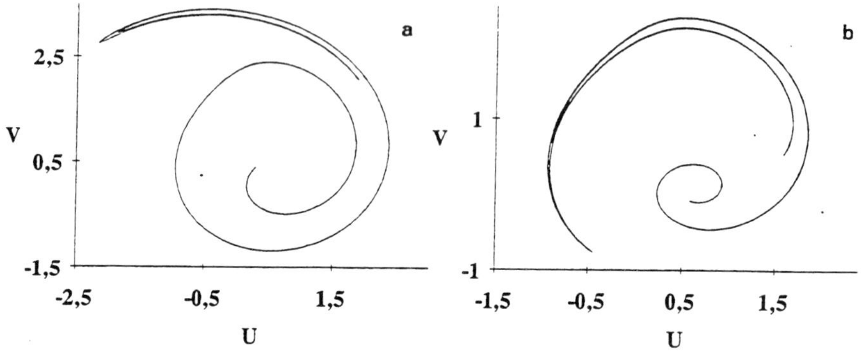

Fig.6. Unstable manifolds of periodic unstable orbits S1(a) and S2(b).

influence of noise and can completely deteriorate the stability of strange attractors. It

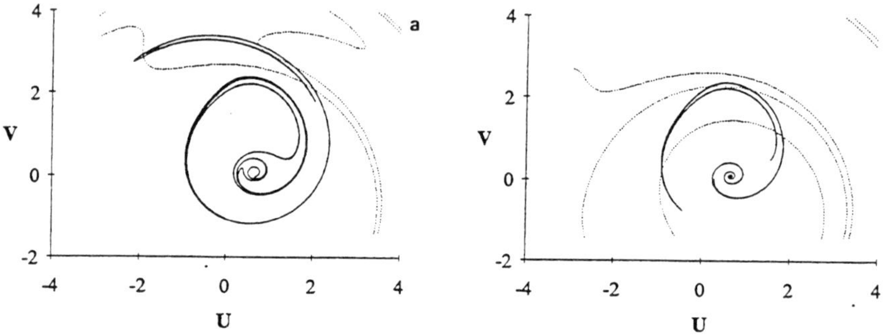

Fig.7. Stable (dotted line) and unstable (solid line) manifolds of unstable orbits S3(a) and S4(b)

becomes evident by decreasing the dissipation parameter δ, when both strange attractors are being destroyed. This situation is depicted in Fig.8 where the solid heavy line denotes the collapse of chaotic oscillations. Below this line only periodic attractor exists. It should be noted here, that there exists a threshold in dissipation value for chaos to arise in this system. We believe that such phenomenon is typical of nonlinear oscillators, and it means that there is no gradual transition from Hamiltonian chaos to dissipative one in passive systems. The phenomenon of chaotic attractors deterioration due to decrease in dissipation was demonstrated also in [9].

In the present work we describe the mechanism for such event to occur: the crisis of basin boundaries - stable manifolds of new saddle points. Since the formation of the new saddles, either in result of a period-doubling or saddle-node bifurcation, is common in dynamical systems, the systematic study of their role is of fundamental importance.

5. PARAMETRIC AMPLIFIERS

A variety of microwave and optical devices fall into a class of parametric amplifiers wherein a signal amplification is due to a time modulation of some nonlinear reactance. These amplifiers, as it follows from the conventional theory of parametric devices, provide low noise amplification because they consist of an entirely reactive circuit and they utilize an ac power supply. At the same time, there are a lot of experimental evidences that parametric amplifiers posses low stability, and their output noise level is in some cases much more higher compared to the theoretical expectations. We have found that the chaotic instabilities may give rise to such a behavior. It is of great interest that exactly the reactive nature of the device and the application of a pumping oscillation for power supply are the main factors responsible for the unstable operation of the amplifiers.

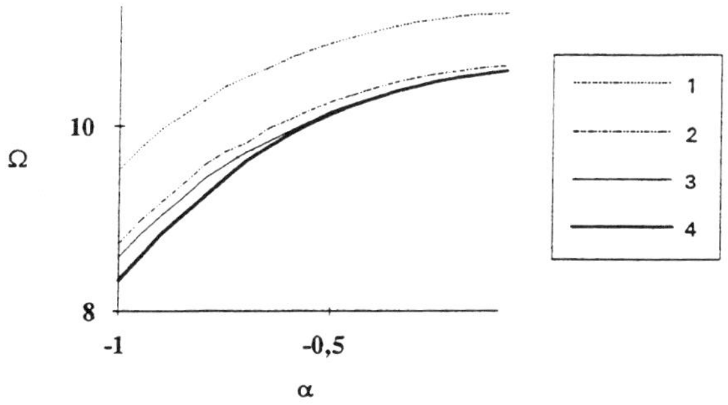

Fig.8. Phase plane of system (6) at the same parameter values as Fig.4. Curve 1-first period doubling, 2-second period doubling, 3-strange attractor appearance, 4-chaos collapse.

The dynamics of parametric amplifiers is adequately described within a scope of mode approach. For single-mode amplifiers the equations used can be reduced to the following system

$$\frac{d^2E}{dt^2} + \omega_0^2 E = -\varepsilon\left[E\omega_0^2 M\cos(pt) + 2\alpha_0\frac{dE}{dt} - \alpha E^2 + \gamma E^3 - A\cos(\omega_s t)\right] \qquad (8)$$

where E is n-th mode amplitude; ω_0 , mode natural frequency; $\varepsilon << 1$, small parameter; M, p, amplitude and frequency of parametric excitation; α_0, dissipation parameter; α, γ, parameters of quadratic and cubic nonlinearities; A, ω_s, amplitude and frequency of external signal.

We will examine the case of primary parametric resonance, i.e. $p = 2\omega$; $\omega \approx \omega_0$. Since $\varepsilon << 1$ the eq.(1) by using the transformation

$$E = -u\cos(\omega t) + v\sin(\omega t)$$

and averaging up to terms of order ε can be reduced to

$$\frac{du}{d\tau} = -\alpha_0 u - \left[\Delta - m + \beta\left(u^2 + v^2\right)\right]v - P\sin(\Omega\tau) \qquad (9a)$$

$$\frac{dv}{d\tau} = -\alpha_0 v + \left[\Delta + m + \beta\left(u^2 + v^2\right)\right]u + P\cos(\Omega\tau) \qquad (9b)$$

where $\Delta = \dfrac{\omega_0 - \omega}{\varepsilon}$; $m = \dfrac{M\omega}{4}$; $\beta = \dfrac{3\gamma}{8\omega}$; $P = \dfrac{A}{2\omega}$; $\Omega = \dfrac{\omega_s - \omega}{\varepsilon}$

Eqs. (9) at P = 0 are rather well investigated [4]. The case P ≠ 0 is now intensively examined by numerous researchers, since it corresponds to many real situations in microwave, electronic, optic, and other devices.

This system permits to study the bifurcations of the two-dimensional tori of the governing physical system via investigation of bifurcations of periodic orbits in averaged equations. It should be noted that when the external signal vanishes the system (9) exhibits regular behavior only.

In order to find the analytical conditions of chaos arising, the Melnikov method was used. System (9) is nonlinear and nonautonomous, thus being the good candidate for demonstrating chaotic behavior, as well as multistability and hysteresis. Specific difficulties are known to occur in applying the well-known Melnikov technique to systems of averaged equations [5]. However, as it was recently shown [6, 7], in the cases of resonant excitation the pointed out problems are not valid. So, following the Melnikov method, which permits to obtain analytical conditions of chaos onset, we start from the Hamiltonian counterpart of system (9). Setting $P=\alpha_0=0$, we have the Hamiltonian

$$H=-\frac{\beta}{4}\left(u^2+v^2\right)^2-\frac{1}{2}\left[(\Delta-m)v^2+(\Delta+m)u^2\right] \tag{10}$$

Here we will consider the case $|\Delta|>m$, which is of particular interest, since the system (10) has multiple singular points if this condition is met. It is demonstrated in Fig.9 where the corresponding phase portrait is given (with notation $\delta_{1,2}=(\pm m-\Delta)/\beta)$).

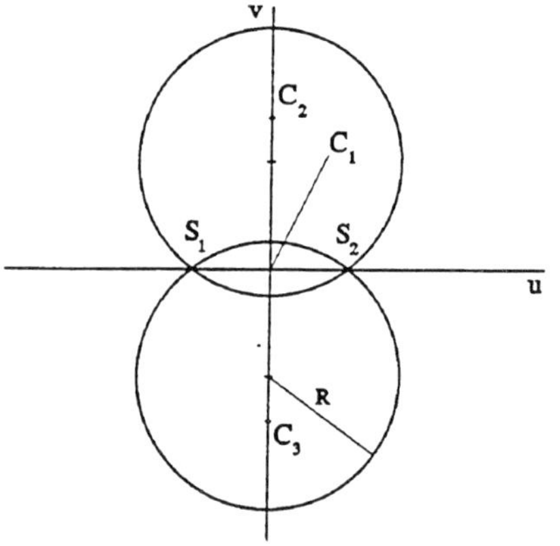

Fig.9. Separatrixes of the Hamiltonian system (10) - two intersecting circles with radii $R=\sqrt{(\delta_1+\delta_2)/2}$ centered at $u=0,\ v=\pm\sqrt{(\delta_1-\delta_2)/2}$. $S_{1,2}$ - saddle points with coordinates $u=\mp\sqrt{\delta_2}$, v =0. $C_{1,2}$ - center-type singular points at $u=0,\ v=\pm\sqrt{\delta_1}$, C_3 - center-type singular point at $u=0$, v =0.

Then we obtain the solution of the system on separatrixes,

$$u=-\frac{\sqrt{\delta_2}(\delta_1+\delta_2)\mathrm{sh}(\tau')}{(\delta_1+\delta_2)\mathrm{ch}(\tau')-q_\pm(\delta_1-\delta_2)}$$

$$|v|=\frac{\delta_2 q_\pm\sqrt{2(\delta_1-\delta_2)}}{(\delta_1+\delta_2)\mathrm{ch}(\tau')-q_\pm(\delta_1-\delta_2)}$$

where

$$q_\pm=\pm\sqrt{\frac{\delta_1+\delta_2}{\delta_1-\delta_2}};\quad \xi=\sqrt{2\delta_2(\delta_1-\delta_2)};$$

$$\tau'=\beta\xi(\tau-\tau_0);\quad \theta_\mp=\arccos(-q_\pm)$$

and different signs correspond to the small and large separatrix loops respectively.

Introducing the perturbation, i.e. setting $P\neq0$, $\alpha_0\neq0$, we calculate the condition of intersection for perturbed separatrixes and formation of homoclinic structure in the phase space of system (9)

$$\frac{P}{\alpha_0} \geq \frac{1}{2\pi\Omega}\left[\xi \pm (\delta_1 + \delta_2)\theta_{\mp}\right]\exp\left(\pm\frac{\Omega\theta_{\mp}}{\beta\xi}\right) - \beta \mathrm{sh}\left(\frac{\pi\Omega}{\beta\xi}\right)\sqrt{2(\delta_1 - \delta_2)} \quad (11)$$

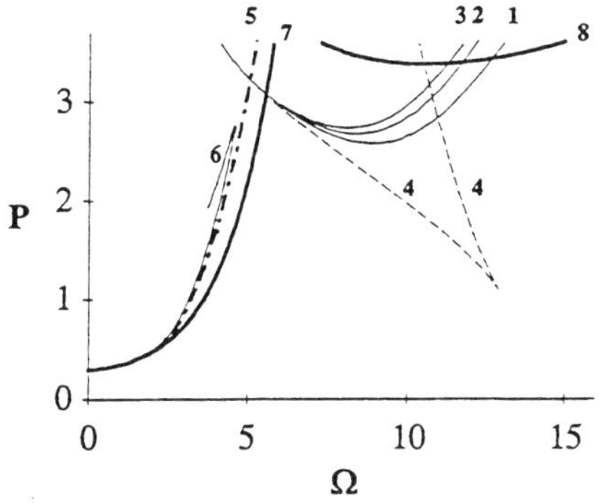

Fig.10. Phase diagram of system (9) obtained numerically at $\alpha_0 = 1$; $\beta = 1$; $\Delta = -6$; $m = 5$. Curves 1-4 correspond to attractors in the vicinity of centers $C_{1,2}$ (see, Fig.9). 1 - first period doubling, 2 - second period doubling, 3 - attractor crises, 4 boundary of multistability, or hysteresis, region. Curves 5,6 correspond to attractors in the vicinity of C_3 center. 5 - symmetry breaking, 6 - first period-doubling. Curves 7,8 - boundaries of homoclinic structure existence areas according to Melnikov criterion (11).

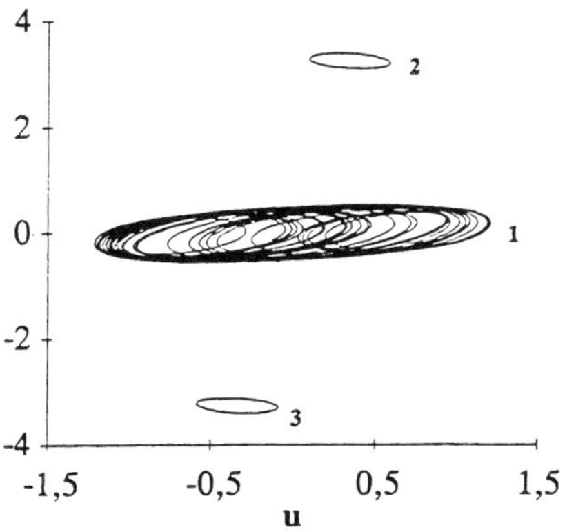

Fig.11. All coexisting attractors of the system (9) at $\Omega = 3.95$, $P = 1.9$.

Relation (11) constitutes the main analytic result of this chapter. It indicates where the areas of complicated behavior of solutions are situated in the parameter space. However, it should be noted that this condition is neither necessary nor sufficient for strange attractor to appear in the phase space. It will be demonstrated below, that homoclinic structure formation does not guarantee the stability of chaotic regime.

5.1. Numerical experiments

Prior to describing the results of computer simulation, we'd like to note that system (9) is invariant with respect to the transformation: $u \to -u$; $v \to -v$; $\tau \to \tau + \pi/\Omega$. It means that two types of attractors can exist in the phase space - symmetric ones (with respect to the point $u = 0$; $v = 0$) and pairs of unsymmetric solutions. Consequently, all bifurcation phenomena observed for one of unsymmetric attractors take place simultaneously for the second one. In Fig.10 the phase diagram of system (9) is depicted.

The best way to explain all possible behavior of the system at hand is to start from the Hamiltonian case, i.e. Fig.9, and then put $\alpha_0 \neq 0$. We obtain three stable foci instead of three centers, and all trajectories will eventually come to one of them, depending on initial conditions. Let us consider now what happens if P is increased at different Ω. At small P the foci turn to stable periodic orbits, or attractors, and the three of them will coexist in the phase space. Herefrom we will for definiteness use the superscripts of C_1 notation in Fig.9 to distinguish them. Attractor 1 undergoes symmetry breaking crises at the line 5 of Fig.10, and to the left of pointed out line four attractors (1, 1', 2, 3) coexist in phase space. At the line 6 both attractors 1 and 1'

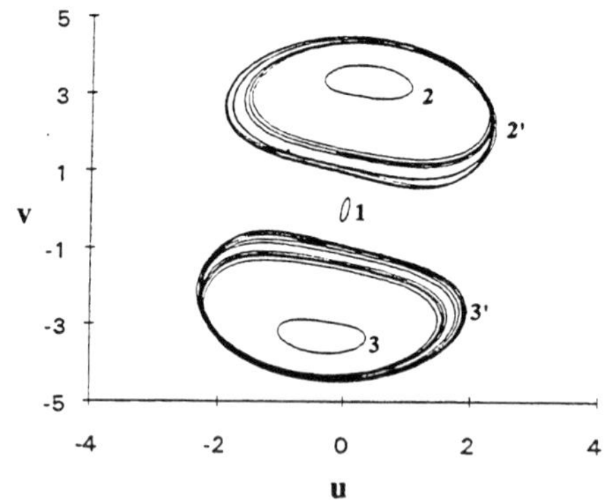

Fig.12. Five coexisting attractors of the system (9) at $\Omega = 9.2$, $P = 2.8$.

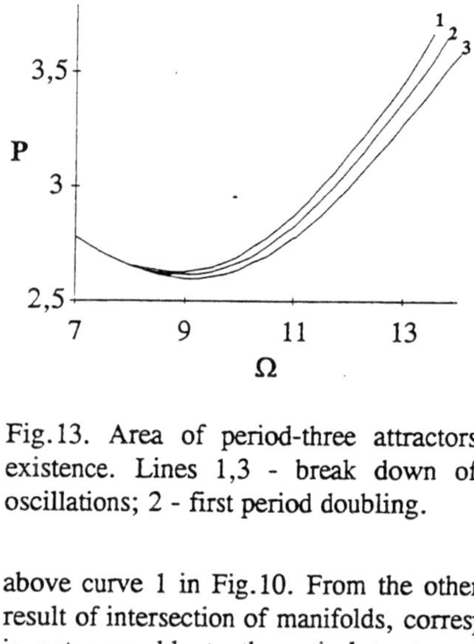

Fig.13. Area of period-three attractors existence. Lines 1,3 - break down of oscillations; 2 - first period doubling.

undergo period doubling bifurcation and inside the region bounded by line 6 strange attractors appear through period doubling cascade. In Fig.11 the situation is given, when they merge and the unified strange attractor arises. However, it turns out that this attractor is very sensitive to variation of parameters and exists only in very narrow band adjacent to line 6. Obviously, its basin of attraction is small and the majority of initial conditions lead to two stable attractors coexisting with it. For this attractor Melnikov criterion (line 7) works well and good predictions can be developed on its basis.

As $P \geq 1.1$ in the vicinity of attractors 2,3 new periodic orbits, 2' and 3', appear within frequency band limited by lines 4 in Fig.10. Orbits 2' and 3' transform to strange attractors through period-doublings, starting from line 1. Finally, two strange attractors may coexist with three regular ones, as is shown in Fig.12. It is clear from Fig.10, that the Melnikov method does not predict the appearance of strange attractors 2', 3'. One of possible explanations to this effect may be other mechanism of strange attractors formation. The pointed out attractors may appear due to intersection of stable and unstable manifolds of other periodic orbits (not originating from saddles $S_{1,2}$ in Fig.9. Such saddle periodic orbits are present in phase space, in the region of parameter space limited by curves 4 or above curve 1 in Fig.10. From the other hand, the homoclinic structure may appear as a result of intersection of manifolds, corresponding to different loops in Fig.9. Such situation is not amenable to theoretical treatment by Melnikov method and can be detected and studied only numerically.

The extreme sensitivity of strange attractors 2', 3' to small fluctuations also takes place in this case. The chaotic oscillations exist only in a narrow area near period doublings, and any phase trajectory is attracted to periodic orbits after chaos break down.

It should be also noted, that the described above phenomena are not all the possible regimes, that can exist in this system. Period-three motion was detected in the area shown in Fig.13, between lines 1 and 3. Moving from line 3 to line 1, we observe period doubling bifurcation at curve 2 followed by a cascade of period-doublings and formation of chaotic attractors. The resulting situation is depicted in Fig.14, where all 7 coexisting attractors are presented.

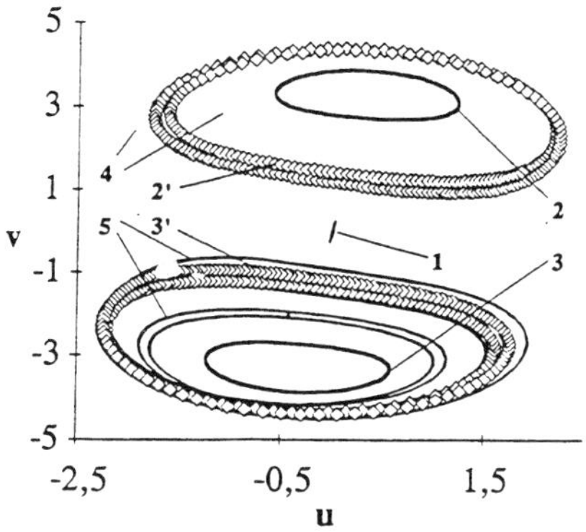

Fig.14. "Period-three" strange attractors (4,5), stable periodic period-one (1,2,3) and period-two (2', 3') orbits (marked by squares to discriminate between them and attractors 4,5). Parameter values: $\Omega = 10$, $P = 2.69$.

5.2 Concluding remarks

The stable operation mode of parametric devices under the influence of a harmonic external signal does not insure its stability for real multifrequency excitation. What's more, a passage from single frequency forcing to even two-frequency one leads to considerable lowering of chaos arising threshold. Unfortunately, the Melnikov method can be applied in very few occasions, so that the numerical experiment is the only universal tool for both finding the chaotization conditions and verifying the predictions made on a basis of Melnikov criterion.

Processing from the condition (4) and from the results of numerical and experimental studies we can conclude:

1. The chaotic oscillations excitation in the weakly nonlinear mode of amplifier operation results solely from the pumping oscillation interaction with the signal to be amplified.
2. The instability threshold with respect to the signal amplitude P is much more low compared with other types of amplifiers (say, transistor ones).
3. The increase in both nonisochronism of oscillations and the quality factor of the resonant circuit leads to the lowering of stochastic instability threshold.
4. The regions in the control parameters space with chaotic behavior occur intermittently with the regions of regular behavior.
5. The typical feature of the system is the possibility for several (more than two) attractors to exist simultaneously in the phase space, i.e. its multistability. We observed up to seven attractors (regular and chaotic) existing simultaneously.

6. SUPERCONDUCTING QUANTUM INTERFEROMETER

We examined the one-contact SQUID which consists of the high-quality circuit inductively coupled with a superconducting ring. The circuit is excited by an external HF pumping oscillator. Such SQUIDs attract constant attention due to a possibility of achieving the sensitivity approaching the Plank limit. However, the expected high-sensitivity level has not been reached in practice. We have shown that the stochastic instability arising in the weakly nonlinear limit can impose a restriction on the sensitivity of SQUIDs.

In the framework of conventional theory of SQUIDs the following equation is used

$$\frac{1}{\omega_0^2}\frac{d^2\phi}{dt^2} + \alpha\frac{d\phi}{dt} + \phi + \ell \sin\phi = \Phi_e + B\cos(\omega_p t)$$

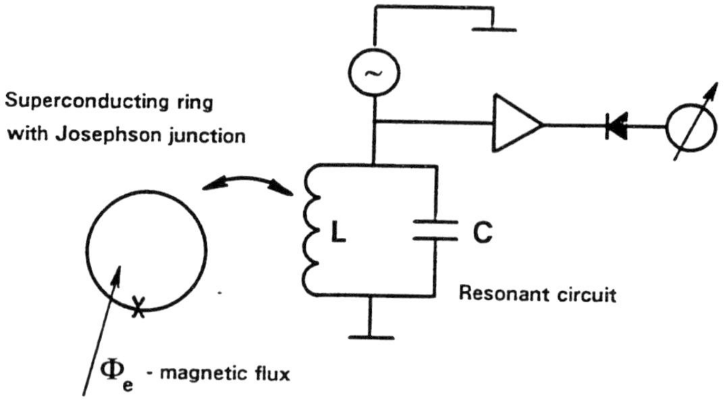

Fig.15. Schematic of the one-contact SQUID.

where ϕ - is the phase of wave function, ω_0, the natural frequency of the interferometer, α, the damping, ℓ, the dimensionless inductance

The adequate description of the SQUID dynamics appears to be possible within the framework of the following averaged equations for the amplitude a and phase γ in the resonant circuit [8]

$$\frac{da}{d\tau} = -a - b\sin\gamma$$

$$a\frac{d\gamma}{d\tau} = -\Delta a + RJ_1(a)\cos\phi_e - b\cos\gamma$$

Here the following dimensionless parameters are used: b is pumping amplitude, $\Delta = 2Q(\omega - \omega_0)/\omega$, the pumping frequency shift, k^2, the coupling coefficient of the tuned circuit with the superconducting ring $Q >> 1$, the Q-factor of the resonant circuit, L, the ring inductance, $J_1(a)$, the first order Bessel function.

We have taken into account the slow time variation of the external magnetic flux ϕ_e, which looks like

$$\phi_e = \phi_0 + \phi_1\cos(\Omega\tau)$$

where ϕ_0 and ϕ_1 are the constant and alternating flux components, Ω, the dimensionless frequency.

The results of investigations indicate that the chaotic instability manifests itself even in the case of adiabatic, equilibrium and non hysteretic processes in a superconducting interferometer closed by a weak link. Such a mode of SQUID operation has been considered as the most promising one for the achievement of the magnetometers high sensitivity limit. There are two main causes of the chaotic oscillations arising. The first one is the finite variation speed of magnetic flux (signal measured or bias flux). The chaotic instability appears if $\Omega \geq 1$. The second cause is the reactive nonlinearity induced in a SQUID circuit by the superconducting interferometer. The chaos arising has the following threshold in the nonlinearity parameter R and the amplitude of pumping b: $R_{th} = 4$; $b_{th} = 1.25$. The results obtained allow to assume that limited SQUID sensitivity observed in some experiments with $R \approx 10$ may be explained by the appearance of chaotic instability. The analysis has shown that avoiding of chaos and retaining the high sensitivity levels requires the adequate value of R from the interval $1 \leq R \leq 4$ to be chosen.

References

[1]A.B.Belogortsev, D.M.Vavriv and O.A.Tretyakov, "On involvement of a passive mode into turbulent motion", Sov.Phys. JETP, Vol.65, pp.737-741, 1987.

[2] O.A.Tretyakov, "Mode basis method," Radiotekhnika i Elektronika, Vol.31, N 6, pp.1071-1082, 1986.

[3] A.B.Belogortsev, "Stochastic oscillations in resonators with nonlinear medium", Ph.D. thesis, Kharkov, 1987.

[4] A.E.Kaplan, Yu.A.Kravtsov, V.A.Rylov, Parametric Generators and Frequency Dividers, Sov. Radio, Moscow, 1966.

[5] S.Wiggins, Global Bifurcations and Chaos: Analytical Methods, NY: Springer-Verlag, New York, 1987.

[6] K.Yagasaki, "Second order averaging and chaos in quasi periodically forced weakly nonlinear oscillator", Physica D, Vol.44, pp.445-458, 1990.

[7] V.B.Ryabov, and D.M.Vavriv, "Conditions of quasiperiodic oscillation destruction in the weakly nonlinear Duffing oscillator", Phys. Lett. A, Vol. 153, N 8,9, pp.431-436, 1991

[8] S.A.Bulgakov, V.B.Ryabov, V.I.Shnyrkov, and D.M.Vavriv: "Effect of the magnetic-flux variations on SQUID stability"; J. Low Temp. Phys., Vol. 83, N 5/6. P.241-255 1991.

[9] G.M.Zaslavsky, R.Z.Sagdeev, Introduction to nonlinear physics, Nauka, Moscow, 1988.

Low Temperature and
General Plasmas
M. Milosavljevic and Z. Petrovic

ISBN 1-56072-399-8

INFLUENCE OF METASTABLES ON THE BREAKDOWN PROBABILITY AND GLOW DISCHARGE IN GASES

Dj.A. Bošan
Faculty of Electronic Engineering, University of Niš
P.O. Box 73, 18001 Niš, Yugoslavia

Abstract. Atoms and molecules in metastable states in processes preceeding and following the electrical breakdown and discharge as well as in discharge formation and in sustaining stationary currents in glow discharge in permanent and some other gases, vapours and their mixtures, mostly at pressures between 1 and 100 *mbar*, were investigated and some partinent results presented in this paper. Metastables have been detected by various methods such as measuring the time delay of electrical breakdown t_d, measuring the contribution of metastables to the discharge current strenght and measuring the secondary electron emission current caused in metastable–metal surface interactions by a special conduction chamber. Particularly the dependence of the breakdown time delay t_d on various parameters has been looked upon, the most important ones being the afterglow period τ, the work function of the cathode material, kind of the gas or the mixture, the pressure, the applied voltage U_w, irradiation of the diode, temperature of the electrodes, size of the interelectrode gap, size of the diode bulb (volume), gas flow rate etc. A part of the study was devoted to the statistics of the stochastically scattered values of t_d. A model approach was applied in order to establish the time–space distribution of concentration of metastables in the discharge tubes, and the influence of concentration of metastables on t_d, on the discharge formative time t_f, on the static breakdown voltage U_s and on other parameters were determined.

1. Introduction

The subject of investigations presented here is the role of metastable atoms and molecules in breakdown mechanisms and probabilities, discharge formative processes, stationary currents and afterglow processes under glow discharge conditions in gases under pressures mostly between 1 and 100 *mbar*.

The problems treated here may be classified from several viewpoints, listed in the following:

1. Presence of metastables: generation, diffusion and deexcitation.
2. Methods of detection: time delay measurements, contribution to the stationary current and using the secondary emission chamber.
3. Transport phenomena of metastables: gas in hermetized diodes, diffusion and gas flow.
4. Character of the process: stationary and nonstationary processes.
5. Processes sensitive to concentration of metastables: time delay of breakdown, discharge formative time, static breakdown voltage, stationary processes, self–quenching processes etc.

In what follows is a selection of results of investigations of electrical breakdown and glow discharge in gas–filled diodes with a special bearing towards the role of atoms and/or molecules in neutral metastable states. This investigations were initiated in 1953 at the laboratories of the Elektronska Industrija in Niš Yugoslavia (former: Zavodi RR) and later, since 1975, at the Faculty of Electronic Engineering in Niš.

The first motivation for this research was a practical one aimed at the industrial development of gas–filled diodes with cold metallic electrodes, including solution of production problems of surge arresters for telephone and telegraph lines. At that time, around 1954, a number of European post–offices placed regulations so that the response of this arresters, i.e. the time delay of the electrical breakdown should be shorter than 1 μs when the applied voltage reaches twice the static breakdown value: $U_w = 2U_s$. Thus t_d become one of the fundamental characteristics of these diodes, initiating a broad investigation of the time delay t_d and related parameters on which it depends.

Here we are going to describe the most important physical magnitudes used in this study:

P – gas pressure,
d – interelectrode gap,
n – concentration of particles in metastable states in the diode,
U_w – working voltage, i.e. the DC voltage applied to the diode, measured directly on the power supply (Figs. 1 and 2),
U_s – static breakdown voltage,
τ – afterglow period, i.e. the time elapsed between the cessation of the previous discharge in the diode, and the next application of the voltage U_w ($U_w > U_s$). During the period τ the voltage on the diode equals zero (Fig. 2),
t_s – statistical time delay, i.e. statistically variable time elapsed from the instant of application of the working voltage U_w on the diode (T_1, Fig. 1) up to the instant when, within the active volume between the electrodes, sufficient free electrons appear that can initiate a new breakdown (T_2, Fig. 1). The size of this active volume is not very much affected by the applied voltage U_w used in this work. The question of the number of electrons necessary for initiation of the breakdown in nitrogen, under different conditions, is elaborated in the monograph [1],
t_f – the discharge formative time spans the instant when these electrons appear within the active volume and the instant when the full glow discharge is developed (T_3), i.e. when the diode current reaches the value i_s (Fig. 1) limited by the circuits impedance. In fact, t_f is most often defined as the time interval between the instants when the diode current reaches 10% and the 90% of its saturation value,
t_d – total time delay (Fig. 1), i.e.:

$$t_d = t_s + t_f \ ,$$

$\bar{t}_d$ – mean value of t_d obtained in a series of measurements under constant conditions. One series usually contains between 100 and 1 000 measurements, or even more.

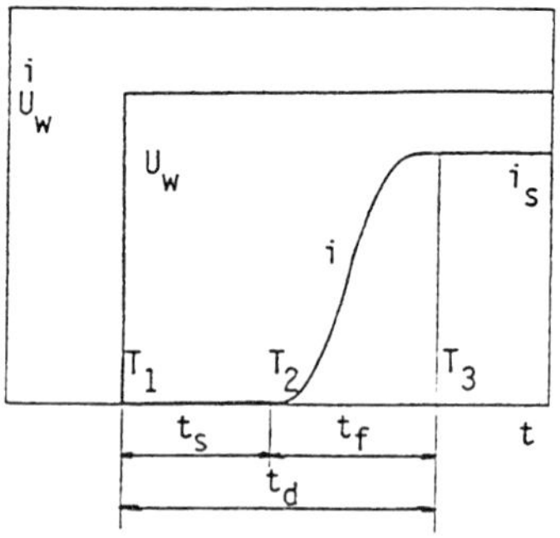

Fig. 1. Time delay of the electrical breakdown in gases:
t_d – total time delay,
t_s – statistical time delay,
t_f – formative time of discharge.

The time intervals t_s and t_f and consequently t_d are stochastic. The dispersion of the data obtained in measurements is large, the values often cover many orders of magnitude (Fig. 23). The research is based on a statistical analysis of these data.

Under conditions in most of the presented experiments, i.e. when $\tau > 1$ s, t_f is for some orders of magnitude shorter than t_s ($t_f \ll t_s$) and it can be considered that

$$t_d \approx t_s \ .$$

Consequently one may measure t_d instead of t_s with sufficient accuracy, which is much simpler in practice.

2. Time delay t_d and the memory curve $\bar{t}_d = f(\tau)$

Experiments on the electrical breakdown in glow discharge were done mostly under pressures between 1 and 100 $mbar$ in the following media: He, Ne, Ar, Kr, Xe, H_2, N_2, O_2, Cl_2, H_2O, Hg, SF_6 and some of their mixtures including air.

Time delay t_d depends on diode properties and on measurement conditions. On kind, pressure, flow rate and temperature of the gas, on diode bulb material, volume and surface, on electrode material, surface properties, shape, gap and temperature. It also depends on the applied voltage, frequency, diode current, glowing time, illumination of the diode, afterglow period etc.

It has been shown that the afterglow period τ parameter is of extreme influence on the value of t_d, but it was neglected, most of the time, in previous investigations. That is why timing schedule for measurements of t_d dependence on the afterglow period τ, $\bar{t}_d = f(\tau)$, the so-called "memory curve" has been worked out (Fig. 2). The figure shows the cycle of measuring t_c. Voltage U_w on the diode is disconnected at instant A (and so is the current i) and the preset afterglow period τ begins. At instant B, U_w is applied again, period τ is completed and t_d begins. At instant C, breakdown occurs in the diode, t_d is completed and the preset glowing time t_i started. At the following instant A, U_w is disconnected and t_i is completed, and one cycle of measuring t_c is finished.

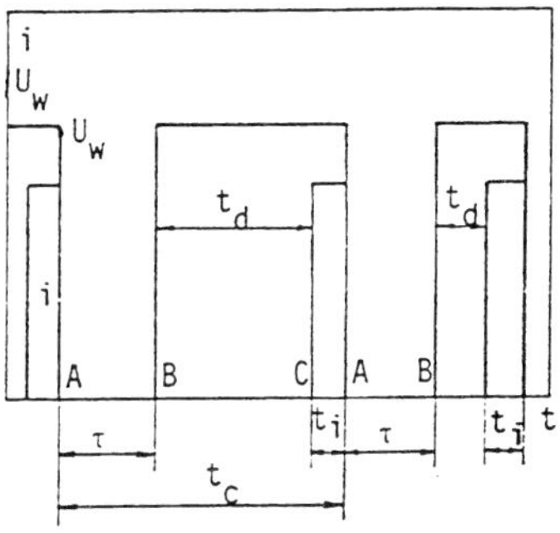

Fig. 2. Time–interval sequences in recording memory curves. The applied voltage U_w and diode current i as a function of time t.

The study of the breakdown time delay was initiated by Jaumann[2], who measured t_d dependence on the overvoltage. However, he did not observe, or at least did not refer on the stochastic character of this phenomenon. Warburg [3 and 4] measured the dependence of time delay on the illumination of the gap by UV light. A correct set of measurements was performed by Zuber [5] and the first statistical approach to the treatment of data was done by von Laue [6].

Afterglow processes and the recovery of dielectric properties in the air, SF_6 and some other insulating gases (under 1 bar and higher pressures) were investigated in detail and the results and data can most often be found in the publications on electroenergetics, e.g. [7, 33, 61, 62 and 72].

That process is continued in order of magnitudes of ns or μs, but in the light of this paper, it is important that always $\tau < 1$ s.

Data on measurements of t_d as a function of τ in glow discharge under 100 $mbar$ are very rare.

The increase of $\bar{t}_d$ with τ up to $\tau = 5$ min was found out by Büge [8], in measurements on glow lamps.

Shatokhin et al. [9] measured $\bar{t}_d$ with τ, ranging from 10^{-3} to 10^{3} s, which will be described in detail later in this paper.

The memory curve $\bar{t}_d = f(\tau)$, in Ar for 3 $s \leq \tau \leq 7$ days was obtained by measurements of Bošan [10] (Fig. 3) and the corresponding memory effects were reported later [11]. Memory curves in Ar, Xe and N_2, and a typical data distribution were given in the paper by Bošan [12].

As it is shown in Fig. 3, the increase of $\bar{t}_d$ spans almost 5 orders of magnitude and can be measured within the range of preset τ from 3 s to 7 days. The shape of this curve could be explained to be due to a process taking place in the diode, within the same range of time. This could be explained exclusively in terms of variation of concentration (n) of neutral metastables of argon atoms, which deexcite during that period and are capable of initiating electrical breakdown by means of secondary emission, provided the voltage U_w on the electrodes is higher than the static breakdown voltage U_s.

This assumption is based, of course, on the thesis that the increase of t_d is the consequence of the reduction of probability of electrical breakdown in the gas.

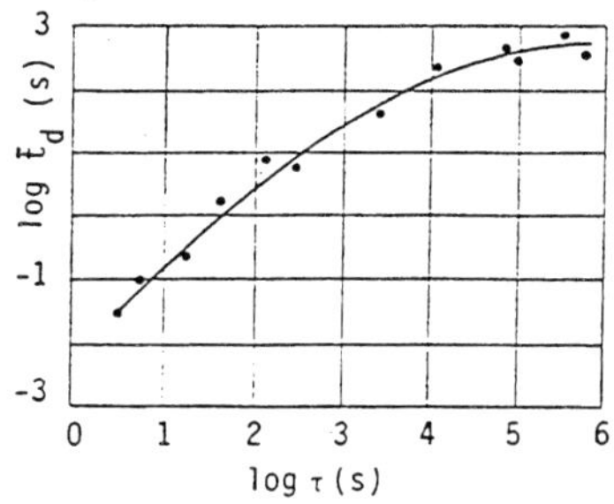

Fig. 3. Memory curve in Ar. $P = 26$ $mbar$.

After the discharge cessation, ions neutralize quickly by recombination and the resonant excited states deexcite by emission of light quanta, thus at $\tau > 1$ s after this cessation only atoms and/or molecules in neutral metastable states remain from the previous discharge.

Gas–filled diodes memorize, in a specific way, the instant when the previous discharge was interrupted, i.e. the duration of the afterglow period τ. The memorizing process lies in the fact that the breakdown probability at every instant (provided all other conditions were constant) depends on total concentrations of free electrons, ions and atoms or molecules in metastable states formed in the previous discharge, or created in some afterglow processes without any external influence, each of which contains enough energy to initiate a new breakdown in a direct or indirect way. On the other hand, these concentrations are functions of time, i.e. they depend on the afterglow period τ. It is supposed that during this period no production of such particles takes place by external sources. The functions, the so–called "memory curves", encompass these phenomena.

Memory curves are functions (Figs. 3, 4, 5, 7, 8, 9, 11, 12, 14, 17, 20, 23 and 24):

$$\bar{t}_d = f(\tau) .$$

As it has been shown, in the period $\tau > 1$ s, the memory processes are linked to neutral metastable states of atoms and/or molecules in the gas filling of the diode.

These considerations are valid when U_w and, consequently the field E, have the values which are not sufficiently high to induce the autoelectronic emission from the cathode.

All processes concerning this memory are stochastic and the data obtained by measurements are scattered yielding specific distributions. Thus, the general characterization of these processes has to be referred to the mean values of measurement data.

In the presented works $\bar{t}_d$ is the mean value of at least 200 experimentally established values of t_d obtained under strictly controlled identical conditions for each prechosen value of τ.

Memory curves have also been recorded under conditions of 10^{-3} $s < \tau < 1$ s [13, 76, 125 and 159]. During this short period new breakdowns in the diode (N_2, 1.3 and 13.3 $mbar$) are initiated both by metastables and ions (Fig. 4). In the diode (Cu, N_2, 6.6 $mbar$), containing auxiliary electrodes for the removal of ions during the period $\tau > 3 \cdot 10^{-4}$ s [14], memory curves have been obtained, thus providing the possibility to separate and estimate the contribution of ions and metastables on the initiation of breakdown in the cited interval τ.

Important problems on connections between the breakdown time delays and the memory processes in gases have been discussed in the work [17]. Between 1953 and 1993, Bošan with collaborators also investigated various influences on $\bar{t}_d$ and consequently on the entire memory curves. The influence of more than 20 experimental parameters on $\bar{t}_d$ have been investigated working with 9 kinds of gases, 3 kinds of vapours and some of their mixtures. Memory curves have been established under different conditions, as it is shown in the papers [10, 12, 17, 18, 25, 34, 40, 42 - 54, 66, 67, 69, 70, 71, 73, 79, 83, 84, 90, 93, 94, 98, 101, 108, 111, 113 - 115, 117, 132 and 162]. Many of the results mentioned are reported or cited in the papers [15, 17, 18 and 124].

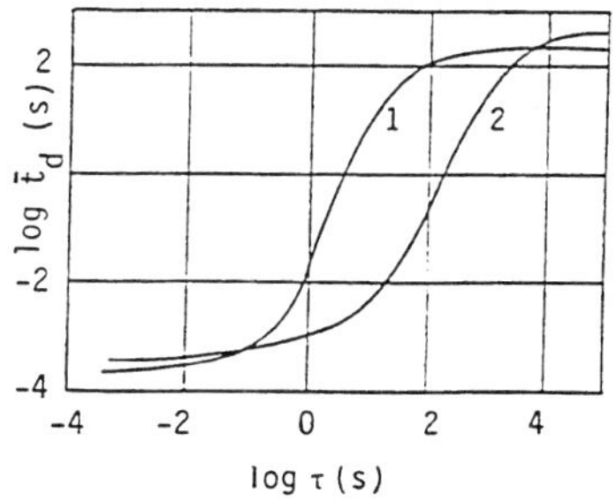

Fig. 4. Memory curves in N_2 under pressure of:
1- 13.3 *mbar*, 2 - 1.3 *mbar*.

All of these measurements have been performed in glow discharges, mainly under pressures below 100 *mbar*.

The described phenomenon of memory is characteristic for different kinds of breakdowns in gases. Van Brunt and Kulkarni [19] elaborated this problem concerning Trichel–pulses in $N_2 + O_2$ and $Ne + O_2$ mixtures under the atmospheric pressure. It has been shown that the time interval between the successive pulses and their amplitudes is related to memory processes. Since in that case $\tau < 1\ ms$, the dominant role in these processes is played by the space charge formed in the previous discharge.

3. Time delay t_d and the concentration n of metastables

A method determining how to obtain information on the change of metastable concentration (n) during the afterglow period (τ) in gas–filled diodes is elaborated in detail in the work [15].

The $\bar{t}_d$ values, as already shown by Meek and Craggs [16], are linked to the electron liberation rate P in the active diode volume as $\bar{t}_d = K/P$. Since this rate (especially for $\tau > 1\ s$) is proportional to the concentration of metastables n, one can also state that $\bar{t}_d = k/n$, where K and k are corresponding constants that depend on the experimental set up. For $\tau > 1\ s$, n refers only to neutral metastables that remain from the previous discharge or were created in the post–discharge processes in the diode. Thus, using the equation mentioned, it is possible to derive, from the memory curve, the reciprocal curve $n = F(\tau)$ being in log – log system symmetric in shape, concerning the straight line $\bar{t}_d = \text{const.}$ ($A - A$, Fig. 5), which satisfies the condition:

$$\frac{d\bar{t}_d}{\bar{t}_d d\tau} = -\frac{dn}{nd\tau} .$$

Both of these curves reach saturation at τ, when the breakdown probability linked to metastables produced in the previous discharge drops significantly below the breakdown probability caused by cosmic rays and radioactive background.

Due to the results obtained and listed above, the measurement of $\bar{t}_d$ values can be used for detection of neutral metastable states in gases.

A frequently used metastable detection method at high metastable concentrations above $5 \cdot 10^6\ cm^{-3}$ is the absorption spectrometry of light [37]. In the case of smaller concentrations it is more effective to detect electrons produced by the secondary emission caused by interactions of metastables with metal surfaces [38]. According to [39] this method may be 10^8 times more sensitive than optical methods.

At the beginning of the curve $n = F(\tau)$, for low values of τ (Fig. 5), the value n is close to initial saturation value of metastable concentration n_0 at instant $\tau = 0$.

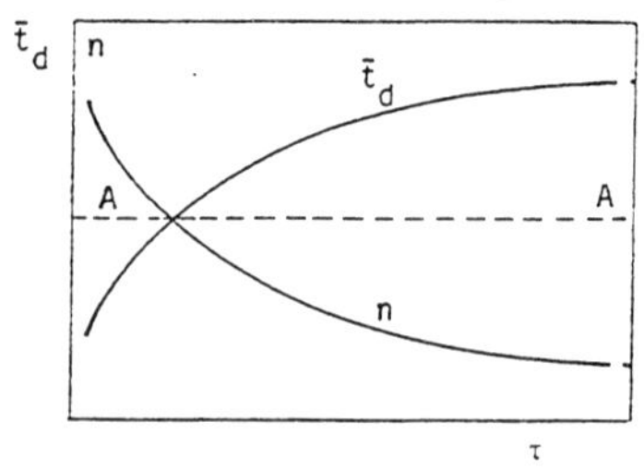

Fig. 5. Time delay $\bar{t}_d$ and the metastable concentration n in function of τ.

In all types of electrical discharges, many kinds of inelastic collisions occur, and consequently new ions and particles in excited states, both resonant and metastable ones, are generated.

Comprehensive bibliography of these processes is cited in works [16, 20, 139 and 147]. Our investigations [21] in nitrogen under gas flow conditions, showed the presence of metastables both upstream and downstream up to 4 m from the production place.

Concentration of metastables in nitrogen plasma can reach the value of $10^{14}\ cm^{-3}$ as presented in [22 and 23]. These monographs contain a lot of information on this subject.

We cite as an example that the metastable concentration of helium immediately after the cessation of the discharge may reach the value of about $10^{13}\ cm^{-3}$ at total concentration of helium atoms of $10^{17}\ cm^{-3}$ and exceed that of electrons being 10^{11} up to $10^{12}\ cm^{-3}$ [22].

The equilibrium (saturation) concentration of N_2 metastables in nitrogen glow discharge depends on the gas pressure and the electric current.

Important data on the concentration of nitrogen molecules in the $N_2(A^3\sum_u^+)$ state at pressures of 0.13 and 2.7 $mbar$, at current from 0 up to 30 mA, have been obtained by optical methods [24].

The decreasing process of n, starting from the value n_0 up to the sensitivity limit of the detection method by measurement of t_d, represents an important subject of this review.

We have had to repeat many of the previously performed t_d measurements of various authors, since they either did not cite or, possibly, even did not consider the afterglow periods in their investigations.

4. Conditioning of the diodes before measurements

In order to obtain reliable results of measurements, it was necessary to stabilize the characteristics of the diodes. Initial measurements on freshly produced diodes, show unstable characteristics. For that reason preconditioning was applied. A certain current density should flow through the diodes, during certain period of time, or a set of current pulses should be applied successively. Thus, in an "artificially aged" diode (Fe, H_2, 15.7 $mbar$) [26], after about 2 500 definite pulses applied, U_s increased for about 3%, but $\bar{t}_d$ decreased for two orders of magnitude. After that conditioning characteristics become stable, and the systematic measurements could be initiated.

All the diodes measured in the way shown in this review were subjected to the conditioning methods described above, i.e. to the stabilization of characteristics. That process encompasses sputtering of the material from the electrodes, until the stationary quality of the surfaces and its interaction with gas environment is attained [71, 106, 108 and 127].

5. Statistical approach to data handling

The stochastic character of magnitude t_d could clearly be seen from its recorded values, under properly controlled conditions. Two methods have mainly been used in order to interpret their distribution. The first one is Laue's distribution [6].

The dispersion of experimental data is evident from Fig. 23. Individual t_d values are scattered within more than five orders of magnitude and in some cases even wider. However, Laue's distribution is shown to be valid in most cases. As it was shown in [27, 28 and 29], Laue's approach is shown to be applicable to afterglow time values from 10^{-3} up to 10^5 seconds, that means also to τ of order of milliseconds when space charge is dominant in initiation of the new breakdown, and to time delay values from 10^{-3} to 10^3 s.

The form of Laue's distribution is as follows:

$$\frac{N_t}{N} = \exp\left[-\frac{t_d - t_f}{\bar{t}_s}\right] ,$$

where

N_t – number of results longer than t_d,
N – total number of measurements,
t_d – individual time delay values,
t_f – discharge formative time,
$\bar{t}_s$ – mean value of statistical time delay.

The results of formative time measurements [30], applying also Laue's distribution, are in accordance with the previous statements. In a diode (Fe, N_2, 0.4 $mbar$) sets of t_d values have been measured at 1%, 3% and 9% overvoltages (Fig. 6). As it can be seen, not only the dependence of distribution on the applied overvoltage has been obtained, but also the discharge formative time t_f, depending on the same magnitude.

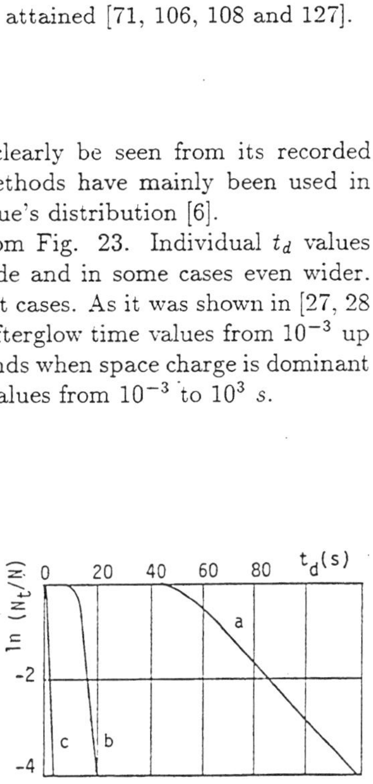

Fig. 6. Laue distribution in the diode (Fe, N_2, 0.4 $mbar$) measured with overvoltages:
a – 1%, b – 3%, c – 9%.

The value of t_f is given by time interval ranging from zero up to the intersection of the obtained curve with the abscissa.

The same method has been used for evaluating the ignition lag of discharge cells in AC plasma displays [31]. Laue's distribution has also been used for processing of data obtained under the conditions typical for electroenergetics, where the values of t_d are of order of ns and μs [72].

Another analytical method has been used for describing the distribution of t_d values measured in the diode (Fe, N_2, 13.3 $mbar$) [27]. It was shown there that the distribution is easily identifiable after about 500 successively measured t_d values, and that its form may be estimated even after about 100 measurements. These distributions were fitted using the least–squares polynomial approximation method. It was found that the best approximation is obtained by the seventh–order polynomial.

Further development of statistical approach can be seen in [16 and 166 – 169].

For some of the measurements, the statistical treatment of t_d values is presented in papers [12, 53 – 59, 65, 77, 92, 102, 108, 112, 115, 116, 146, 151, 153, 154 and 156].

This problem has also been treated and the corresponding literature is cited in [16 and 40].

6. Dependence of time delay on the kind and pressure of the gas and the interelectrode gap

Measurements in Ar and Xe (Fig. 7) [12] and in He and Kr (Fig. 8) [34], were in accordance with previous measurements of t_d in various gases [35], this time within the range of τ up to about 10^5 s. As Braunbek [36] has shown, breakdown probability can be expressed in the following way:

$$W \approx K \cdot e^{\alpha d} ,$$

where

W – breakdown probability,

K – constant,

α – first Townsend coefficient,

d – electrode gap.

Since α increases with the atomic number of elements, W also increases, and the decrease of $\bar{t}_d$ is caused. It has been confirmed in this case within the mentioned range of afterglow period τ.

Especially illuminating is [25 and 47] to compare the results of measurements of t_d and memory curves up to τ about 10^3 s in N_2 and in the laser mixture $CO_2 - N_2 - He$, both under pressure of 25 $mbar$ [25]. In accordance with the known mechanism, deexcitati-

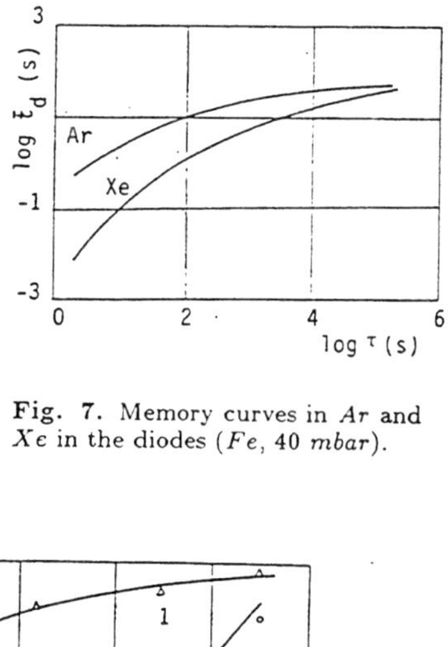

Fig. 7. Memory curves in Ar and Xe in the diodes (Fe, 40 $mbar$).

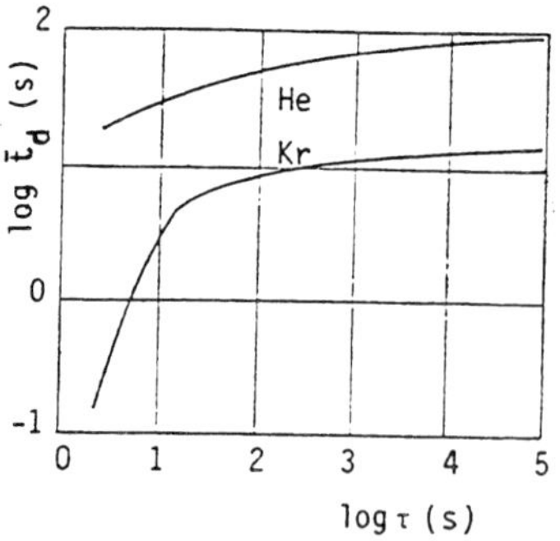

Fig. 8. Memory curves in He and Kr in the diodes (Fe, 40 $mbar$).

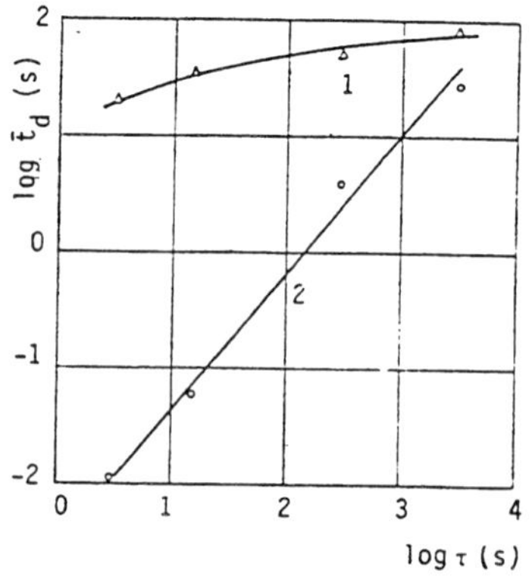

Fig. 9. Memory curves in the diodes (Fe, He). The pressure is: 1– 40 $mbar$, 2 – 1.3 $mbar$.

on of metastables of nitrogen in laser mixture is more intensive than in pure N_2, the concentration n is lower and $\bar{t}_d$ considerably higher, as it has been shown by the measurements.

Some results of the measurements performed in this region are given in the papers [46, 63 –65, 69, 98 and 110].

A theoretical approach to the problem of t_d dependence on gas pressure is presented in [35]. An example of recorded memory curves in He is shown in Fig. 9 [67]. This problem has also been approached in the papers [66, 68 – 71, 108 and 132].

In order to investigate the $\bar{t}_d$ dependence on the electrode gap, a diode has been made (Fe, N_2, 3 $mbar$), with a mechanism for changing the interelectrode gap (d) continuously, by spinning an iron "butterfly", by means of a magnet situated externally to the diode [54]. Since all other conditions have been identical, with the change of d, U_s has also changed, so that all the measurements have been performed with the overvoltage of 2%. With d increasing from 1.5 to 12 mm, the decrease of $\bar{t}_d$ has also been obtained in accordance with previous theoretical considerations [35], but this time it has been the shift of the complete memory curve. This problem has also been treated in the works [50, 51, 94, 105, 111 and 113].

7. Dependence of the time delay on overvoltage and on the work function of cathode material

The time delay of electrical breakdown in gases, depending on overvoltage, under the conditions typical for electroenergetics (pressure mostly 1 bar and high overvoltage) has been investigated in detail, with corresponding t_d values usually between ns and ms [72].

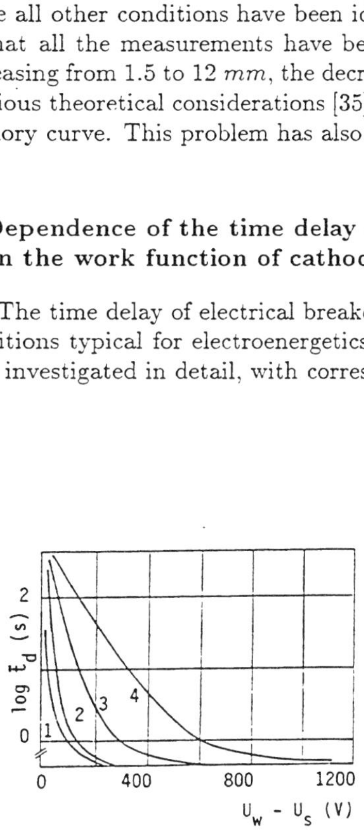

Fig. 10. $\bar{t}_d$ dependence on U_w in argon-filled diodes with electrodes made of: 1 – Ba (evaporated on Cu), 2 – Al, 3 – Cu, 4 – Fe.

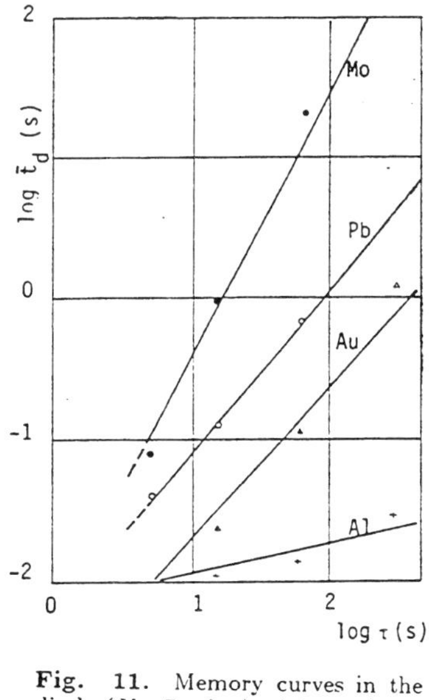

Fig. 11. Memory curves in the diode (N_2, 7 $mbar$) with 4 pairs of electrodes of different materials.

In the reviewed papers, these phenomena have been investigated under pressures below 100 $mbar$ in the regime of glow discharge. Typical curves $\bar{t}_d = f(\Delta U)$, obtained under these conditions, where $\Delta U = U_w - U_s$, in argon, for cathodes made of Ba, Al, Cu and Fe respectively, are shown in Fig. 10 [10].

In order to maximize reliability of measuring t_d dependence on cathode material, and to eliminate differences in quality and pressure of the gas, a nitrogen–filled diode has been made, under pressure of 7 $mbar$, with 4 pairs of electrodes, where each pair has been made of different materials: Al, Pb, Au film and Mo [44]. The results are shown in Fig. 11.

In all the cases it has been shown that, under the same other conditions, $\bar{t}_d$ increases together with the increase of work function of cathode material, due to the decrease of coefficient γ of secondary electron emission caused by metastables. There is an exception with gold which has the coefficient γ higher than it should be according to the sequence of work functions, and that is why the curves for Pb and Au in Fig. 11 have replaced each other.

The dependence of t_d on overvoltage has also been treated in the papers [10, 12, 44, 45, 48, 60, 73, 83, 84, 93 and 114].

The dependence of t_d on cathode materials has been discussed in the papers [10, 12, 42, 44, 45, 49, 50, 73, 74, 105 and 109].

8. The dependence of t_d on temperature

Breakdown time delays t_d and complete memory curves were obtained by measurements on diodes in which the electrode temperature could be varied. Thus a diode was manufactured with a tungstene wire cathode that could be electrically heated. A thermocouple was soldered to the active part of the cathode in order to measure the temperatures. For that reason the diode glass bulb had 4 electrical lead–throughs for the cathode system.

The diode has been filled with nitrogen at 4 $mbar$ [43]. Some results of these measurements are presented in Fig. 12. This problem was also treated in [48, 79, 80, 87, 93, 111 and 115].

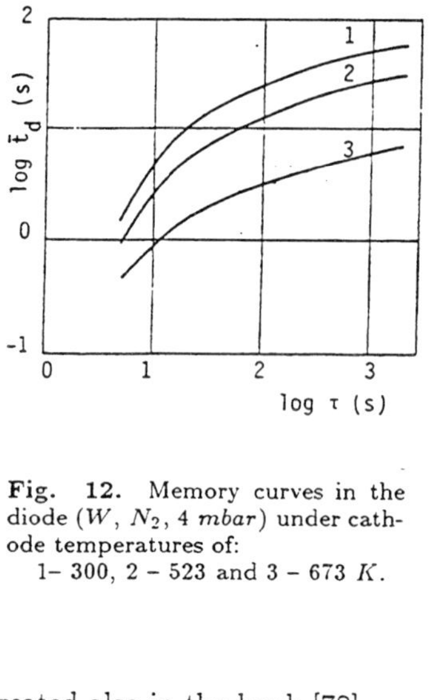

Fig. 12. Memory curves in the diode (W, N_2, 4 $mbar$) under cathode temperatures of: 1- 300, 2 - 523 and 3 - 673 K.

From the cited papers, particularly in Fig. 12 one can see that t_d decreases with the rise of temperature, this can be ascribed to the increase of the deexcitation coefficient γ of metastables on metallic surfaces with the rise of temperature. Principally this is in accordance with the Richardson–Dushman formula. This topic is treated also in the book [78].

9. The dependence of t_d on electromagnetic radiation

The influence of electromagnetic radiation on t_d was investigated, and so visible light, laser light of selected wavelenghts and radioactive gama radiation were applied to the diodes. Here we present the influence of white light and that of gama rays. The effects are primarily due to the "positive" optogalvanic effect [81], and under certain conditions partly on the photoelectric effect on electrode surfaces. The positive optogalvanic effect is the consequence of a selective, resonant light apsorption by atoms or molecules in the gas leading to an increase in concentration of metastables. Therefore there is a rise of the breakdown probability and a reduction of the breakdown time delay.

The dependence of $\bar{t}_d$ on U_w is shown in Fig. 13 measured on mass-produced indicator glow-lamps held permanently in darkness or exposed to daylight. The measurements were done each after one day afterglow period, ($\tau = 24\ h$), i.e. one measurement of t_d per day. As can be seen reducing the applied voltage, when U_w approaches the U_s value, the $\bar{t}_d$ values tend to infinity. This could be considered to be a very reliable valid definition of the static breakdown voltage [45].

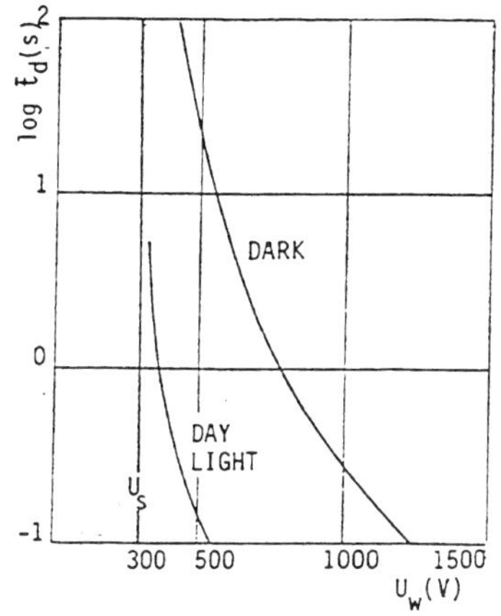

Fig. 13. Influence of light and the applied voltage U_w on $\bar{t}_d$ in signal glow lamps. $\tau = 24\ h$.

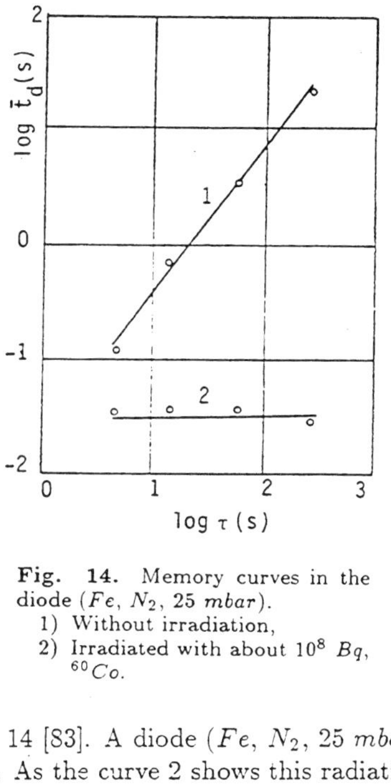

Fig. 14. Memory curves in the diode (Fe, N_2, 25 $mbar$).
1) Without irradiation,
2) Irradiated with about 10^8 Bq, ^{60}Co.

A similar influence of radiation is shown in Fig. 14 [83]. A diode (Fe, N_2, 25 $mbar$) is irradiated by approximately 10^8 Bq ^{60}Co γ source. As the curve 2 shows this radiation maintains a constant value of n in the diode and consequently the constant value of $\bar{t}_d$ during the time τ.

Memory curve were also established at small τ values ($\tau < 1\ s$), for the diode (Cu, N_2, 6.6 $mbar$) [85] that was either irradiated by a 10^6 Bq ^{60}Co source or protected from radiation. In this case one can count on a certain influence of the photoelectric effect on the electrodes.

From this experiment it can be seen that in conditions of the presence of space charge, the influence of the external irradiation of the diode on t_d can be neglected in comparison with the influence in the period $\tau > 1\ s$.

Effects of this type were investigated and reported in [45, 60, 82, 84, 90, 138, 142 and 144].

10. Transport of metastables due to diffusion and gas flow

Results of measurements of t_d in a gas flow diode (Fe, N_2, 1 bar) at flow rates between 0 and 100 $Ndm^3 \cdot min^{-1}$ are shown in Fig. 15 [12]. There was no auxiliary electrical discharge in the gas flow, i.e. no ion production or source of metastables. As can be seen, at preset values of $\tau = 3\ s$, the curve has a pronounced maximal value around 10 $Ndm^3 \cdot \min^{-1}$ corresponding to a minimal concentration of metastables. Similar results were obtained in Ar and N_2 at 80 $mbar$.

Up to now there have been no plausible explanations for this behaviour.

Fig. 16 shows the results of t_d measurements in the system [21] with an auxiliary pair of electrodes for generation of ions and metastables. The auxiliary current $i_a = 150\ \mu A$. Nitrogen has flowed through a glass tube of inner diameter of 10 mm, with the velocity of 10 $cm \cdot s^{-1}$, under pressure of 80 $mbar$. t_d has been measured in the diode with Fe electrodes, on condition that $\tau = 5\ s$. The ratio on the ordinate has been $\bar{t}_{d0}/\bar{t}_{di}$, where $\bar{t}_{d0}$ represents the $\bar{t}_d$ value obtained at $i_a = 0$, and $\bar{t}_{di}$ at the mentioned auxiliary current. The abscissa represents the distance between the auxiliary and the measuring diode.

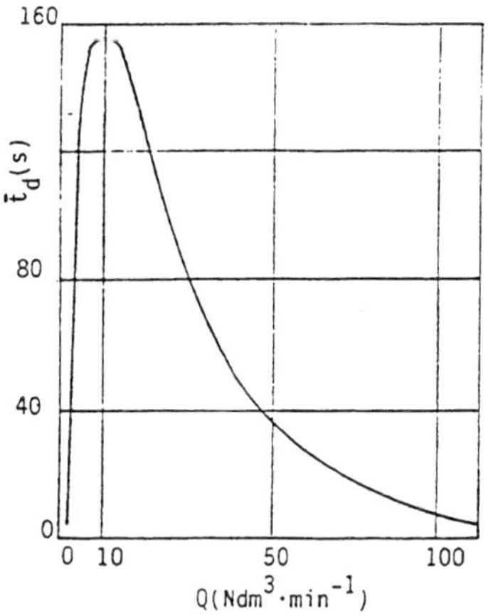

Fig. 15. Dependence of $\bar{t}_d$ on the gas flow rate Q in the diode (Fe, N_2, 1 bar) without auxiliary discharge. $\tau = 3\ s$.

The measurements have been performed upstream and downstream the gas and the results (Fig. 16) show that the transport of metastables could be recorded by this method, in a described system, up to 4 m. Curve b, recorded on the measuring diode which, in that case, has been placed upstream the gas flow (gas flows from the measuring toward the auxiliary diode), shows that a number of metastables reaches the measuring diode, against the gas flow,by diffusion through its own gas. Similar problems are presented in papers [70, 86, 87, 91, 111 and 133].

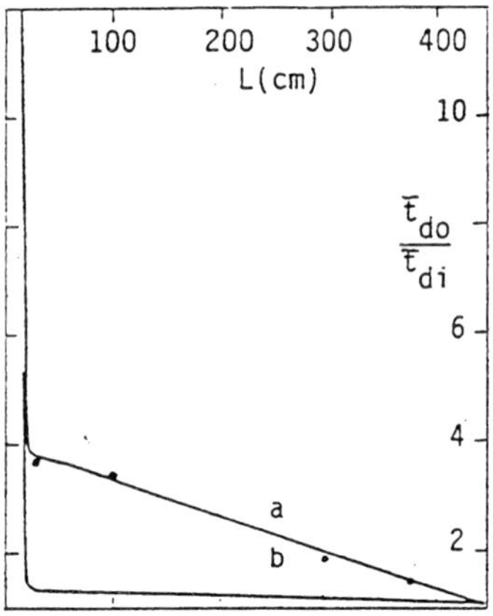

Fig. 16. Transport of metastables through the diode (Fe, N_2, 80 $mbar$). Gas flow spead: 10 $cm \cdot s^{-1}$. Gas is flowing a) toward the measuring diode, b) toward the auxiliary diode. $\tau = 5\ s$. Auxiliary current $i_a = 150\ \mu A$. L – interdiode distance.

The transport of metastables by diffusion has been studied in the diode with two pairs of electrodes (Fe, Ne, 13.3 $mbar$) [88]. In a glass tube of diameter 70 mm, the diodes have been 42 cm apart. Memory curves have been established up to $\tau = 300\ s$, with the current through the auxiliary diode as a parameter. For the auxiliary current equal to zero, the expected curve has been obtained. With the current flowing through the auxiliary diode, at the increase of the current, t_d decreased and the curves shifted appropriately (Fig. 17). This points out the generation of metastables in the discharge on the auxiliary diode and their transport to the measuring diode by diffusion.

Results on investigations of diffusion of charged particles and metastables in nitrogen are shown in Figs. 18. and 19. For these measurements a special 40 mm wide glass tube was manufactured about 1 m long (Fe, N_2, 13.3 $mbar$) with two pairs of electrodes. The measuring diode was incorporated at one end of the tube while the auxiliary one was movably mounted in a glass sleeve that could be repositioned along the main tubing by external means without changing the gas conditions [89]. The distance between the diodes (L) could be varied up to 0.5 m as presented in Fig. 18. By this set-up differences in quality and pressures of the gas were eliminated which would have otherwise appeared inevitably if measurements have been performed on a set of tubes with different interdiode distances.

Between the two diodes a constant DC voltage could be applied. The measuring diode was operated under G.M. counter conditions, thus it was possible to measure the time t of the active particle diffusion from one to the other diode, i.e. to establish the time span between the start of the current flow through the auxiliary diode and the appearance of the signal at the measuring diode. The interval between two succesive measurements was kept at 30 s. Each point on the curve $t = f(L)$ is the mean value of 100 measurements.

The obtained diagrams show two pronounced inflection points one at $L = 1.3$ cm and the other at $L = 8$ cm, which correspond to the ranges of diffusion of electrons and ions in the tube. Each section of the curve corresponds to particles having the dominant influence on the initiation of breakdown in the measuring diode. Part I for electrons, part II for ions and III for metastables whose role gets manifested at time intervals exceeding 10^{-3} up to 10^{-2} s or at distances between the diodes between 8 and 50 cm.

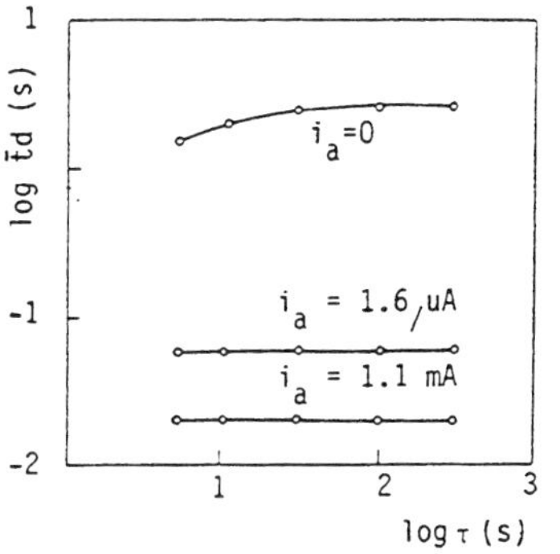

Fig. 17. Memory curves in the diode (Fe, Ne, 13.3 $mbar$). The currents i_a in the auxiliary diode are given on the curves. Interdiode distance is 42 cm.

The obtained values for coefficients of metastable diffusion have been higher for about two orders of magnitude than those cited in literature. This is attributed to the exceptional sensitivity of the method applied [39]. The diffusion coefficient and field E are given in Fig. 19. Problem of diffusion of metastables has been elaborated also in [109, 115, 116 and 145].

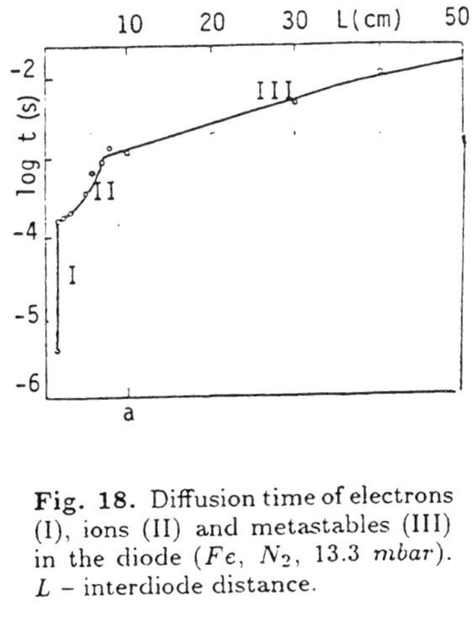

Fig. 18. Diffusion time of electrons (I), ions (II) and metastables (III) in the diode (Fe, N_2, 13.3 $mbar$). L - interdiode distance.

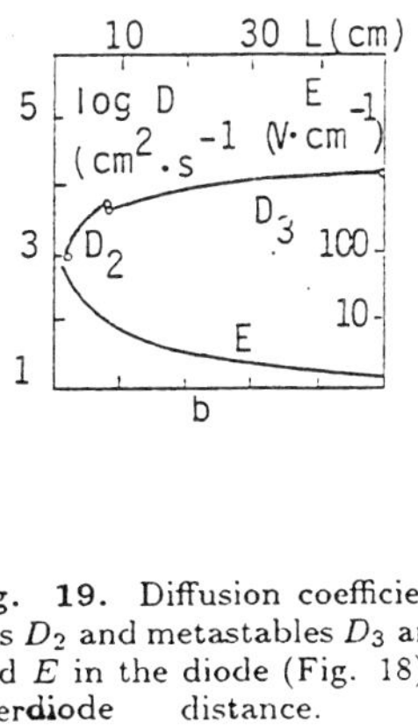

Fig. 19. Diffusion coefficients of ions D_2 and metastables D_3 and the field E in the diode (Fig. 18). L - interdiode distance.

The results of the experiments cited so far could be explained by the hypothesis that, under given conditions, electrical breakdown is initiated by secondary electrons emitted from the electrodes, in processes induced by metastable atoms or molecules generated in electrical discharge or in some afterglow processes. In other words, the process of memorizing in gas filled diodes is explained by the decrease of the concentration of metastables in the afterglow period. There has been no other acceptable explanation. The following experiment has provided the exact proof for this hypothesis.

11. Detection and laser spectral analysis of metastables – memory carriers in neon

In order to establish exactly which kind of the particle is responsible for the memory in neon a glass tube with two pairs of 5 μm gold plated electrodes was manufactured that was constructed so that the gas and diode gap could be illuminated by laser light beams of a tunable dye laser. The interdiode distance was 10 *cm* and the tube was filled at 13.3 *mbar* of spectroscopially pure neon [90].

With no radiation and no current at the auxiliary electrodes memory curve 1 in Fig. 20 was obtained as expected. At 1 *mA* at the auxiliary diode, without radiation or illumination, curve 2 was obtained. The shape of curve 2 could be explained to be due to the diffusion of metastables generated at the auxiliary electrodes that were in a steady state constant concentration at the measuring diode. In other words at constant auxiliary current there is constant concentration of metastables at the measuring diode and the average value of the breakdown time delay $\bar{t}_d$ does not depend on the afterglow time τ. Measurements at zero current through the auxiliary diode but illuminating the measuring diode by laser light of 614.3 *nm* yielded curve 3.

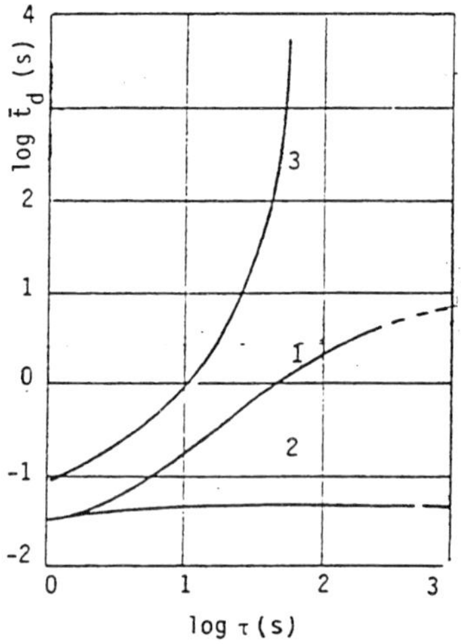

Fig. 20. Memory curves in the tube with two diodes (*Au*, *Ne*, 13.3 *mbar*). Interdiode distance – 10 *cm*.

1. No radiation, auxiliary current $i_a = 0$.
2. No radiation, $i_a = 1$ *mA DC*.
3. Irradiation with laser light. $\lambda = 614.3$ *nA*, $i_a = 0$.

This phenomenon is explained by the negative optogalvanic effect [81]. The metastable atom of neon in the $1S_5$ state absorbs resonantly the light of the mentioned wavelenght, so that this state is deexcited and the concentration of metastables drastically reduced tending to zero at $\tau < 100$ *s* and $\bar{t}_d$ tending to infinity, under given experimental conditions. The confirmation of this thesis is supported by the resonant characteristic of this process, because irradiation by somewhat lower or higher wavelenghts than the mentioned one, did not influence $\bar{t}_d$ and the memory curve. The resonance has been so sharp, that the applied wavelenght of dye laser had to be permanently controlled during the experiment by the optogalvanic method displayed osciloscopically.

Thus, the thesis that the cause of memory process in gases lies in the process of the decrease of the concentration of metastable states has been supported. In the case of neon metastable atoms were in the $1S_5$ state.

12. Metastables among fragments formed by the disintegration of SF_6 in glow discharge

A glass tube of 20 mm diameter for measurements in gas flow, containing three diodes with electrodes made of chemically pure iron, has been provided in order to detect metastable states of atoms and molecules, which are formed as fragments during disintegration of SF_6 in glow discharge. The first diode in the flow passage has been an auxiliary one for generation of fragments, the second one serving to eliminate ions, and the third being the measuring one. Distance between the diodes has been 15 cm each. SF_6 has flowed under pressure of 5 $mbar$ with flow rate of 110 $bar \cdot cm^3 \cdot s^{-1}$ [18].

Having eliminated the possibility that electrons, ions or photons from the auxiliary discharge could reach the measuring diode, it was found that electrical current in the auxiliary diode (i_a) influence the current through the measuring diode (i_m) operated at constant voltage. There was a rise of i_m increasing i_a. Measurements have been done up to $i_a = 3\ mA\ DC$. The phenomenon has been interpreted in terms of transport of neutral metastable fragments of SF_6 by gas flow, and partly by diffusion, from the auxiliary toward the measuring diode. This process has been investigated in detail in the work [95]. The kind of this metastable state has not been identified yet.

It has also been possible to record memory curves after glow discharge in SF_6, which supports previous conclusions [52].

Some additional informations on glow discharge in SF_6 are given in [57, 126, 134, 137 and 155].

13. Dependence of metastable generation rate on the current in the diode

Cernogora has shown [24] by the method of light absorption, that the generation rate of metastables in glow discharge in nitrogen, and their concentration depend on the intensity of electrical current and have maximum at certain current strenghts. The same phenomenon has been found in experiments to be described, under considerably different conditions, mostly by the method of metastable detection via t_d measurements.

A typical example of measurement in N_2, from the paper [68], is shown in Fig. 21, together with some results from the paper [24].

The minimums of curves B and C correspond to the maximal generation rate of metastables, as well as the maximum on curve A. The extremes on these curves, of course, appear at different current strenghts, due to considerably different conditions of experimenting.

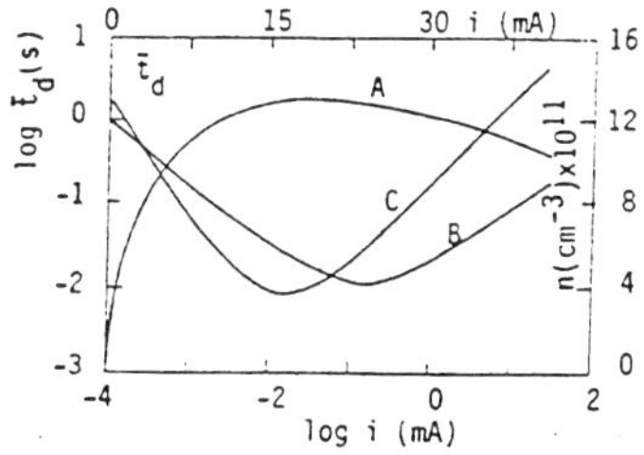

Fig. 21. A – Dependence of n on diode current (between 0 and 40 mA) in N_2 under pressure of 1.3 $mbar$.
Time delay dependence on the diode current (between 10^{-4} and 10^2 mA) in N_2 under pressure of: B – 1.3 $mbar$, C – 4.0 $mbar$.

In the papers cited in this work, these phenomena have been found under very different conditions by the methods of: a) t_d measurements, b) measurements of metastable contribution to stationary current, c) measurements by means of a chamber for secondary electron emission caused by metastables, which is going to be described at the end of this paper. The results of this investigations are given also in the papers [47, 63, 74, 98 – 100, 103 and 113].

14. Dependence of t_d on the diode bulb volume

In order to investigate diffusion of neutral metastables produced in the active interelectrode space, a glass diode [96] has been made having two volumes of unequal size separable by means of spherical ground glass stopper. While in vertical position, the sphere separated the large spare volume so that the volume of the diode was 31 cm^3, with the inner surface of the glass bulb of 54 cm^2. When in horizontal position, the rolling sphere opened up the spare volume, so that the total volume of the diode was now 5818 cm^3 having a surface of 2379 cm^2. The tube has been helium–filled under pressure of 1 $mbar$.

The anode has been made of pure iron, and the cathode has been copper–made, with 5 μm thick galvanic layer of gold. At constant voltage on the diode; the current in large volume has been, depending on the intensity of the current, even 60% lower than the current in the small volume.

When the passage between the two parts of the diode was open, neutral metastables diffused into the enlarged volume, and their contribution to the intensity of the current in the diode become negligible. They were deexcitated on the walls of the bulb, as well as in the gas by metastable–metastable collisions (self–quenching process). The probability for electrons and ions to leave the electrical field between the electrodes is several orders of magnitude lower that for metastables, so that the migration of charged particles could be neglected. (The role of metastables in the discharge current strenght is elaborated in [16, 32, 75 and 170].)

A similar nitrogen–filled diode with "two volumes", of 1.3 $mbar$ [97], has been used for corresponding investigations by t_d measurements. Thus, under other identical conditions, two comparative memory curves have been obtained, where the curve obtained for the "large" volume has completely been shifted towards higher values of $\bar{t}_d$.

Opening of the passage between the two parts of the diodes in both cases made possible the diffusion of neutral metastables generated in the discharge, considerably reducing their concentration in the active space of the diode and thus causing the effects mentioned above.

Similar measurements are described in papers [46, 101, 107 and 130].

15. Distribution of n in the gas–filled diode

A model concerning physical processes in the gas–filled diode, whose volume is considerably larger than the zone of discharge, i.e. the active volume between the electrodes has been worked out. It has been supposed that metastable $N_2(A)$ molecules, during the discharge, dominantly diffuse out of the active volume in the bulb, where they deexcite, but exist in considerable concentrations, even after the discharge cessation.

The change in the concentration (n) has been treated by solving Boltzmann's kinetic equation in order to find the distribution of n in the diode bulb in stationary state [112]. The same has been done in Kr, Fig. 22 [146]. Considering the most important deexcitation mechanisms of metastable states, solving a system of equations mentioned, the distribution of n in the bulbs was also obtained in papers [146, 151 and 153].

The decrease of n of $N_2(A)$ metastables during τ is also obtained. The results mentioned show that after $\tau \approx 1$ s the active diode volume becomes sinks of metastables, because the retrodiffusion of the metastables from the entire bulb toward the active volume become the dominant diffusion process. Consequently the effective life of metastables in the active volume is prolonged and the concentration is kept higher than in case of absence of retrodiffusion. This is confirmed by the existence of a certain plateau on the concentration decrease curve given in papers [117, 145 and 154].

Using this model a relation between the increase of t_d and the decrease of n during the period τ has been established [124].

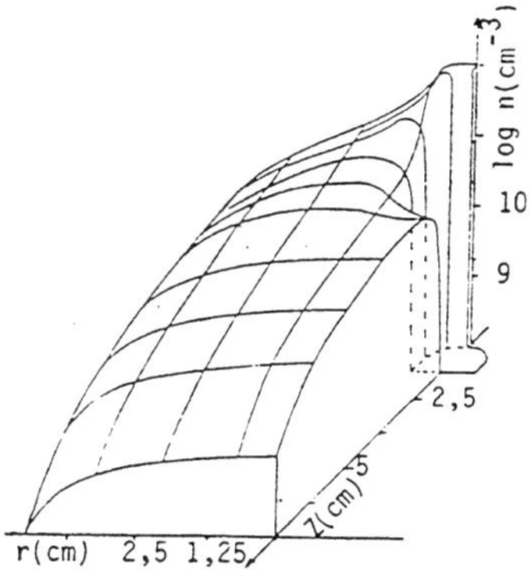

Fig. 22. Distribution of metastable concentration n of Kr 5^3P_2 states in a cylindrical diode bulb. $P = 2.7$ $mbar$.

16. Automation of the measurements

In order to perform t_d measurements in a faster and more reliable way, as well as in a wider range of τ, starting from about 10^{-2} up to 10^4 s, an automatic system with computer data acquisition and processing has been assembled. The measurements are performed in time cycles given in Fig. 2. Before starting with the measurements, the following parameters were preset: time τ, time t_i, voltage U_w, current through the diode i during t_i, as well as the number of measuring cycles to be repeated. Then, the system is initiated and the expected number of t_d measurements is automatically performed after which it switchess of automatically too. At the end of a series of measurements the stored data are displayed including the number of measurement cycles completed, the minimal and maximal t_d values obtained in the series, as well as the sum of t_d values. The data are, of course, ready to be used in further handling. This system is described in the papers [15] and [114].

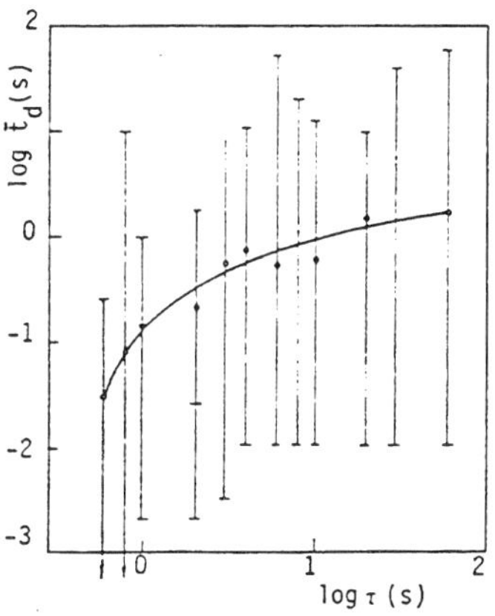

Fig. 23. Memory curve with dissipation of t_d values in the diode (Fe, N_2, 13.3 $mbar$).

In this way a statistically relevant mass of data for further processing is obtained. One of the cases is shown in Fig. 23 obtained by a nitrogen-filled diode with pure iron electrodes, under pressure of 13.3 $mbar$. The dispersion of the obtained t_d values is obvious in this figure.

In the paper [15] the question has been risen whether the experimentally obtained curve $\bar{t}_d = f(\tau)$ could provide reliable data from which the $n = F(\tau)$ dependence can be estimated, i.e. the variation of concentration of metastables in the given system in time. The dilemma arises from the fact that U_w considerably influences the t_d values, so that the proper overvoltage needed for the measurements is not known. The question is whether the change of U_w influences the slope of the memory curve, or not. If the answer is positive, this method could not be used for obtaining the function $n = F(\tau)$, and in case of the negative answer, such a method could be used.

In the paper[15] recorded memory curves on the diode (Fe, N_2, 40 $mbar$) operated successively at overvoltages of 15, 20 and 25% are presented. As it has been shown at these overvoltages 3 curves were obtained parallelly shifted along the ordinate, i.e. with considerably changed t_d values, but the slopes of these curves ($d\bar{t}_d/d\tau$, where $d\bar{t}_d$ is given in percents) have remained unchanged for each τ value.

Analogous results have been obtained with neon, under the same pressure of 40 $mbar$.

Such an influence of the voltage U_w on the memory curves most probably has the following explanation. The time delay t_d is a function of the probability of electrical breakdown in the gases. The probability of the breakdown in the afterglow period $\tau > 1$ s is related to two processes. One is the release of the initial electron, usually from the cathode, by secondary emission caused by metastables, and the other is the formation of the discharge due to this electron. Each of these two processes has its own probability.

Because these metastables are neutral particles, the secondary emission induced by them takes place with a probability that does not depend on the applied voltage, i.e. on the electric field E between the electrodes. Because of this, the slope of the memory curve does not depend on U_w or E, but only on the change of the metastable concentration n.

The process of discharge formation is, however, a function of the field E, which results in the change of $\bar{t}_d$ value.

From the above, it can be concluded that, within a specific range of relevant parameters, memory curves can be applied to measure the change in the metastable concentration of gas atoms or molecules in gas-filled diodes during the afterglow period, i.e. the ratio $dn/d\tau$.

17. Modeling of the deexcitation process of metastables in the afterglow period

In order to give a physical interpretation of the $n = F(\tau)$ dependence, that is taking care of the fact that the concentration n (especially for the small τ values) decreases much faster than according to a simple exponential law of independent metastable deexcitations, that would yield a constant value of the half deexcitation time independent on τ, a model is proposed considering, in the first approximation, only the most influential deexcitation mechanisms.

There are three neutral metastable states ($A^3\sum_u^+$, $W^3\Delta_u$ and $a'^1\sum_u^-$) in molecular nitrogen having sufficient energy [150, 157 and 171], a long enough life–time [41, 158 and 171] and probability of being produced in glow discharge that can generate a secondary electron emission from the cold metallic cathode and thus trigger the breakdown in a nitrogen–filled diode [124].

We consider that the breakdown at long afterglow times ($\tau > 1$ s) is due to the $N_2(A^3\sum_u^+)$ metastables within the active volume of the diode.

Various investigations [118 – 120] indicate that the $N_2(A^3\sum_u^+)$ metastable states are, most probably, the initiators of secondary emission and of electrical breakdown in nitrogen under the conditions described. This has recently been confirmed by work [121]. In spite of the fact that we have mainly been referring to the $N_2(A^3\sum_u^+)$ state in nitrogen, this research includes events due to joint effects of all neutral metastables of atoms and molecules present in diodes, including those of impurities, created in previous glow discharges as well as in decaying afterglow processes [122].

At the instant of cessation of the stationary glow discharge of a sufficient duration (t_i), there is a saturation concentration of metastables n_0 established in the active diode volume. Since the breakdown probability and consequently the time delay in afterglow period depend on the concentration (n) of these metastables, there is a question what effects n depend upon during this period ($\tau > 1\ s$) in the active interelectrode volume. One may suppose the influence of two groups of deexcitation mechanisms and a third process are related to the metastables diffusion.

The metastable deexcitation mechanisms of group I are described by a concentration independent coefficient B in the differential equation proposed, meaning that the deexcitation of one metastable particle in the next second has a constant probability, i.e. it is independent on the metastable concentration n and thus on the afterglow time τ, similar to the independent decay of radioactive nuclei.

Mechanisms of group II (described by the coefficient C in the same equation), refer to self–quenching processes via which the deexcitation probability of a metastable particle depends on n and thus decreases in τ.

We consider that in group II processes, the dominant role can be attributed to the binary metastable – metastable collisions by which one or both excited molecules can deexcite. In the differential equation the constant $C = 1.1 \cdot 10^{-9}\ cm^3 \cdot s^{-1}$ is used [123].

The third process is diffusion. During the glow discharge with constant current in the diode, the constant number of metastables is generated in each second. The active volume between the electrodes is a source of metastables that fill up the entire diode bulb, due to their dominant diffusion toward the walls of the bulb where their deexcitation is carried out, as well as partly in the gas filling of the diode.

That is why a stationary distribution of the metastable concentration has been established [146] with the constant concentration gradient in each part of the diode volume.

After the discharge cessation, the generation of metastables is stopped and in the period $\tau > 0.6\ s$ the active volumes are transformed from the metastable source to their sinks, because of the significant difference between the deexcitation coefficients on the electrodes and on the glass bulb ($\gamma_E = 5 \cdot 10^{-3}$ and $\gamma_G = 3 \cdot 10^{-5}$).

Because of that the concentration n is decreased in active space in relation to the concentration in the entire diode.

Consequently the retrodiffusion of the metastables from the entire diode bulb towards the electrodes, becomes an important diffusion process.

In the equation it is described by the member which contains a diffusion coefficient D.

A differential equation describing the time variation of metastable concentration is proposed:

$$-\frac{\partial n}{\partial \tau} = D\Delta^2 n + Bn + Cn^2$$

where

n – concentration of $N_2(A^3\sum_u^+)$ metastables in the diode,
τ – afterglow period,
D – diffusion coefficient.

The constant B in the equation contains constants typical for deexcitation processes with the values from the literature:

– mean radiative life–time of free isolated $N_2(A^3\sum_u^+)$ metastables [41 and 158],

- mean concentration of nitrogen atoms during τ considered [24],
- deexcitation rate constant in the collisions with nitrogen atoms [104],
- impurity concentration,
- deexcitation rate constant in the collisions with the impurities [163],
- deexcitation rate coefficient on the electrodes [147], and
- deexcitation rate coefficient on the glass bulb [164].

The solution of the differential equation gave a theoretical curve $n = F(\tau)$ shown in Fig. 24 (with dashed lines), which is in the first approximation in accordance with the experimental one (full lines in the same figure). Beside that, the range of values of initial (saturation) metastable concentration has been obtained

$$n_0 > 1 \cdot 10^{11}\ cm^{-3} ,$$

which is in accordance with the data given in literature, i.e. between 10^{10} and 10^{14} cm^{-3}, obtained by other methods [16, 20, 22, 24, 139, 163 and 165]. The numerical value of the constant k from the equation $\bar{t}_d = k/n$ has also been obtained.

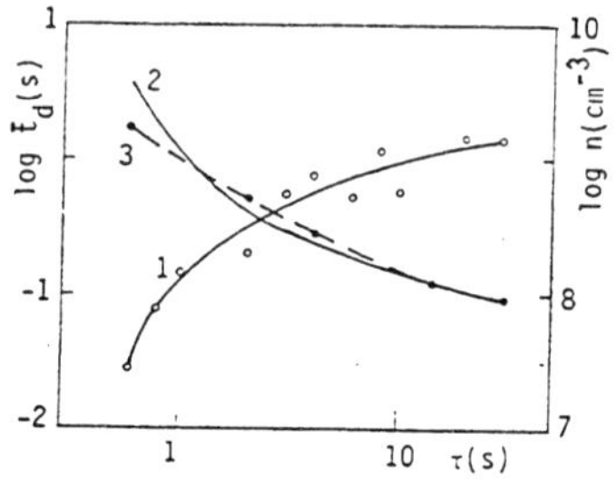

Fig. 24. 1 – $\bar{t}_d = f(\tau)$, 2 – $n = F(\tau)$ – experimental curve, 3 – $n = F(\tau)$ – theoretical curve in the diode (Fe, N_2, 13.3 $mbar$).

In addition to this, the two other significant curves have been obtained, and they are: the half–deexcitation time ($T_{1/2}$), i.e. the time interval in which the concentration of metastables drops to one half of its initial value by deexcitation. The curve, very different from that of the radioactive decay case, increases in time and tends to infinity close to the value of τ, where $\bar{t}_d$ tends toward the saturation constant. This phenomenon is also threated in [143]. The second curve shows the ratio of the deexcitation rate by the self–quenching processes (Cn^2) to the processes independent of n (Bn). It can be seen from the curve that self–quenching processes dominate during deexcitation up to about above 1 s after the cessation of the discharge.

Details on this problem are given in the paper [124].

18. Another view of the memory process in the diode

In the paper [9] three memory curves are given, recorded up to $\tau = 10^3$ s, in Ar with Mo, Cu and ($Ba - Sr - Ca$)–oxide electrodes, respectively. The authors presume that after $\tau = 1$ s there are no metastables in the diode and that memory is a consequence of a temporal change of activity of "emission centres" on the electrode surface. In other words, the role of a phenomenon from the domain of the solid state physics is presumed.

According to the results described in the present paper, this interpretation cannot be accepted for several reasons, the most important ones being:

1. The electrical field applied is several orders of magnitude smaller than needed for autoelectronic emission.
2. The experiments with the diode having mercury electrodes at room temperature have given standard memory curves and the mentioned effect cannot be attributed to the Hg surface.

3. The experiment with the laser light has identified the kind of metastable which is the carrier of the memory.
4. The hypothesis of surface effect is denied by the results of the memory curves recorded on the diode with "two volumes".

Etc.

19. Dependence of discharge formative time on metastable concentration

Discharge formative time in the gas (t_f), besides the characteristics of the diode, considerably depends on the measurement conditions. The measurements on the diode (N_2, 1.3 *mbar*) with the overvoltage of 1.6% in a circuit with a serial resistance of 34 $M\Omega$, have shown the formative time of subnormal glow discharge of 40 *min*. During this time, the increase of the effective value of the current from 0.1 up to 1 μA has been recorded. The discharge has been selfsustaining during the whole of this period and it consisted of the pulses, which appeared aperiodically, always starting with the current $i = 0$. However, the mean time interval between the pulses has been shortened until the current has reached the saturation value i_s.

Another diode (N_2, 1.3 *mbar*) operated at 1% overvoltage in a circuit with a serial resistance of 5.5 $M\Omega$ showed a steady rise of the current from 0 to 5 μA during about 150 seconds, however, in the first 10 seconds there were sporadic current pulses that can be observed on a scope. These sporadic pulses had the base line at zero current and after that initial period become periodic riding on a steadily rising DC base till a saturation level is reached that marked the end of the formative process.

These experiments suggest the increase of the metastable concentration after each pulse and, with it, the probability of appearance of new pulses, and the contribution of metastables to the electrical current increase.

These formation processes are observable only in diodes that rested at least 48 hours between measurements. At shorter resting times the t_f values were of the order of milliseconds, indicating the presence of great concentration of metastables that survived from previous discharges and are contributing to the short discharge formation.

This hypothesis is confirmed by direct measurements of the dependence of t_f on τ in a diode (N_2, 6.7 *mbar*). The formative times t_f have been measured at an overvoltage of 20% with a serial resistance of 24 $M\Omega$. The voltage has been applied at different values of τ, from 10^0 to 10^5 *s*, with t_f ranging from 10^0 to 10^2 *s*. Thus, the relation between t_f and the metastable concentration has been confirmed. Other investigations on t_f are presented in papers [135, 149, 160 and 161].

20. Dependence of static breakdown voltage on metastable concentration

On the diode (Ar, 7 *mbar*) the static breakdown voltage U_s has been measured by initiation of breakdown with short–lasting illumination of the diode by UV light. The diode which has not been plugged in for 48 hours, showed a certain value of U_s. After breakdown a current flow of order 0.1 μA through the diode was sustained and U_s has been measured every five minutes, after the short interruption of the current. Until reaching the saturation value n_0, U_s decreased for about 8%. This phenomenon has been found in many gases, but always at pressures below 10 *mbar*.

Breakdown voltage has been investigated under different conditions and the results are presented in [128, 129, 131, 136, 138, 140 - 142, 144, 148 and 152].

21. Measurements of metastables by a secondary electron emission chamber in gas flow

A detection chamber for metastables operating similar to an ionization chamber was placed in a tubular system for measurements in gas flow that was assembled from glass tubes of an inner diameter of about 65 mm. The flow rate and gas pressure could be regulated in the range from 0.5 to 1 000 $mbar$ by a system of valves and pumps. Upstream from the detection chamber an auxiliary diode was operated at glow discharge creating photons, electrons, ions and metastables at discharge current i_a. The repositionable metastable detection chamber was placed at a sufficient distance L so that metallic grids or mashes and light mazes could be placed between it and the glow discharge source. Between this elements voltages, i.e. electric fields could be established stopping the transport of charged particles and photons and enabling only neutral metastables to reach the detection chamber with the gas flow and by diffusion.

The detection chamber consists of two insulated coaxial 150 mm long aluminium cylinders. The inner diameter of large cylinder is 54 mm, and the gap 7 mm. Thus, the field is near-homogeneous. The gas flow under pressure of 0.5 $mbar$ passed between the two cylinders. At current $i_a = 2\ mA\ DC$ and $L = 30\ cm$ the $U - I$ characteristic of the detector has been recorded. The plateau extended from 15 V up to 450 V with a slope of 15%/100 V with the current in the detector approximately 1 nA.

At the operating voltage in the detector of about 50 V, the dependence of the detector current on i_a has been obtained. The results show that the source of the current in the detector is exclusively secondary electrons released there by neutral metastables created in the auxiliary discharge and that there is no charge multiplication in the gas within the chamber.

First measurements in He under pressure of 0.5 $mbar$, by means of this detector have shown that the generation rate of metastables reaches its maximum at $i_a \approx 2\ mA\ DC$ being higher than $10^{10}\ s^{-1}$.

22. Conclusion

The phenomenon of "memory" after electrical discharges in gases, which is manifested through the increase of the time delay of electrical breakdown ($\bar{t}_d$) with the increase of time passing from the instant of the cessation of previous discharge (τ), has been investigated in a large number of gases and vapours. Magnitude $\bar{t}_d$, as well as "memory curves", $\bar{t}_d = f(\tau)$, have been investigated under various conditions and it has been concluded that the increase of $\bar{t}_d$ with τ, i.e. the memory process, in many cases lasts even 10^5 s and more. This phenomenon is explained by the presence of long-living metastable states in the gas and by the process of their deexcitation, i.e. the decrease of the concentration (n) of neutral metastable states of atoms and/or molecules in the gas during the time τ.

In some cases the kind of metastables has been identified by the laser spectroscopic analysis, and in some other it has been theoretically assumed. From the experimental curves $\bar{t}_d = f(\tau)$, by the formula $\bar{t}_d = k/n$, the curve $n = F(\tau)$ has been obtained. Modeling deexcitation processes during τ and solving a suitable differential equation, the theoretical curve $n = F(\tau)$ has been obtained, which is in accordance with experimental data. By solving this equation, numerical data about the process described have been obtained. The curves $T_{1/2} = \varphi(\tau)$ have been plotted, where $T_{1/2}$ represents the half-deexcitation time, which increases in time and tends to infinity, nearing the instant τ, where

$\bar{t}_d$ reaches saturation. The curve $Z = \Phi(\tau)$ has also been obtained, where Z represents the ratio between the deexcitation rate by self–quenching processes (metastable–metastable collisions) and the deexcitation rate by other processes. It has been shown that self–quenching processes dominate in the first second of the afterglow period τ.

Some other magnitudes and phenomena have also been investigated, which, beside $\bar{t}_d$, depend on n, and they are: discharge formative time, static breakdown voltage, contribution of metastables to the intensity of the current of self–sustaining discharge, etc.

A chamber for detection of metastables in the flowing gas under pressure of the order of *mbar* has been made operating by measuring the current of electrons formed by secondary emission caused by metastables, and in which there is no multiplication of charged particles.

Thus, in addition to $\bar{t}_d$ measurements and the measurements of the contribution to the stationary current of the discharge, this chamber represent the third method for detection of metastable states in experiments described, confirming their contribution to discharge processes in gases.

Acknowledgment

The author would like to thank the Ministry of Science and Technology of the Republic of Serbia who contributed significantly to the realization of this investigation by financial support.

I would also like to express my thanks to the Faculty of Electronic Engineering in Niš, specially to the dean the Faculty Dr Ninoslav Stojadinović, for his understanding and considerable support in my work.

My special thanks go to Dr Djura Krmpotić, professor of the Faculty of Physics in Belgrade for his great help while I was working on this paper as well as for his unselfish and considerable support and help through all these forty one years of my work.

References

List of abbreviations

SPIG	– Symp. of the Physics of Ionized Gases
ESCAMPIG	– Europhysics Sectional Conference on the Atomic and Molecular Physics of Ionized Gases
ETAN	– Conf. of Electronics, Telecommunications, Automation and Nuclear Engineering
SD	– Zbornik referatov simpozija o elektronskih sestavnih delih in materialih
JUKEM	– Jugoslovenski simpozijum o mjerenjima i mjernoj opremi
SUSEJ	– Stručni seminar studenata elektrotehnike Jugoslavije

[1] Raether H 1964 *Electron Avalanches and Breakdown in Gases* (London: Butterworths)
[2] Jaumann G 1895 *Annalen der Chemie und Physik (Wiedemann's Annalen)* **55** 656
[3] Warburg E 1896 *Annalen der Chemie und Physik (Wiedemann's Annalen)* **59 9** 1
[4] Warburg E 1897 *Annalen der Chemie und Physik (Wiedemann's Annalen)* **62 11** 385
[5] Zuber K 1925 *Annalen der Physik* **76** 231
[6] von Laue M 1925 *Annalen der Physik* **76** 261
[7] Maller V N and Naidu M S 1981 *Advances in high voltage insulation and arc interruption in SF_6 and vacuum* (Oxford, New York: Pergamon Press)
[8] Büge M 1928 *Archiv für Elektrotechnik* **19** 480
[9] Shatokhin V L, Milovanova R A and Vasilevsky D V 1991 *Proc. XXth Int. Conf. on Phenomena in Ionized Gases* (Pisa) *Contrib. Papers* **3** 681–2
[10] Bošan Dj A 1956 *Industrial R. and D. Report* TD–P02 Nov–Dec 1956 Comp. Elektronska industrija, former: Zavodi RR (Niš)

[11] Bošan Dj A 1962 *Abstracts of 1th Symp. on Phys. of Ionized Gases* (Beograd) F-5 Published by the Federal Nuclear Energy Commission of Yugoslavia (Beograd)
[12] Bošan Dj A 1978 *Proc. 5th Int. Conf. on Gas Discharges* (Liverpool) IEE Conf. Publ. no 165 273–6 Published by the Institution of Electrical Engineers (GB)
[13] Mijović B J 1985 *Doct. Thesis* Fac. of Electronic Eng. Univ. of Niš
[14] Pejović M, Marković V and Milosavljević Č 1991 *J. Phys. D: Appl. Phys.* **24** 677–80
[15] Bošan Dj A and Jovanović T V 1992 *J. Phys. D: Appl. Phys.* **25** 436–41
[16] Meek J M and Craggs J D 1978 *Electrical Breakdown of Gases* (New York: Wiley)
[17] Bošan Dj A, Pejović M M and Vujović M V 1980 *Acta Phys. Acad. Sci. Hungaricae* **49** **1–3** 23–38
[18] Bošan Dj A, Zlatić V Ž and Mijović B J 1990 *J. Phys. D: Appl. Phys.* **23** 304–7
[19] Van Brunt R J and Kulkarni S V 1990 *Physical Review* **A42** **8** 4908–32
[20] Smirnov B M 1982 *Vozbuzdennie Atomi* (Moskva: Energoizdat)
[21] Bošan Dj A and Lukić M M 1980 *Proc. 6th Int. Conf. on Gas Discharges and their Applications* (Edinburgh) IEE Conf. Publ. no 189 part **2** 46–7 Published by the Institution of Electrical Engineers (GB)
[22] Smirnov B M 1974 *Ioni i vozbuzdennie Atomi v Plazme* (Moskva: Atomizdat)
[23] Radcig A A and Smirnov B M 1986 *Parametri Atomov i Atomnih Ionov* (Moskva: Energoatomizdat)
[24] Cernogora G, Hochard L, Touzeau M and Matos Ferreira C 1981 *J. Phys. B: At. Mol. Phys.* **14** 2977–87
[25] Pejović M and Bošan Dj A 1980 *Proc. SD-80* 353–9
[26] Bošan Dj A and Simonović D 1984 *J. Phys. D: Appl. Phys.* **17** L1–L4
[27] Pejović M M, Bošan Dj A and Nikolić Z 1982 *J. Phys. D: Appl. Phys.* **15** 867–72
[28] Pejović M M, Mijović B J and Bošan Dj A 1983 *J. Phys. D: Appl. Phys.* **16** 1953–7
[29] Pejović M, Marković V and Mekić S 1991 *J. Phys. D: Appl. Phys.* **24** 779–81
[30] Radović M K and Bošan Dj A 1987 *J. Phys. D: Appl. Phys.* **20** 639–41
[31] Roepcke J and Zahn R J 1990 *Contrib. Plasma Phys.* **30** 511–22
[32] LLewellyn-Jones F 1967 *Ionization Avalanches and Breakdown*, Science Paperbacks and Methuen and Co. LTD
[33] Chunikhin A and Zhavoronkov M 1989 *High voltage switchgear – analysis and design* (Moscow: Mir Publ.)
[34] Bošan Dj A, Mijović B and Pejović M 1981 *Elektrotehnika* (Beograd) **30** **4** 659–60
[35] Kiselev Yu V 1965 *Proc. 7th Int. Conf. on Phenomena in Ionized Gases* (Beograd) 838–43
[36] Braunbek W 1926 *Ztschrf. Phys.* **39** 6
[37] Bolden R C, Hemsworth R S, Shaw M J and Twiddy N D 1969 *Proc. 9th Int. Conf. on Phenomena in Ionized Gases* (Bucharest) *Contrib. Papers* 27
[38] Martišovitš V and Košinár I 1973 *Proc. 11th Int. Conf. on Phenomena in Ionized Gases* (Prague) *Contrib. Papers* 485
[39] Lancaster B M and Nygaard K J 1976 *Proc. 12th Int. Conf. on Phenomena in Ionized Gases* (Eindhoven) *Contrib. Papers* 91
[40] Bošan Dj A 1975/76 *Doct. Thesis* Fac. of Physics Univ. of Beograd
[41] Rayzer Yu P 1987 *Fyzika Gazovogo Razryada* (Moskva: Nauka)
[42] Bošan Dj A and Pejović M 1978 *Proc. SD-78* 147–151
[43] Bošan Dj A and Pejović M M 1979 *J. Phys. D: Appl. Phys.* **12** 1699–702
[44] Pejović M and Bošan Dj 1979 *Proc. XXIIIth ETAN* I.403–I.409
[45] Bošan Dj A 1962 *Industrial R. and D. Report* 19–II–1962 Comp. Elektronska industrija, former: Zavodi RR (Niš)
[46] Bošan Dj A, Nalbani B and Pejović M M 1980 *Proc. XXIVth ETAN* I.269–I.274
[47] Pejović M M, Bošan Dj A and Dimitrijević B 1980 *Proc. XXIVth ETAN* I.359–I.366
[48] Pejović M M and Bošan Dj A 1980 *Proc. Vth ESCAMPIG Contrib. Papers* 133
[49] Bošan Dj A and Pejović M M 1980 *Fizika* (Yugosl.) **12** **1** 371–4
[50] Pejović M M, Bošan Dj A and Nallbani B A 1982 *J. Phys. D: Appl. Phys.* **15** L31–L34
[51] Pejović M M and Bošan Dj A 1982 *Proc. JUKEM-82* 971–80
[52] Bošan Dj A, Radović M K and Jakšić M B 1984 *Proc. XXVIIIth ETAN* II.249–II.254
[53] Marković V and Pejović M 1990 *Proc. XXXIVth ETAN* XIII.125–XIII.130
[54] Pejović M M, Bošan Dj A and Stoiljković B 1980 *Elektrotehnika* (Beograd) **2** 260–2
[55] Pejović M, Miljković M and Milošević S 1982 *Elektrotehnika* (Beograd) **31** 437–40
[56] Pejović M, Mijović B and Nikolić Z 1983 *Elektrotehnika* (Beograd) **32** 471–4
[57] Bošan Dj A and Jakšić M B 1985 *Proc. XXIXth ETAN* X.107–X.114
[58] Pejović M and Ristić D 1987 *Proc. SD-87* 299–303
[59] Bošan Dj A, Radović M K and Janković D S 1987 *Proc. XXXIth ETAN* X.165–X.170
[60] Bošan Dj A 1956 *Industrial R. and D. Report* 23–X–1956 Comp. Elektronska industrija, former: Zavodi RR (Niš)
[61] Christophorou L G (Editor) 1980 *Gaseous Dielectrics II, Proc. IIth Int. Symp. on Gaseous Dielectrics*, Knoxville, Tennessee, USA (New York, Oxford: Pergamon Press)

[62] Christophorou L G (Editor) 1982 *Gaseous Dielectrics III, Proc. IIIth Int. Symp. on Gaseous Dielectrics*, Knoxville, Tennessee, USA (New York, Oxford: Pergamon Press)
[63] Pejović M and Dimitrijević B 1981 *Proc. SD-81* 205–10
[64] Pejović M M and Krstić G 1989 *J. Phys. D: Appl. Phys.* **22** 235–7
[65] Pejović M and Krstić G 1989 *Proc. XXXIIIth ETAN* X.151–X.156
[66] Mijović B, Pejović M and Bošan Dj 1982 *Proc. XXVIth ETAN* I.449–I.456
[67] Pejović M, Mijović B and Bošan Dj 1982 *Proc. SD-82* 139–43
[68] Pejović M M and Dimitrijević B 1982 *J. Phys. D: Appl. Phys.* **15** L87–L90
[69] Pejović M M, Mijović B J and Bošan Dj A 1983 *J. Phys. D: Appl. Phys.* **16** L149–L151
[70] Mijović B, Pejović M and Bošan Dj 1983 *Proc. SD-83* 75–9
[71] Pejović M M, Mijović B J and Bošan Dj A 1984 *J. Phys. D: Appl. Phys.* **17** 351–5
[72] Gänger B 1953 *Der elcktrische Durchschlag von Gasen* (Berlin: Springer)
[73] Pejović M M, Bošan Dj A and Krmpotić Dj M 1981 *Plasma Physik* **21** **3** 211–5
[74] Pejović M, Bošan Dj and Golubović S 1982 *Proc. SD-82* 145–50
[75] Newton R R 1948 *Phys. Rev.* **73** **6** 570
[76] Pejović M and Mijović B 1988 *Zh. Tekh. Fiz.* (USSR) **58** **11** 2124–9
[77] Mekić S 1991 *M. Sc. Thesis* Fac. of Electronic Eng. Univ. of Niš
[78] Eichmeier J 1981 *Moderne Vakuumelektronik* (Berlin: Springer)
[79] Pejović M, Golubović S and Radović M 1984 *Proc. SD-84* 79–83
[80] Pejović M M and Radović M K 1989 *Acta Phys. Hungarica* **65** **1** 19–23
[81] Labat J 1993 *Resonance optoeffects in gas discharges, Proc. XVIth SPIG, Invited Papers*
[82] Filipović R D 1987 *M. Sc. Thesis* Fac. of Electronic Eng. Univ. Niš
[83] Pejović M M (Mentor), Milošević S and Miljković M 1981 *Proc. XXIth SUSEJ* 7 138–145
[84] Bošan Dj A and Radović M 1981 *Proc. SD-81* 201–204
[85] Marković V and Pejović M 1990 *Zbornik radova Fil. Fak.* (Niš), *Serija fizika* 1 45–50
[86] Bošan Dj A, Lukić M and Pejović M 1981 *Proc. XXVth ETAN* I.555–I.561
[87] Bošan Dj A and Pejović M M 1986 *Acta Phys. Hungarica* **59** **3–4** 239–45
[88] Bošan Dj A, Radović M and Pejović M 1983 *Proc. XXVIIth ETAN* V.383–V.390
[89] Radović M and Bošan Dj 1983 *Proc. SD-83* 81–5
[90] Bošan Dj A, Radović M K and Krmpotić Dj M 1986 *J. Phys. D: Appl. Phys.* **19** 2343–9
[91] Bošan Dj, Lukić M and Pejović M 1980 *Proc. Vth ESCAMPIG Abstr. of Inv. Talks and Contrib. Papers* 105
[92] Pejović M M, Stefanović M and Bošan Dj A 1980 *Proc. VIIth Kongr. Mat., Fiz. i Astr. Jugosl. Abstr.* II.35
[93] Pejović M M and Bošan Dj A 1981 *J. Phys. D: Appl. Phys.* **14** 693–8
[94] Pejović M, Bošan Dj A and Nalbani B 1981 *Proc. XXVth ETAN* I.355–I.362
[95] Zlatić V Ž 1990/91 *Doct. Thesis* Fac. of Physics Univ. of Beograd
[96] Bošan Dj A, Zlatić V Ž and Mijović B J 1993 *Proc. XVIth SPIG Contrib. Papers* 242–4
[97] Bošan Dj A and Jovanović T V 1993 *Proc. XVIth SPIG Contrib. Papers* 245–7
[98] Pejović M, Mijović B, Bošan Dj and Golubović S 1983 *Proc. XXVIIth ETAN* V.375–V.382
[99] Stojanović S R, Bošan Dj A and Radović M K 1985 *Proc. XXIXth ETAN* X.121–X.126
[100] Pejović M and Marković V 1992 *J. Phys. D: Appl. Phys.* 1217–20
[101] Pejović M, Krstić G and Andrejić Lj 1984 *Proc. SD-84* 85–9
[102] Pejović M M (Mentor), Miljković M and Milošević S 1981 *Proc. 11th SUSEJ* 7 192–199
[103] Pejović M and Dimitrijević B 1982 *Proc. XXVIth ETAN* I.441–I.448
[104] Young R A and St. John G A 1968 *J. Chem. Phys.* **48** 895–7
[105] Pejović M (Mentor), Miljković M S and Milošević S M 1982 *Proc. 22th SUSEJ* 8 13–20
[106] Pejović M, Bošan Dj and Golubović S 1983 *Proc. SD-83* 69–73
[107] Pejović M (Mentor) and Andrejić Lj 1983 *Naučni podmladak, Tehničke nauke*, Univ. Niš **1–2** 13–21
[108] Pejović M, Golubović S and Dimitrijević B 1984 *Proc. XXVIIIth ETAN* II.229–II.236
[109] Radović M K and Bošan Dj A 1984 *Proc. XXVIIIth ETAN* II.237–II.242
[110] Krstić G 1989 *M. Sc. Thesis* Fac. of Electronic Eng. Univ. of Niš
[111] Pejović M M 1980 *Doct. Thesis* Fac. of Electronic Eng. Univ. of Niš
[112] Radović M K 1985 *Proc. XXIXth ETAN* X.127–X.132
[113] Pejović M M, Bošan Dj A and Golubović S M 1986 *Acta Phys. Hungarica* **59** 273–8
[114] Jovanović T V 1992 *M. Sc. Thesis* Fac. of Electronic Eng. Univ. of Niš
[115] Radović M K 1987 *Doct. Thesis* Fac. of Electronic Eng. Univ. of Niš
[116] Bošan Dj A, Stojanović S R and Radović M K 1986 *Proc. XXXth ETAN* X.77–X.82
[117] Radović M K, Bošan Dj A and Mijović B J 1986 *Proc. XXXth ETAN* X.83–X.89
[118] Folkard M A and Haydon S C 1973 *J. Phys. B: At. Mol. Phys.* **6** 214–26
[119] Haydon S C and Williams O M 1973 *J. Phys. B: At. Mol. Phys.* **6** 227–31
[120] Haydon S C and Williams O M 1976 *J. Phys. D: Appl. Phys.* **9** 523–36
[121] Ernest A D 1992 *Experimental Studies of Metastable Particle Effects in Weakly Ionised Nitrogen and Neon, Doct. Thesis* (Univ. of New England)

[122] Marković V Lj, Pejović M M and Petrović Z Lj 1993 *Proc. XVIth SPIG Contrib. Papers* 251–4
[123] Hays G N and Oskam H J 1973 *J. Chem. Phys.* **59** 1507
[124] Bošan Dj A, Radović M K and Jovanović T V 1994 *J. Phys. D: Appl. Phys.* **27** 286–95
[125] Marković V 1990 *M. Sc. Thesis* Fac. of Electronic Eng. Univ. of Niš
[126] Bošan Dj and Prokić M 1982 *Proc. XXVIth ETAN* I.437–I.440
[127] Simonović D 1983 *M. Sc. Thesis* Fac. of Electronic Eng. Univ. of Niš
[128] Radović M, Pejović M and Živković B 1985 *Proc. SD-85* 157–60
[129] Radović M K, Pejović M M and Bošan Dj A 1986 *Proc. SPIG-86 Contrib. Papers* 439–42
[130] Nalbani B A 1981 *M. Sc. Thesis* Fac. of Electronic Eng. Univ. of Niš
[131] Mijović B and Pejović M 1986 *Proc. XXXth ETAN* X.114–X.120
[132] Mijović B 1981 *M. Sc. Thesis* Fac. of Electronic Eng. Univ. of Niš
[133] Lukić M M 1980 *M. Sc. Thesis* Fac. of Electronic Eng. Univ. of Niš
[134] Sejdiu Š 1991 *M. Sc. Thesis* Fac. of Electronic Eng. Univ. of Niš
[135] Pejović M, Dimitrijević B and Golubović S 1986 *Proc. XXXth ETAN* X.99–X.106
[136] Pejović M and Djordjević V 1986 *Proc. XIIth JUKEM-86* 701–708
[137] Kučani N 1986 *M. Sc. Thesis* Fac. of Electronic Eng. Univ. of Niš
[138] Pejović M and Filipović R 1987 *Proc. SD-87* 277–281
[139] Polak L S, Sergeev P A and Sloveckij D I 1977 *Teplofyzika Vysokih Temperatur* **15** 15–23
[140] Radović M K and Marković V Lj 1987 *Proc. XXXIth ETAN* X.159–X.164
[141] Marković V Lj and Radović M K 1987 *Zbornik radova Fil. Fak. u Nišu Knjiga XI* 131–6
[142] Pejović M, Filipović R and Radović M K 1987 *4 Jugosl. Simp. o primeni fizike Proc.* 289–92
[143] Bošan Dj A and Radović M K 1988 *Proc. 14th SPIG* 397–400
[144] Pejović M and Filipović R 1988 *Proc. XXXIIth ETAN* X.187–X.193
[145] Bošan Dj A and Radović M K 1988 *Proc. XXXIIth ETAN* X.195–X.202
[146] Radović M K 1988 *Zbornik radova Fil. fak. u Nišu Serija fizika i hemija* 1 47–52
[147] Maller V N and Naidu M S 1974 *J. Phys. D: Appl. Phys.* **7** 1406–11
[148] Pejović M M and Filipović R D 1989 *Int. J. Electronics* **67** **2** 251–6
[149] Pejović M M (Mentor) and Andrejić Lj 1982 *Proc. 22th SUSEJ* **8** 6–12
[150] McGowan J Wm (Editor) 1975 *Advances in Chemical Physics* Vol XXVIII The Excited State in Chemical Physics Part 1 (New York London: Wiley)
[151] Radović M K 1989 *Proc. XIXth Int. Conf. on Phenomena in Ionized Gases* (Belgrade) *Contrib. Papers* **4** 1006–7
[152] Pejović M and Filipović R 1989 *Svetotehnika* (USSR) **10** 14–6
[153] Radović M K and Manić V M 1989 *Zbornik radova Fil. fak. u Nišu Serija fizika i hemija* **2** 57–64
[154] Radović M K 1990 *Proc. XVth SPIG Contrib. Papers* 225–226
[155] Bošan Dj A, Zlatić V Ž and Sejdiu Š I 1990 *Zbornik radova Fil. fak. u Nišu Serija fizika* 1 33–8
[156] Radović M K 1990 *Zbornik radova Fil. fak. u Nišu Serija fizika* **1** 39–44
[157] McGowan J Wm (Editor) 1981 *Advances in Chemical Physics* Vol XLV The Excited State in Chemical Physics Part 2 (New York London: Wiley)
[158] Brömer H H and Spieweck F 1967 *Planet Space Sci.* **15** 689–92
[159] Pejović M, Marković V and Milosavljević Č 1990 *Proc. XXXIVth ETAN* XIII 131–137
[160] Pejović M, Marković V and Mekić S 1991 *Proc. XXXVth ETAN* 83–87
[161] Pejović M, Živković J and Ristić G 1993 *Proc. XVIth SPIG Contrib. Papers* 248–250
[162] Marković V Lj, Petrović Z Lj and Pejović M M 1993 *Proc. XVIth SPIG Contrib. Papers* 255–58
[163] Csernysheva N V and Ionih Yu Z 1979 *Optyka i Spektroskopya* **47** 67–72
[164] Janča J 1972 *Scr. Fac. Sci. Nat. UJEP Brunensis Physica* **2** 75–93
[165] Kenty C 1967 *J. Chem. Phys.* **47** **8** 2545–51
[166] Wijsman R A 1949 *Phys. Rew.* **75** **5** 833
[167] Loeb L B 1948 *Rev. of Modern Phys.* **20** **1** 151
[168] Tagashira H 1965 *Proc. 7th Int. Conf. on Phenomena in Ionized Gases Contrib. Papers* 277
[169] Tagashira H 1969 *J. Phys. D: Appl. Phys.* **2** **2** 1193
[170] Granovski V L 1971 *Električeski tok v gaze – ustanovivšisja tok* (Moskva: Nauka)
[171] Werner H J, Kalcher J and Reinsch E A 1984 *J. Chem. Phys.* **81** 2420

Low Temperature and
General Plasmas
M. Milosavljevic and Z. Petrovic

ISBN 1-56072-399-8

RESONANCE OPTO EFFECTS IN GAS DISCHARGES

J.M. Labat
Faculty of Physics, University of Belgrade
P.O.B. 560, Belgrade Yugoslavia

1. INTRODUCTION

Among a numerous effects connected with the resonance absorption of radiation by a gas discharge, only two are known to alter some of the electrical properties of the discharge. These two effects, called optogalvanic and optovoltaic, will be considered and their experimental evidence, physical modelling and possible application discussed.

Optogalvanic effect (OGE) was first described by Penning [1] as a variation of the electrical impedance of a neon discharge when irradiated by a light from another neon discharge. The effect is clearly based on the pumping effect of the radiation on the metastable levels, which are critically involved in the ion-electron production mechanism in the discharge.

Extensive study and practical application of OGE had to wait a considerate period of time until the tunable lasers have been introduced. Actual development of OGE as a spectroscopic tool began with the work of Green et al [2]. Since then, a wide range of applications to OGE has been reported, such as small concentration detection, isotopic analysis, laser calibration and wavelength stabilization, sub Doppler spectroscopy and others. OG technique is now considered as an alternative to standard absorption and fluorescence techniques. Exceptionally good review article on optogalvanic spectroscopy including extensive list of references was published by Barbieri et al [3].

Optovoltaic detection as a new probe for laser spectroscopy was mentioned recently by Brandenberger [4]. Optovoltaic effect was described as an optically induced emf's, rather than changes in impedance. The emf appears on the two auxiliary electrodes that are at some distance from a RF discharge, whenever the neutral gas close to one of the electrode is resonantly illuminated by a laser. The effect seems to be also a promising spectroscopic tool, although its possibilities and physical model are still to be elaborated in more details.

2. OPTOGALVANIC EFFECT

2.1. Experimental evidence

The optogalvanic (OG) effect consists basically in the change of a gas discharge current, whenever the discharge itself is illuminated by a light of the wavelengths corresponding to the plasma atomic or molecular transition. Although the effect was found on the great number of gases, it was predominantly investigated in discharges in noble gases, neon in particular. The reason for this preference is the fact that majority of spectral lines emitted by neutral neon atom is situated in the wavelength range of Rhodanine 6G dye. This dye, due to its high efficiency is the must popular dye used in dye lasers. In terms of modelling, the neon is also convenient because the mentioned transitions terminate on one of the four lowest lying levels. Two of these levels are metastable ($2s_5$ and $2s_3$), one is quasiresonant ($1s_4$) and one ($1s_2$) is resonant level. The upper levels belong to the system of ten $2p$ levels (the Pashen notation of levels will be used in this report). For these reasons the neon will also be considered here, pure and also with the minute impurities of argon.

Typical experimental set-up for optogalvanic measurements is given in Fig.1.

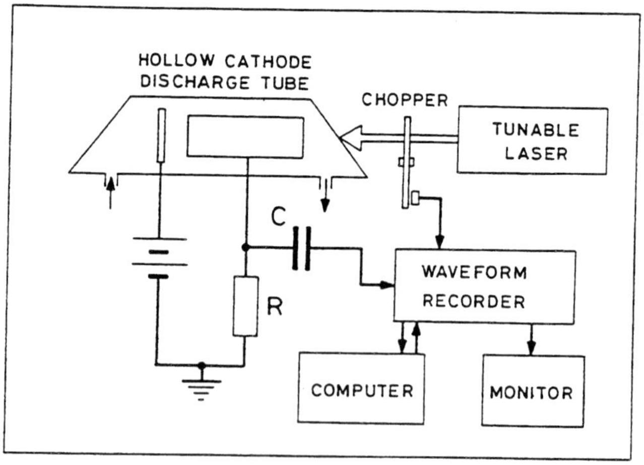

Figure 1.

The plasma of a discharge is illuminated by the beam of tunable dye laser. The optogalvanic signal is detected as a change of the voltage drop on a series discharge resistor. The plasma of a hollow cathode discharge is preferably used because of exceptionally low noise of this type of discharge. The signal is averaged by the lock-in system or the waveform recorder interfaced to computer. Since the transient effect are lost in the first case, the second set-up is regarded as being more versatile. Optogalvanic signals recorded during continuous scanning of the tunable laser in 580 - 650 nm wavelength range are given in Figure 2. The discharge illuminated was driven in pure neon at the pressure of 600 Pa and discharge current was 25 mA.

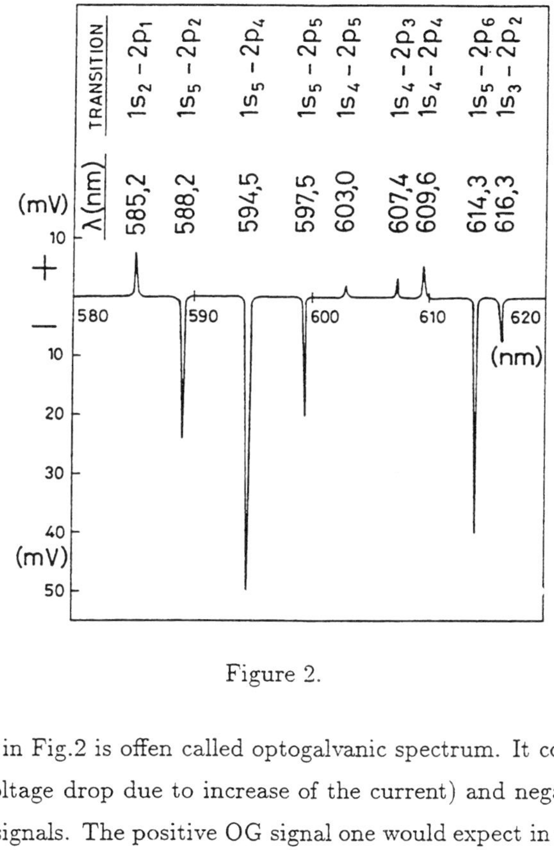

Figure 2.

The record shown in Fig.2 is offen called optogalvanic spectrum. It consists of positive (increase of the voltage drop due to increase of the current) and negative (decrease of the voltage drop) signals. The positive OG signal one would expect in the case the laser energy is used to transfer bound electrons to some levels that are preferably ionized, what consequently leads to additional production of free electron. If the laser irradiation causes depletion of some levels, predominantly metastable, that play principal role in the ionization process, the density of free electrons will be reduced, leading to a negative OG signal. However, this interpretation is oversimplified and cannot be used as a general rule. Collisional coupling of the adjacent levels and radiational transfer may change this basic idea. Furthermore, the ionization process may be also greatly influenced by minor impurities, with atoms of low ionization energy. For example, optical pumping of 614.3 nm neon transition with the metastable lower level ($1s_5$) should always produce negative OG signal. Experiments show that minute quantities of argon will cause the signal to be positive in a relatively large range of neon pressure and discharge current [5].

Therefore, experimental results, particularly concerning the sign of the OG signal are disputable and simplified theories cannot be trusted in the general case.

However, for pure neon, as was the case of OG spectrum in Fig.2, the positive lines are always consequence of pumping the transitions orinating from $1s_4$ resonant lower level, while negative signals correspond to transitions with $1s_5$ metastable lower level.

2.2. Origin of the OG signal

Referring to figure 1. the discharge circuit itself consists of the source, discharge tube and series resistor.

The voltage drop (U_R) on the discharge series resistor (R) will be changed when the discharge current (I) is changed due to alternations in the plasma conductivity (w). The optogalvanic signal is actually a change in the voltage drop ΔU_R that can be expressed as [6]

$$\Delta U_R = I \frac{R \cdot R_d^2}{R + R_d} \Delta w \tag{1}$$

where $R_d = dU_d/dt$ is the equivalent discharge resistance. The change in conductivity is a function of $T_e, n_e, \Delta T_e$ and Δn_e. In the case of low density plasma the relative change of electron density is much larger than the relative change of electron temperature, hence the optogalvanic signal U_{OG} is equal to

$$U_{OG} \equiv \Delta U_R = I \frac{R \cdot R_d^2}{R + R_d} \cdot \Delta n_e \tag{2}$$

Change of the electron temperature that also contributes to optogalvanic effect was found to have for a few orders of magnitude smaller influence on the OG signal [7,8], for the range of parameters met in this experiment. It is therefore justified to that the OG signal as being consequence of the change of electron density only.

Typical experiment OG signals obtained using chopped laser light are presented in fig. 3.

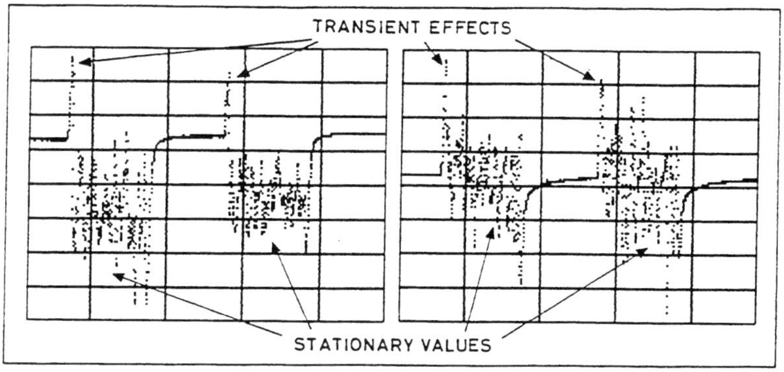

Figure 3.

The noisy parts of the traces are actual OG signals. The stationary values were obtained after elimination of the noise by the standard averaging methods. These values are found to be directly proportional to the laser intensity up to 50 mW. This indicates that just a weak absorption occurs [9] what enables certain simplification of the theoretical model. In general, perturbation of the discharge characteristics induced by the laser radiation is small in most cases, so that the OG signal is directly proportional to the number of photons absorbed [10].

2.3. Modelling of the optogalvanic effect in neon

Usual way to model the OG effect is to write down a kinetic equations including those levels whose population is greatly perturbed by the laser radiation and the equation for electron (ion) density [10]. An attempt to obtain qualitative as well as quantitative comparison with experimental data, requires knowledge of the plasma parameters, namely, temperature and density of electrons, and also dynamical plasma resistance. This kind of model was successfully applied to neon positive column OGE [11].

Modelling of the OGE in neon that will be presented here is based on the similar kinetic approach, that uses the experimental values of the electron temperature and density measured in a central, field-free region of the hollow cathode discharge and calculated stationary levels populations. Special attention is paid to include all the relevant collisional and radiative processes and also the terms involving impurities. Model includes six levels (ground level included): five 1s levels, while 10 2p levels lumped together are treated as one. Influence of collisional and radiative processes to the population of each of these levels in described by a kinetic equation. Values of electron temperature and density and also the temperature and pressure of the gas (neon) are parameters of the system. Processes included are given in figure 4. [12].

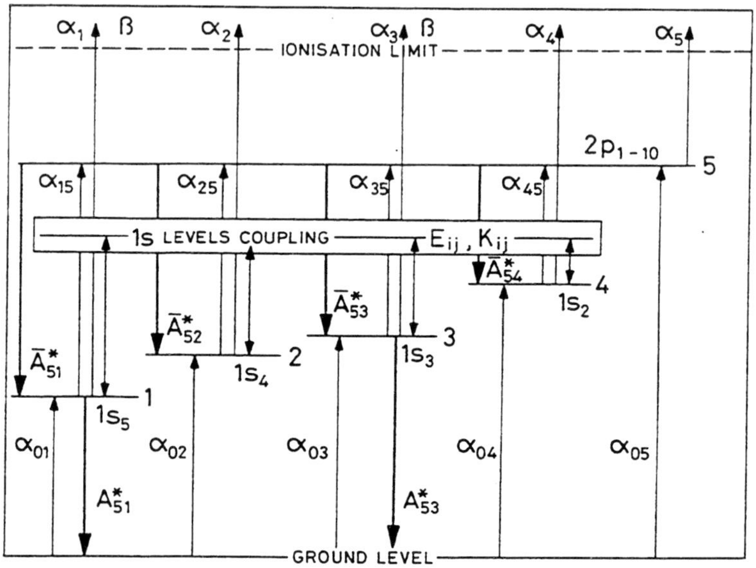

Figure 4.

System of equation for stationary case $\frac{dn_i}{dt} = 0$ and no laser radiation is of the form:

$$\frac{dn_i}{dt} = \sum_{j=1}^{5} n_j a_{ij} + b_i; \qquad i = 1, \ldots, 5 \tag{3}$$

where coefficients of the system are as follows:

$$a_{ii} = -[n_e(\alpha_i + \alpha_{i5} + E_i) + (n_1 + n_3)\beta_i + n_0 K_i + \frac{Da}{n_0 \Lambda^2} + A_i^*]$$

$$a_{ij} = n_e E_{ij} + n_0 K_{ji} \qquad\qquad a_{55} = -\overline{A^*} - n_e \alpha_5$$

$$a_{i5} = \overline{A^*}_i \qquad\qquad b_i = n_0 n_e \alpha_{0i} \tag{4}$$

$$a_{5i} = n_e \alpha_{i5} \qquad\qquad b_5 = 10 n_0 n_e \alpha_{05}$$

The other quantaties are: n_i, n_e – populations of atoms excited to level i and electrons, α_i, α_{ij} – rate coefficients for collisional ionization of level i and excitation from i to j by electrons, β_i – coefficient of ionization in collisions of metastables, E_{ij}, K_{ij} coefficients of excitation transfer by electrons and neutrals and A_i^* and $\overline{A_i^*}$ are effective radiative transition probabilities from level i and from level 5 to level i (averaged over all 2p levels), $\overline{A^*}$ is total effective radiative deexcitation of level 5 ($\overline{A^*} = \sum_{i=1}^{4} \overline{A_i^*}$) and D_a is the coefficient of ambipolar diffusion. The system of five nonlinear algebraic equation has been solved by an iterative method. The values of coefficients for neon were taken from [13] [14] and [15].

Presence of impurities can be included in the model by introduction a term $n_{Ar}w_{pi}$ where n_{Ar} is the density of argon atoms and w_{pi} is the rate coefficient for Penning ionization [5]. The argon was chosen because its ion has three $4s^4P$ levels that are energetically close to 1s levels of neon, enabling very efficient Penning ionization. By inclusion of this effect one obtains population densities that are slightly lower than in the case of pure neon. As expected, densities of $1s_5$, $1s_4$ and $1s_2$ atoms are particularly reduced at higher pressures.

Further step is modelling of the OG effect on the basis of calculated population densities of the five levels. Since the OG signal depends on the change of electron density, an additional equation is included describing the rate of change of the electron concentration. Resonance pumping by the laser radiation will perturb directly only the upper and lower levels of the transition. Additional term of the form $\sigma I(n_i - n_j)$, is included in the two corresponding rate equations where σ is photoexcitation cross section, I is the laser beam intensity and n_i and n_j are lower and upper levels of the transition. The electron equation is off the form:

$$\frac{dn_e}{dt} = -n_e^2 B - \frac{Da}{\Lambda^2} n_e + n_e \sum_{i=1}^{4} n_i \alpha_i + n_e n_5 \alpha_5 + n_e \sum_{i=1}^{10} p_i \alpha_5 + \frac{1}{2}(n_1 + n_3)^2 \beta + n_{Ar} \sum_{i=1}^{4} n_i w_p \tag{5}$$

where B is the recombination rate coefficient and p_i is number density of the $2p_i$ level. Meanings of the other values were already introduced. Laser pumping will indirectly produce changer also in densities of other levels and the electron density, denoted by Δn_i and Δn_e, respectively. By subtracting respective equation for the unperturbed case from those including laser perturbation, one obtains the set of equations for transition

$1s_5 - 2p_6(614.3\ \mathrm{nm})$ pumped by the laser:

$$\frac{d}{dt}(\Delta n_5) = \sigma I(n_1 - n_5) - \frac{\Delta n_5}{\Theta_5}$$

$$\frac{d}{dt}(\Delta n_1) = -\sigma I(n_1 - n_5) - \frac{\Delta n_1}{\Theta_1} + \Delta n_5 \overline{A_{51}^*}$$

$$\frac{d}{dt}(\Delta n_2) = -\frac{\Delta n_2}{\Theta_2} + \Delta n_5 \overline{A_{52}^*} + \Delta n_1 P_{12}$$

$$\frac{d}{dt}(\Delta n_3) = -\frac{\Delta n_3}{\Theta_3} + \Delta n_5 \overline{A_{53}^*}$$

$$\frac{d}{dt}(\Delta n_4) = -\frac{\Delta n_4}{\Theta_4} + \Delta n_5 \overline{A_{54}^*} \qquad (6)$$

$$\frac{d}{dt}(\Delta n_e) = -\frac{\Delta n_e}{\Theta_e} + \sum_{i=1}^{5} \Delta n_i \lambda_i$$

where

$$\frac{1}{\Theta_i} = [n_e(\alpha_i + \alpha_{i5} + E_i) + (n_1 + n_3 + n_i)\beta_i + n_0 K_i + \frac{D_a}{n_0 \Lambda^2} + A_i^*]$$

$$\frac{1}{\Theta_5} = \sum_{i=1}^{4} A_{5i}^* + n_0 \sum_{j=1}^{10} H_{5j}$$

$$\frac{1}{\Theta_e} = 2n_e B + \frac{D_a}{\Lambda^2}$$

$$\overline{A_{5i}^*} = A_{5i}^* + \sum_{j=1}^{10} m_j A_{ji}^* \qquad (7)$$

$$P_{12} = n_0 K_{12} + n_e E_{12}$$

$$\lambda_i = n_e \alpha_i + (n_1 + n_3)\beta_i + n_{Ar} w_{pi}$$

$$\lambda_5 = n_e \alpha_5 W$$

$$W = 1 + \sum_{i=1}^{10} m_i$$

2.3.1. Solution of the system for transition $1s_5 - 2p_6(\lambda = 614.3\ \mathrm{nm})$

Optical pumping of $1s_5 - 2p_6$ transition by the laser radiation of 614.3nm wavelength perturbs directly the metastable $1s_5$ lower level (notation:1) and $2p_6$ upper level (notation:5). Since the OG signal in directly proportional to the change of the electron density, the solution of the system (6) is in the form:

$$\Delta n_e = (n_1 - n_5)\sigma I \left[C_{eo} + \sum_{i=1}^{5} C_{ei} e^{-\frac{t}{\Theta_i}} + C_{ee} e^{-\frac{t}{\Theta_e}} \right]. \tag{8}$$

This expression describes temporal dependence of the OG signal during the laser switch-on phase. However, as $t \to \infty$, from (8) follows also a stationary value of Δn_e :

$$\Delta n_e^{st} = (n_1 - n_5)\sigma I C_{eo} = \rho C_{eo}. \tag{9}$$

Since the value

$$\rho = (n_1 - n_5)\sigma I$$

is always positive because of $n_1 \gg n_5$, the sign of the optogalvanic signal depends on the value of the expression for C_{eo} :

$$C_{eo} = \Theta_e[\underbrace{\lambda_5\Theta_5}_{(a)} + \underbrace{\Theta_5 \sum_{i=1}^{4} \lambda_i \theta_i \overline{A_{5i}^*}}_{(b)} - \underbrace{\lambda_1\Theta_1}_{(c)} - \underbrace{\lambda_2\Theta_1\Theta_2 P_{12}(\overline{A_{51}^*}\Theta_5 - 1)}_{(d)}] \tag{10}$$

First (a) and third term (c) describe a contribution to ionization by the lower and upper level directly perturbed by the laser. The second term (b) gives a contribution of other 1s levels that are perturbed by radiative deexcitation of the upper level, while the last term (d) gives a contribution of the neighbouring lower level perturbed by the excitation transfer from the lower level of the transition. By analysing of the expression (10) one may obtain the following conclusion: The effect will be negative if the relative ionization rate of the lower level is higher than that of the higher level of the pumped transition and it will be positive in the opposite case. The relative ionization rate is the quantity $\lambda_i\theta_i$; that is the number of ionization acts per one depopulation.

In a simplified form one can write

$$\Delta n_e^{st} = \rho\Theta_e(\lambda_5\Theta_5^* - \lambda_1\Theta_1^*) \tag{11}$$

where

$$\Theta_5^* = \Theta_5(1 + \sum \frac{\lambda_i}{\lambda_5}\Theta_i \cdot \overline{A_{5i}^*}) \tag{12}$$

and

$$\Theta_1^* = \Theta_1[1 + \frac{\lambda_2}{\lambda_1}\Theta_2 P_{12}(\overline{A}_{51}\Theta_5 - 1)] \tag{13}$$

are the effective relaxation times of upper and lower level directly perturbed.

As a conclusion of this section one can state that, in accordance with the eq. 10, for transition $1s_5 - 2p_6$ and others with the metastable lower level, the OG signal should be always negative in the case of discharge in pure neon.

2.3.2. Solution of the system for transition $1s_4 - 2p_4(\lambda = 609.6$ nm)

Procedure analogous to that used for transition $1s_5 - 2p_6$, leads to a system of equations for $1s_4 - 2p_4$ transition with the resonant lower level. The perturbed levels in this case are 2 and 5, while the level 1 is just indirectly perturbed. However, the metastable level (1) plays also in this case a predominant role in determination of the sign of the OG signal. The reason is strong coupling between levels 2 and 1. For the stationary case an expression for the quantity C_{eo} is identical with eq. 10, except that $\lambda_1 \leftrightarrows \lambda_2$ and $\Theta_1 \leftrightarrows \Theta_2$.

2.3.3. Contribution of processes to the stationary OGE

Contribution of various terms in expressions for C_{eo} eq. 10 for gas pressure of 200 Pa is given in figs. 5a and b versus discharge current for $1s_5 - 2p_6$ transition (fig. 5a)

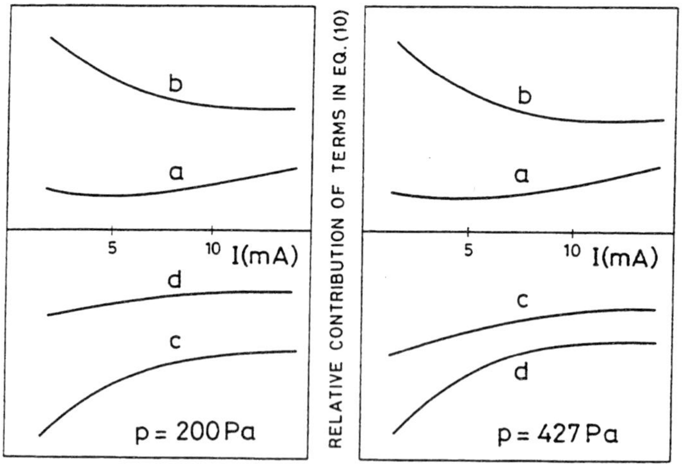

(a) (b)

Figure 5.

it follows that the sign of the effect predominantly depends on the second and third terms (b and c) of expression (10). As shown in section 2.3 the density of excited atom sharply increases for lower discharge currents. Ionization in mutual collisions of metastable atoms becomes a principal electron production mechanism. This process at low discharge current has for two orders of magnitude larger contribution than ionization by electronic collisions.

2.3.4. Transient effects

Temporal dependence of the OG signal or the electron density Δn_e during laser switch-on phase is described by the eq. (8), while that during the switch-off by:

$$\Delta n_e^{off} = \Delta n_e^{st} - \Delta n_e^{on} \tag{14}$$

where Δn_e^{st} is the stationary value of OG signal described by the eqs. (9) or (11).

Supposing the laser pulse lasts much longer than any of the relaxation time (in our experiment 1 ms), in majority of cases one will observe peaks in OG signals during switching on and off phases. Shape of the OG signal is primarily influenced by a relaxation times of particular levels involved. The upper level has shortest relaxation time ($\Theta_5 \simeq 10^{-8} s$) so that its population density will attain a higher value in switching-in much before the other levels densities will be appreciably altered. During this period, the electron production rate will be maximal. Figures 6a and 6b present calculated number densities of various levels involved in OG effect during switch-on phase for $1s_5 - 2p_6$ and $1s_4 - 2p_4$ transition respectively in pure neon.

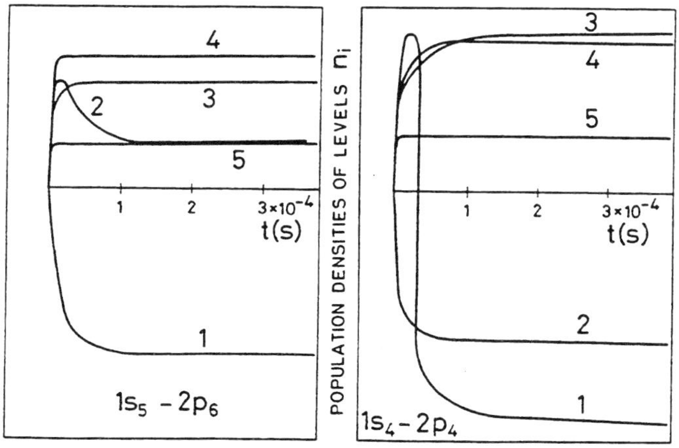

(a) (b)

Figure 6.

Existence of peaks is basically consequence of the ratio of relaxation time of the lower and upper levels, namely it is always: $\Theta_5 \ll \Theta_{1,2}$. Along with this it is also essential that $\Theta_e \ll \Theta_{1,2}$, so that electron density follows the change of upper level and increases before the density n_1, n_2 drops. For the case of $\Theta_e \simeq \Theta_{1,2}$ rise of n_e is too slow and the peak, that corresponds to the maximal ionization, is not present.

2.3.5. Influence of impurities

Presence of minor quantities of argon in the neon discharge due to Penning ioniza-

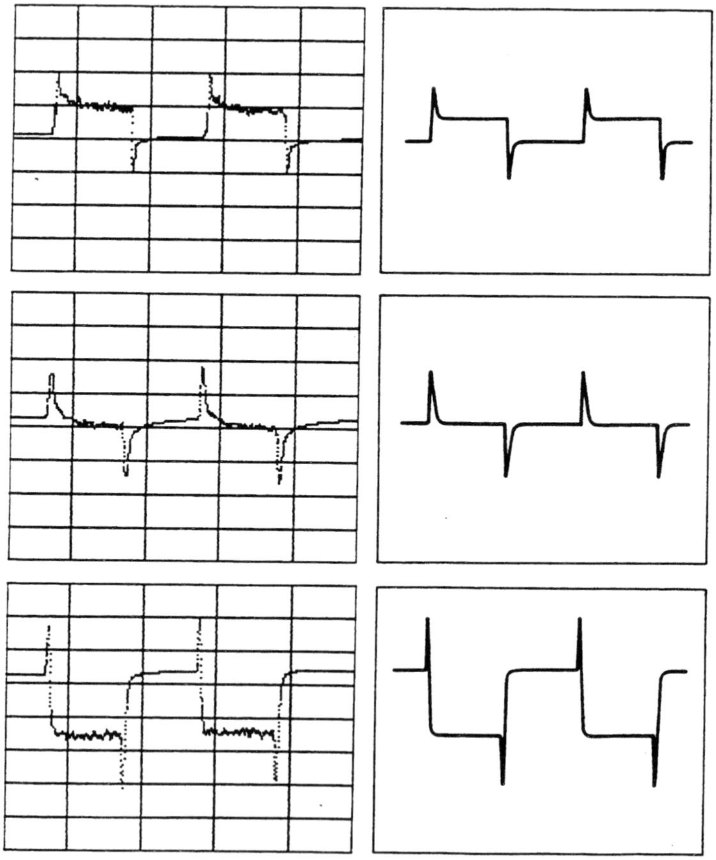

(a) (b)

Figure 7.

tion will change drastically values and sign of the OG signals. The rate coefficients for Penning ionization of lowest neon levels are very large: $w_{p_1} = 1.5 \times 10^{-11} cm^3 s^{-1}$ [16], $w_{p_2} = 6.8 \times 10^{-10} cm^3 s^{-1}$ and $w_{p_4} = 1.4 \times 10^{-9} cm^3 s^{-1}$ [17]). Inclusion of this effect will alter values of OG calculated on the basis of eq. (8). In fig. 7a and b we present experimental (a) and theoretical (b) OG signals for various neon pressures with 0.1% of argon, when transition $1s_4 - 2p_4(609.6nm)$ is pumped by a laser.

Appreciable peaks that were obtained experimentally and theoretically are consequence of the physical conditions in the plasma at relatively low discharge current [5].

Experimental and calculated values of stationary OG signals are presented in figs. 8a and b for various discharge currents and three values of the gas pressure. The transition is $1s_4 - 2p_4$.

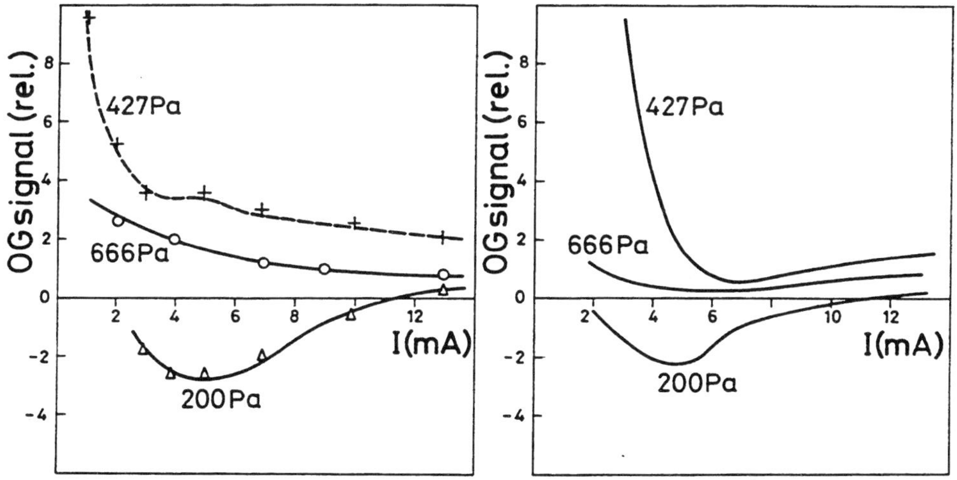

(a) (b)

Figure 8.

Typical features found by the experiment such as negative value for the low pressure range that tends to small positive value at larger currents, is demonstrated also by the model. For higher values of the gas pressure the OG signal becomes positive, what is also confirmed by the experiment. Large values of the signal for lower discharge currents are caused primarily by the high values of the electron temperature under these conditions.

This alters the ratio of the population density of the lower and upper levels of the particular transition and consequently leads to the observed effect.

3. OPTOVOLTAIC EFFECT

3.1. Experimental evidence

Recently, a new effect caused by the resonant absorption of laser radiation by the discharge in krypton, has been described by Brandenberger [4]. RF discharge was induced in a commercial krypton spectral lamp, and the optogalvanic (OV) signal of the order of 0.1V was induced on the lamp electrodes by the laser resonant excitation of 785.5 nm ($1s_3 - 2p_3$) transition in krypton in the region closer to one of the electrodes. The signal was explained by a difference in secondary emission of electrons caused by the metastable atoms bombardment. The difference is a consequence of metastable atoms destruction by the resonant pumping near the one electrode.

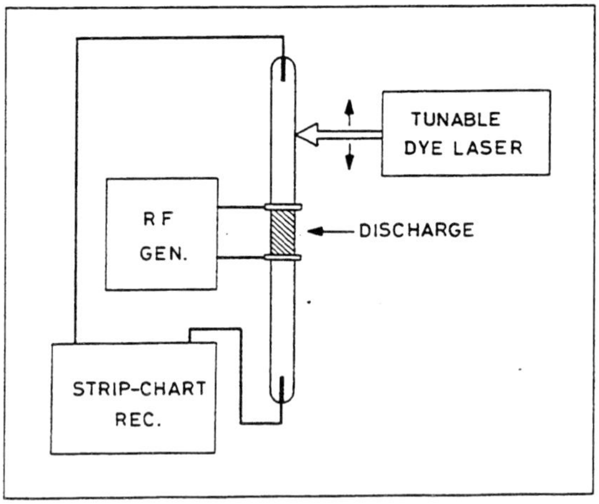

Figure 9.

In order to study the effect in more details we have done a series of experiments in a wider range of the laser wavelengths, using longer glass tubes filled with neon at 20 mbar. The tube of 0.4 cm ID had end-on electrodes 7cm apart. The discharge was induced by a low power (1W) RF generator at 45 MHz in a central part of the tube using two external ring electrodes at a distance of about 1 cm (fig. 9).

The discharge plasma occupied practically only the region between the ring electrodes and there was always a space of unexcited gas between the ring and the tube electrodes. By careful positioning of the ring electrodes it was always possible to obtain equal potentials of the end-on electrodes and the zero reading of the recorder. The unexcited gas close to one of the auxiliary electrode was then illuminated by a CW laser radiation. By scanning continuously the laser wavelength in the range between 580 nm and 620 nm, the spectrum of the OV signal was obtained figure 10 [18].

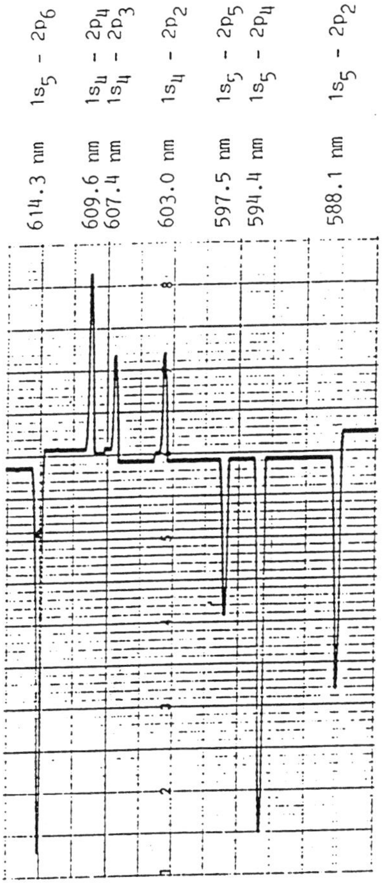

Figure 10.

Individual lines and corresponding transitions were identified as given in the figure. Although unexpected, we have obtained positive as well as negative values of the OV signal similar to OGE. For the laser power of the order of 10 mW, the OV signals were of the order of 1 mV. The negative value of the signal means that the electrode close to the illuminated region is at lower potential in respect to the other. This occurred always for the laser wavelengths corresponding to the transitions with metastable $1s_5$ being the lower level. On the other hand, the signal is positive when the lower level is resonant $1s_4$.

3.2. Discussion

The physical basis of the effect stems from a mechanism of charging the electrodes. The metastable atoms diffuse from the discharge region and finally reach the electrodes. The other particles, charged or excited, do not leave the discharge region, because they rapidly recombine or deexcite. Owing to high energy of the neon metastable $1s_5$ atoms, in interaction with the electrode they cause ejection of the electrons leaving thus the electrode positively charged. This effect was taken into account by Mewe [19] in evaluation of the floating potential of a Langmuir probe.

In the case the RF discharge is positioned in the middle of the interelectrode distance, both electrodes will be charged to the some positive potential. Depletion of metastable atoms by the laser resonance pumping in the region close to one electrode will, consequently, cause this electrode to be less positive, i.e. more negative in respect to the other. Negative OV signals were indeed obtained for all of the transition that have the $1s_5$ metastable lower level (see fig.10). Origin of the positive signal can be found by the following reasoning. Collisions of the ground state neon atoms with the metastables may lead to the excitation transfer of $1s_5$ to the $1s_4$ resonant state. In this way, the population density of atoms excited to the resonant state will build up. This process is very efficient due to very small energy difference between the two levels (0.051 eV). Due to imprisonment of the resonant radiation, density of $1s_4$ atoms may be considerably high. By the resonant pumping of transitions with the resonant lowel level, population of $1s_5$ metastable atoms will increase a a consequence of spontaneous deexcitation of the optically pumped common upper levels. Subsequently, more metastables will reach the electrode, charging it more positive.

Ratios of negative and positive OV signal intensities of transition having the same upper level do not depend on the particular pair of lines chosen and are the same at the particular distance. This was found for 594.5 nm ($1s_5 - 2p_4$) and 609.6 nm ($1s_4 - 2p_4$) line pair, and also for 588.2 nm ($1s_5 - 2p_2$) and 603.0 nm ($1s_4 - 2p_2$) spectral lines. This, most probably, stems from the fact that intensity of the OV signals depends on the population densities of atoms in the two lower levels, that is same in the both cases. Supposing existence of equilibrium, population ratio of the two lowest levels ($1s_5$ and $1s_4$) is determined by:

$$\frac{n_5}{n_4} = \frac{K_{45}}{K_{54}} \tag{15}$$

where quantities n_5 and n_4 are densities of metastable ($1s_5$) and resonant ($1s_4$) state atoms respectively. K_{45} and K_{54} are rate coefficient for collisional excitation transfer between the two levels induced by the ground state atoms. However value of n_5/n_4 ratio calculated on the basis of K_{45} and K_{54} values of Steenhuijsen [13] are far to large in comparison with measured by the laser induced fluorescence method. It has been shown that the ratio does not depend on the gas pressure and is constant along the tube what proves validity of eq. (15). Laser induced fluorescence measurements were done using a pair of spectral lines, such as 588.1 nm ($1s_5 - 2p_2$) and 603.0 nm ($1s_4 - 2p_2$) with the common upper level. Initially, we have pumped 588.1 nm line and observed the fluorescence on 603.0 nm. Further, we have pumped the 603.0 nm line and observed the fluorescence on the other. The same we have repeated also for other pair of lines origination from common upper level $2p_4$ (notation: 2) and having $1s_5$ (notation:5) and $1s_4$ (notation:4) lower levels (594.4 nm and 609.6 nm).

The density ratio follows from:

$$\frac{n_{1s_5}}{n_{1s_4}} = \frac{J_{24}}{J_{25}} \cdot \frac{I_{42}}{I_{52}} \cdot \frac{g_5}{g_4} \cdot \frac{\lambda_{42}^4}{\lambda_{52}^4} \tag{16}$$

where J_{24} and J_{25} are laser intensities at wavelengths corresponding to $1s_5 - 2p_2$ and $1s_4 - 2p_2$ transitions, I_{42} and I_{52} are fluorescence intensities on λ_{42} and λ_{52} spectral line wavelengths, while g_5 and g_4 are statistical weights of $1s_5$ and $1s_4$ levels. From the two mentioned pairs of lines, we have obtained on the basis of the formula (16) the ratio $n_{1s_5}/n_{1s_4} = 1.2 \pm 0.1$ that is for an order of magnitude larger than calculated values from eq. (15). This discrepancy is must probably a consequence of unreliable values for excitation transfer coefficients K_{ij}, that were determined under gas discharge conditions [13].

Further studies of the negative OV signal intensities have shown that proposed model has a correct physical background. Namely, the $1s_5$ level (notation:1) atoms population is reduced by the laser optical pumping for an amount that is proportional to the laser intensity at the particular wavelength (J_{las}), probability of induced transition from $1s_5$ level upwards (B_{52}) to some $2p$ level (notation:2), and the branching ratio (K) due to spontaneous transition downwards:

$$\Delta n_{1s_5} \propto J_{las} \cdot B_{52} \cdot K \tag{17}$$

where

$$K = \frac{\sum A_{2i}(except\ to\ 1s_5)}{\sum A_{2i}}, \qquad (i = 1s_5, 1s_4, 1s_3, 1s_2)$$

and

$$B_{52} = 6.01 \cdot 10^3 \frac{g_5}{g_2} A_{52}(\lambda_{52}(nm))^3.$$

Due to relatively weak radiation field and small population of the upper ($2p$) levels, stimulated emission has been omitted.

As shown in the table, intensities of the OV signal I_{OVS} are indeed proportional to Δn_{1s_5} calculated on the basis of the formula (17). The values have been normalized to 1 for 588.1 nm line.

Transition	λ (nm)	$J_{las} \cdot B_{52} \cdot K$ (rel.units)	I_{OVS}
$1s_5 - 2p_2$	588.1	1	1.0
$1s_5 - 2p_4$	594.4	1.47	1.43
$1s_5 - 2p_5$	597.5	0.32	0.58
$1s_5 - 2p_6$	614.3	0.8	1.56

Table 1.

The found agreement of relative values of the two quantities (for an exeption of 614.3 nm line),justifies the proposed leading role of metastable $1s_5$ atoms in the creation of the OV effect. Due to various channels the $1s_4$ level is coupled to $1s_5$, this simple check is not applicable to positive signals. We conclude by the statement that along with the optogalvanic, the optovoltaic effect can be also used as a simple, high resolution spectroscopic technique with the discharge being at the same time the object and detector. However, it is much to be done until the complete physical basis of the OV effect will be comprehensible.

REFERENCES

1. Penning, F.M.: Physica, 8 (1928) 137
2. Green, R.B., R.A.Keller, G.C.Luther, P.K.Shenck and J.C.Travis; Appl.Phys.Lett. 29 (1976) 727
3. Barbieri, B. and N.Beverini: Rev.Mod.Physics 62 (1990) 603
4. Brandenberger, J.R.; Phys.Rev.A 36 (1987) 76
5. Labat, J.M. and S.Bukvić : J.Phys. D.Appl.Phys. 21 (1988) 1396
6. Bukvić, S. and J.M.Labat: Contr.Papers of SPIG '86 (Šibenik) (1986) 423
7. Maeda, M. Y.Nomiyama and Y.Miyazoe; Opt.Commun. 39 (1981) 64
8. Valentini, H.B.: Opt.Commun. 53 (1985) 313
9. Pinarosa, P., Y.Demers and J.M.Cagne :J.Opt.Soc.Am.B 1 (1984) 704
10. Lawler, J.E.: Phys.Rev.A 22(1980) 1025
11. Doughty,D.K. and J.E.Lawler; Phys.Rev.A 28 (1983) 773
12. Pivalica, S.: Diploma thesis, Univ. of Belgrade (1990) unpubl.
13. Steenhuijsen, L.W.G.: Beitr.Plasmaphys. 21 (1981) 301
14. Drawin, H.W.; Associated Euratom - CEA EUR, CEA FC 383 (1966/67)
15. Wiese, W., M.V.Smith and B.M.Miles; Atomic Transition Probabilities NSRDS - NBS 4, Vol.I (1966) (Washington. D.S.:US. Government Printing Office)
16. Biondy, M.A.: Phys.Rev.88 (1952) 660
17. Mihailov,A., N.Ljepojević (private communications)
18. Labat, J.M. and S.Bukvić; Contr.Papers of XIX ICPIG, J.M.Labat editor.(1989) 366
19. Mewe,R.; Physica 46 (1970) 626

Low Temperature and
General Plasmas
M. Milosavljevic and Z. Petrovic

ISBN 1-56072-399-8

APPLICATION OF FIELD EMISSION ELECTRON SOURCES IN SMITH-PURCELL FEL

Yuri A. Romantosov
Radio Astronomy Institute of Ukrainian Acad. Of Sc.,
4 Krasnoznamennaya Str., Kharkov, Ukraine
(Phone: (7-0572) 44-87-18, Fax: (7-0572) 47-65-06)

I. Smith-Purcell Free-Electron Laser (S-P FEL) concept.

Forty years ago S.J. Smith and E.M. Purcell have found the visible light emission while passing the relativistic electron flow in the vicinity of the diffraction grating [1]. Such phenomenon served as the basis of one of the most interesting millimeter and submillimeter wave-bands free-electron lasers concept. The visit card of S-P FEL is an open resonator containing periodical structure.

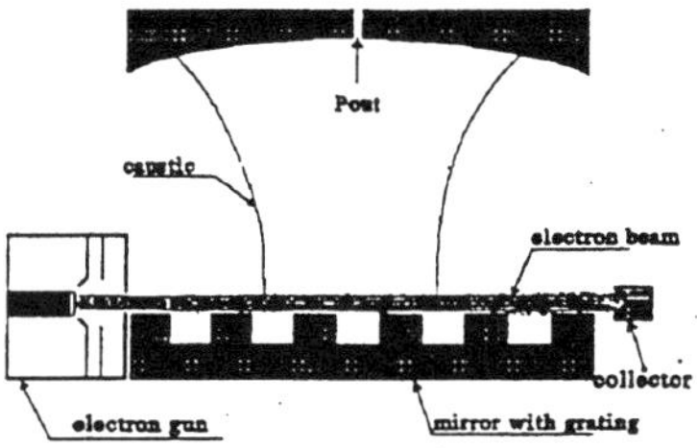

fig 1.

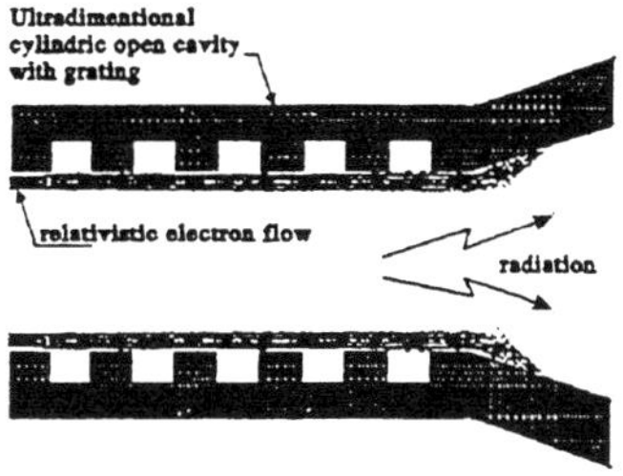

fig.2

Ordinary [2-4] those are quasi-optical oscillatory systems with the diffraction grating on one of the mirrors (fig. 1). In the case of super-high power relativistic S-P FEL [5] electrodynamical system has been fabricated as slightly corrugated cylindrical open cavity whose cross-section dimension sufficiently exceed the wave length of radiation (fig. 2). As an active matter the S-P FEL use thin free-electron flow passing over the periodical structure surface.

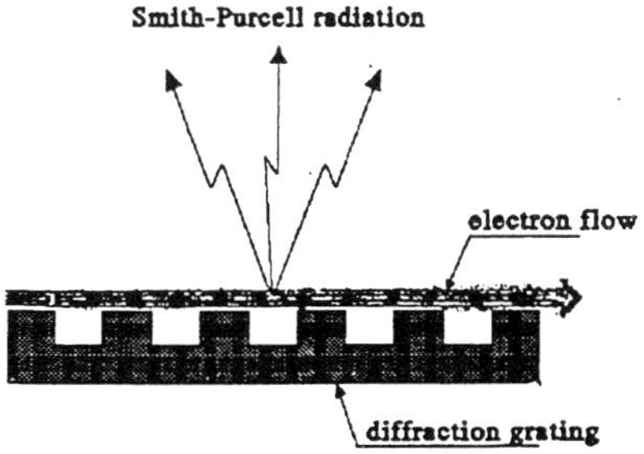

Condition for vertical S - P radiation:

$$1 / \lambda = v_0 / c,$$

where: 1 – period of the grating; λ – wave length
v_0 - electron flow velocity; c - light velocit

fig. 3.

If the conditions of Smith-Purcell radiation (fig. 3) is satisfied here, the electrons excite electromagnetic field both within vicinity of periodical structure - surface field and in the whole volume of open resonator due to the fast space harmonic of S-P radiation. This radiation reflects from upper mirror and falling again on the grating transforms partly into surface harmonics of periodical structure field. If the appropriate condition of synchronism $V_o \geq V_{ph}$ between initial velocity of electron flow V_o and phase velocity V_{ph} of one of the surface field space harmonics are met, the electrons bunching and amplification of the electromagnetic oscillations occurred. Such procedure of feed-back is the characteristic property of S-P FEL by contrast to classical tubes (TWT, BWO, distributed interaction klystrons) where the feed-back is set up through the surface harmonic application. The S-P feed-back allows to excite very effectively the open resonators satisfying the best way the millimeter and short on wave bands. By this time the oscillators of the given type -orotron, ladetron, relativistic radiation generator are known and investigated [2-8]. They have demonstrated successfully the ability to generate coherent and high-stable radiation of the middle and high power in millimeter and submillimeter regions of electromagnetic wave spectrum.

II. Smith-Purcell Amplifiers and Multipliers

Though the potential of S-P FEL, as the explorations shown [9-13], has not been exhausted. The perspective of such-type FEL further development are sufficiently connected with the creation of active devices operating in non autonomic regimes. Above all it means the amplifiers and multipliers, since by now the problem of creation of high-stable sources of millimeter and , especially, submillimeter middle power radiation is not solved completely.

It should be pointed out that the creation of above mentioned devices presents a far more complicated problems compared with autogenerators. So, it could be expected that further efforts in the field of such S-P FEL development should determine the priority orientation both for theoretical and experimental works. The main direction of the efforts is the search of comparatively simple and sufficiently efficient constructions allowing to influence effectively active matter physical characteristics with injected signal.

Now two types of such constructions are known. At first, it is a single-resonator device (fig.4) of reflective type where external signal injected directly into resonator and influence the electron beam through the forced oscillations of electrodynamical system field. Indisputable merit of the single resonator device is its simplicity. However experimental[12] and theoretical [13] investigations shown that the significant amplification of 10÷20 dB can be reached only in strongly regenerative operation mode, and maximum achievable efficiency

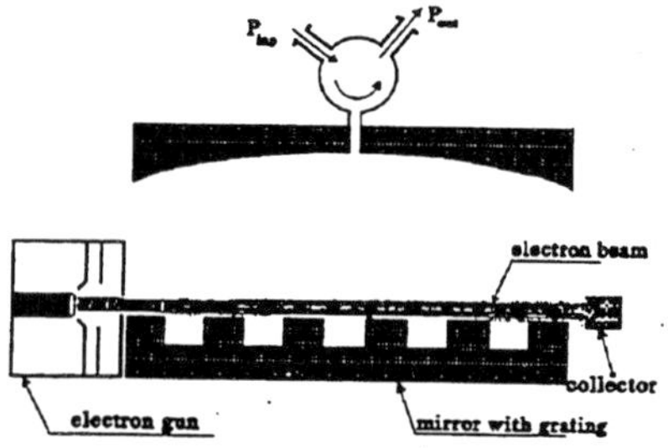

fig 4.

does not surpass the units of percent.

Secondly, numerous studies [10,11] established that the improvement of S-P FEL working characteristics can be reached applying the constructions

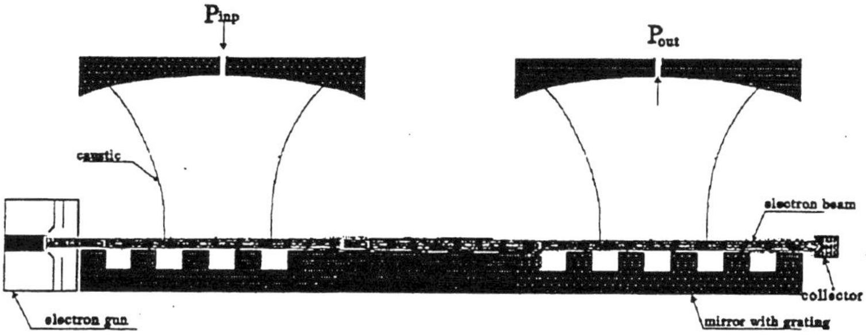

fig. 5

in which the input of signal and its transformation into beam density modulation are spatially separated from output circuit where electron bunches energy is extracted by microwave field. At the expense of this it has been possible to perfect the device operation. As a results, in two-resonator construction (fig 5), similar in type to klystron, the power gain enhance on 10÷20 dB and efficiency may run to 20÷30 percents. In addition the relative amplification band can be expanded to several percents. But because of large spatial extension such S-P FEL design can not be easily realized with three or more resonators.

III. Orofemitron - a new type of S-P FEL

In my point of view, to combine in one design all advantages of S-P FEL mentioned above and at the same time to get rid of their main shortcomings is possible only with the modern technological achievements of vacuum microelectronics. Such originating in the silicon industry are shown now great prospects for novel microwave electronics devices to be created with the applying of low voltage field emission electron sources [14].

Usually they fabricated as gated matrices of conic picks or sharp blades grown on the silicon substratum (fig. 6). The vertical dimensions of working elements are varied from 1 to 15 microns, the current per one pick give the value from 1 to 10 micro ampere,

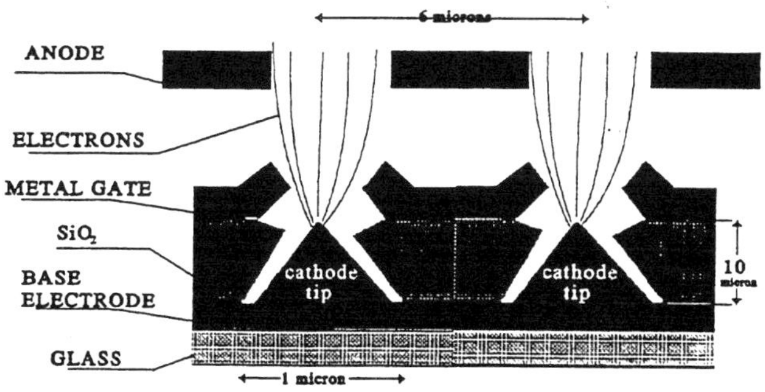

fig. 6.

the pick or blade edge radius being 10÷100 nm and extracting gate voltage of an order 100 V. With a matrix parking density of 10^4 picks per mm^2 such cathodes allow to obtain current of an order 10÷100 mA. It is significance that very small dimensions of such devices active branches make the process of field emission microwave modulation practically inenertionless and ,in addition, the point to emphasize is that matrix field emitter cathodes give of principal possibility to increase the density of thin electron beams. This is very attractive because their effective thickness $a \approx 0.2\lambda * v_0 / c$ may amount the several percent of wave length λ in nonrelativistic case (initial velocity of electrons $v_0 << c$-the light velocity) and thermocathodes can not ensure the current of an order 100 mA for such flow by the reason of technological and physical limitations. These facts in connection with sharp slope of gated field-emitter current-voltage characteristics make such devices rather attractive for creation compact and highly efficient converters of millimeter band signal into modulation parameters of electron beam. The application of such devices can essentially simplify development of S-P FEL amplifiers and multipliers as in this case the total length of the tube, weight and overall dimensions of focusing magnets do not practically increase. The gain and efficiency will be sufficiently high due to preliminary electron bunching by input signal and the possibility to optimize electron-wave interaction in output resonator. In this connection it should be noted that earlier appropriate schemes of amplifiers and multipliers were successfully realized for the centimeter region of electromagnetic wave spectrum. First of all I mean the **femitrons** - devices of klystron type created by W. Dyke [15] and **gigatron** - a new X-band pulse amplifier tube designed for linear accelerator collider application [14]. As for S-P FEL, such method of electron flow density modulation makes their construction more workable in comparison with multiresonator one. One can be convened with these conclusion if to examine the S-P FEL construction named as **orofemitron** [16].

Let us consider the basic elements of the orofemitron construction (fig. 7).

The electrodynamic system consist in miniature open resonator with short focus and plane mirrors. Lower one covered in part by diffraction grating. Sheet electron beam been modulated by input signal passes in the vicinity of the grating. The accelerating voltage (initial beam velocity v_0), period of grating l and natural frequency ω_r of resonator are chosen for S-P vertical radiation condition $\omega_r l / v_0 \cong 2\pi$ and due to that the

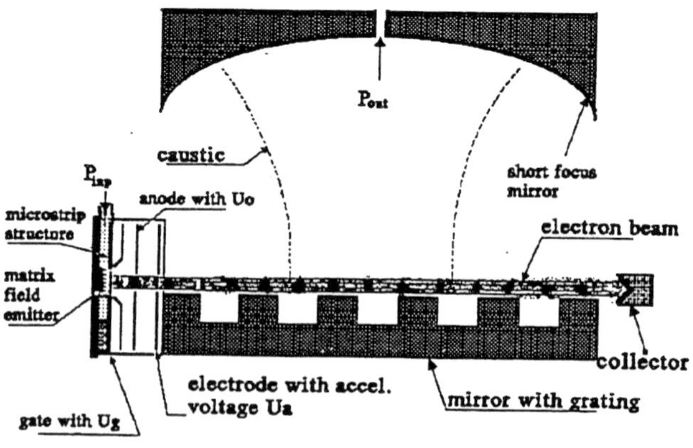

fig. 7.

electromagnetic oscillations are excited. Distributed interaction of electron beam and surface part of resonator field leads to additional bunching of electrons and to effective transformation of their kinetic energy into electromagnetic one. Microwave power is taken out through coupling slope in the center of upper mirror.

The distinctive feature of such type S-P FEL is special electron gun. It is intended both for forming the electron flow given parameters and modulating of the beam current by input signal. The electron source is the gated field emission cathode array (GFECA). Controlling electrode-the gate is under the static positive potential relative to very sharp cathode tips, thereby even with low gate-cathode voltages (~100 V) the field emission from the tips is available. Focusing and accelerating electrodes shape the emitted electrons into thin flow. Its form is preserved on the whole device length due to focusing magnetic field. It should like to note the researches carried out by R.Spindt, P. McIntyre, H. Gray and et.al. They have established [14] that optimum forms of electrodes and static field geometry allow to decrease transverse components of electron velocity and the pulsing of electron flow border.

The gate of field emitters array is coupled with input circuit by high frequency. Significant is that the array is a great reactive load caused by the gate-base capacity. In high frequency the impedance of such load sharply decreases and the input signal transmission to controlling electrode becomes ineffective. The best way out of a difficult condition is to add the inductance in series to the array capacity. In resonant frequency such an oscillatory circuit with the duality factor Q can provide the capacity microwave voltage with Q-times higher amplitude then the input signal one. Depending on practical necessity input circuit with field-emitter cathode array can be made both as a half-wave cavity or lumped-constant resonant circuit on the basis of microstrip and dielectric lines [14].

As it was mentioned above in this devices the microwave modulation of emitted current takes place practically inenertionless. That is why the time-domain dependence and the value of current in the modulator output can be easily obtained with the help of current-voltage characteristic of the emitter array. Such is usually approximated by Fowler-Nordheim formula for the value I of field emission current

$$I_0 \approx A * U^2 / \exp(B / U)$$

with a significant relative slope $(U / I_0) * (dI_0 / dU) = 2 + B / U \approx 10 \div 15$ and high nonlinearity. Here: A and B are constants defined by construction and material of the cathode; U - extracting voltage ~100 V. This points to the possibilities to amplify the signal power and , if it is necessary, to multiply its frequency. These conclusions are illustrated by fig. 8 and fig. 9.

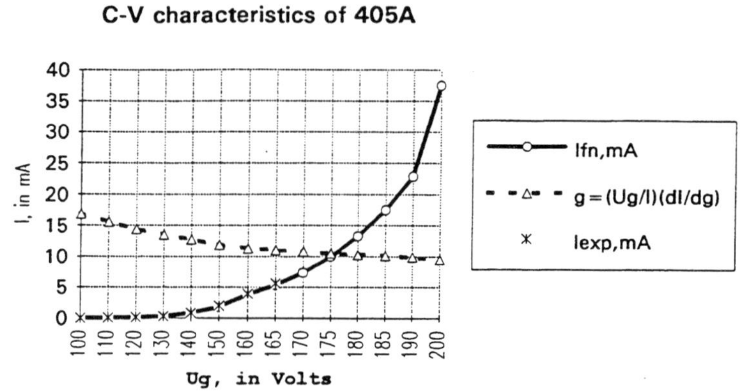

fig. 8.

Fowler-Nordheim formula for the C-V characteristic of the sample 405A [IEEE Trans. on E.D.,vol. 38, N. 3] :

$$I_{FN} = 0.00163 * U_g^2 / \exp(1492 / U_g), \text{ in A.}$$

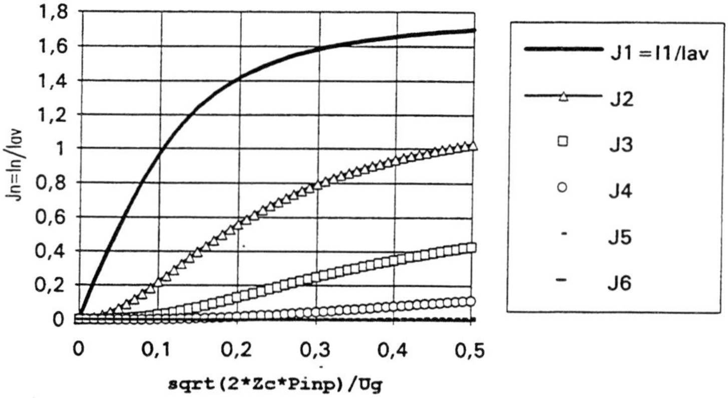

fig. 9. Value of relative amplitude $J_n = I_n/I_{av}$ of current harmonic vs coefficient V_M of gate voltage RF modulation, where I_{av} is average anode current.

Control signal: $U_c = Ug*[1 + V_M * \cos(\omega t)]$;

$$V_M = \sqrt{2 Z_C * P_{inp}} / U_g;$$

$$Z_C \approx 1k\Omega;$$

$$U_g = 160V, \ U_{anode} = 350V$$

The chose of this or that mode is defined by the bias voltage U_g and the value of controlling signal U_c amplitude relation to U_g .

The possibilities mentioned above, have been grounded by W. Dyke groupie in a due time and have made a basis of the femitrons. But 30 years ago the low-voltage field-emitters were not available that is why W. Dyke has used bielectrode modulator construction with extracting voltage about 10 kV. Probably this explain the fact that specific mechanism of the gain enhance connected with the additional acceleration of the pre-formed electron bunches with the voltage being equal to U_a was not used in femitrons. The matter is that in this case the velocity increment needed for electrons effective distributed interaction with resonator field, imparted not in an ordinary way to electron beam with homogeneous density, but to the flow of electron bunches been formed by input signal which voltage $U_c \sim U_g$. Hence, the relation between microwave power being given up to resonator field by the pre-accelerated bunches, and the input signal power spent for the bunches creation, will be the grater, the grater accelerating voltage U_a is compared with the gate voltage U_g . That is why the corresponding increment of the whole device gain should be expected due to such type electrostatic pumping.

IV. Orofemitron theory and numerical experiment results

The validation of proposed S-P FEL advantages have been examined with the help of numerical experiment taking into account characteristics of Stanford field emission cathode array [17] and oscillatory system of miniature orotron (diffraction radiation generator) [18]. The applicable mathematical model depend on joint solution of Maxwell's equations and equations of beam particles motion with the well-known method of eigenfunction expansion [19] adapted for such electrodynamical systems by O. Tret'jakov and D. Vavriv [10]. Using standard presumptions [10]:

1) Form of the Maxwell's Equations solution:

$$\vec{E}\left(\vec{r},t\right)=\vec{\Re}\left(\vec{r}\right)*C(\tau)*\exp[-i\omega t-i\gamma(\tau)]$$

where: $\vec{\Re}\left(\vec{r}\right)$ is the electrical field eigenfunction;

ω - frequency of external signal close to the natural ω_r frequency of the forced mode;

$C(\tau)$ and $\gamma(\tau)$ - the required amplitude and phase of the mode being excited, slow varying functions in comparison with $\exp[-i\omega t]$.

2)

a) The quality factor of loaded resonator $Q >> L\omega/2/V_0 \Rightarrow$ one resonator mode operation and during particle life-time C and γ are const;

b) $V_0 \approx V_{ph}(\omega) \quad \Rightarrow$ one slow-wave interaction of beam and the resonator field.

c) one-dimensional motion of the beam particles $\Rightarrow$the field of synchronous space harmonic can be presented in the following form:

$$\left\langle \Re_y \right\rangle_s \approx f(y / L) * \exp(i * y * \omega / V_{ph}),$$

where $f(y / L)$ is the spatial resonator field distribution along the beam path,

usually $f = (w * \sqrt{\pi / 2})^{-1/2} / \exp((y / L - 0.5)^2 / w^2), w = R / L, R$ – field spot radius on the diffraction grating.

one can obtain the self-consistent model of output resonator field dynamic. The model include the Van der Poll - type equations for field amplitude F and phase γ

$$\frac{dF}{d\tau} = -F + \frac{G_a}{4\pi} \int_0^1 f(\xi) \int_0^{2\pi} J_a(\tau, \varphi - \gamma) * \cos\ [\ \Psi(\xi, \varphi) + n * \varphi\ ] d\varphi\ d\xi;$$

$$\frac{d\gamma}{d\tau} = \Delta - \frac{G_a}{4\pi F} \int_0^1 f(\xi) \int_0^{2\pi} J_a(\tau, \varphi - \gamma) * \sin\ [\ \Psi(\xi, \varphi) + n * \varphi\ ] d\varphi\ d\xi;$$

$$F(0) = F_0 \text{ and } \gamma(0) = \gamma_{in}.$$

and secondly, equation of beam particles motion

$$\frac{d^2\Psi}{d\xi^2} = [\ (1 + \Phi_\phi^{-1} d\Psi / d\xi\ )^2 - \beta_\phi^2\ ]^{3/2} * \{\ \frac{\gamma_o - 1}{2\beta_\phi^2} \Phi_\phi * F(\tau) * f(\xi) * \cos\ [\ \Psi(\xi, \varphi) + n * \varphi\ ] -$$

$$p_{ph}^2 \frac{1}{\pi} \int_0^{2\pi} J_a(\tau, \tilde{\varphi} - \gamma) * \sum_{k=1}^{\infty} \frac{\Gamma_k}{k} \sin(\ k * [\ \Psi(\xi, \tilde{\varphi}\ ) + n * \tilde{\varphi} - \Psi(\xi, \varphi) - n * \varphi\ ]\ )\ d\tilde{\varphi}\ \};$$

$$d\Psi / d\xi = \Phi_s \text{ and } \Psi = 0 \text{ at } \xi = 0 \text{ for } \varphi \in [0, 2\pi].$$

Here: $\tau = t\omega / 2Q$, the "slow" time;

$\xi = y / L$, $(0 < \xi < 1)$, the particle spatial coordinate y normalized by interaction region length L ;

$\varphi = \omega t_0$, the initial phase of particle, must be considered as its' "mark";

$\Psi(\xi, \varphi) = \Phi_\phi \xi - \omega t_L(\xi, \varphi)$, the variation of electron phase on trajectory during its' life time $t_L(\xi, \varphi)$;

$\gamma(\tau)$, the phase of the oscillations excited in the resonator;

$F(\tau) = C(\tau) / E_{rel}$, the field amplitude C normalized by $E_{rel} = U_{rel} / L$;

$U_{rel} = m_e c^2 (\gamma_o - 1) / e$, the value of accelerating beam voltage;

$J_a(\tau, \varphi - \gamma) = I_b(\tau, t) / I_{AV}$, the entrance beam current $I_b(\tau, t)$ normalized by its' average value I_{AV}

$f(\xi)$ – the excited mode spatial distribution in the direction of the beam motion ;

$G_A = \dfrac{2QL^2|I_{AV}|}{\omega_r N_r U_{rel}}$, the parameter of nonlinearity and bifurcations;

$\Delta = 2Q(\omega_r - n\omega)/\omega_r$, the relative detuning of n - harmonic of signal frequency ω and ω_r resonator frequencies;

$\Phi_\Phi = \omega L / v_{ph}$ and $\Phi_0 = \omega L / v_0$ - slow - wave and electron beam transit angles;

$\Phi_S = \Phi_\phi * (v_{ph}/v_o - 1)$, parameter of electron beam - slow wave nonsynchronousity;

$p_{ph} = \omega_p L / v_{ph}$ - space charge parameter; ω_p - beam plasma frequency;

$\beta_\phi = v_{ph} / c$ - relation between slow wave phase velocity v_{ph} and light speed c;

$\gamma_\Phi = 1./\sqrt{(1-\beta_\Phi^2)}$; $\gamma_o = 1./\sqrt{(1-\beta_o^2)}$; $\beta_o = v_0 / c$.

These equations are coupled through amplitude F and normalized entrance current J_a been preliminary modulated by the input signal.

The power gain was calculated with this general formula

$$K = [2Z_c / R_c]*[U_A / U_g]^2 * F^2 / [U_{inp} / U_g]^2,$$

where:

Z_c – coupling impedance of the input circuit;

$R_c = \dfrac{2QL^2}{\omega_r N_r}$ – coupling resistance of output resonator;

U_A – voltage of additional acceleration of electron bunches shaped by input signal with voltage U_{inp};

U_g – the gate voltage;

Note that in this formula amplitude F must be obtained from the self-consistent analysis.

In linear operation mode (input signal voltage U_{inp} << gate voltage U_g) the analytical solution is available and it lead to the direct formula for the maximum gain value

$$K_{max} \approx [Z_C I_{AV} / U_g] \, g_1^2 \, [U_A / U_g] \, [\lambda v_o / c / R][J_O / (1 - J_O)^2],$$

dependent on:

$g_1 = (U_g / I_o(U_g))(dI_o / dU_g) \approx 10 \div 20$ - relative slope of the C- V characteristic of GFEC;

$J_O = I_{AV} / I_{ST.MIN} < 1$ - relation between beam current I_{AV} and minimum value $I_{ST.MIN} \sim 1/Q$ of current of self - excited auto - oscillations;

R – radius of the open resonator field spot on the diffraction grating;

λ - wave length of input signal and beam velocity v_o.

Usually the values of K_{max} are an order of 20÷30 dB.

On figures 10 and 11 one can see some results of calculations, which were carried out for electrodynamical system of miniature orotron [18] using characteristics of Stanford gated field emitter cathode array [17] (C-V characteristic from sample 405A, Ua=2400 V, Ug=160 V, g_1=11.3,Iav=0.1 A, Zc=1430Ω, Jo=0.4÷0.8,λ=4.26 mm,).

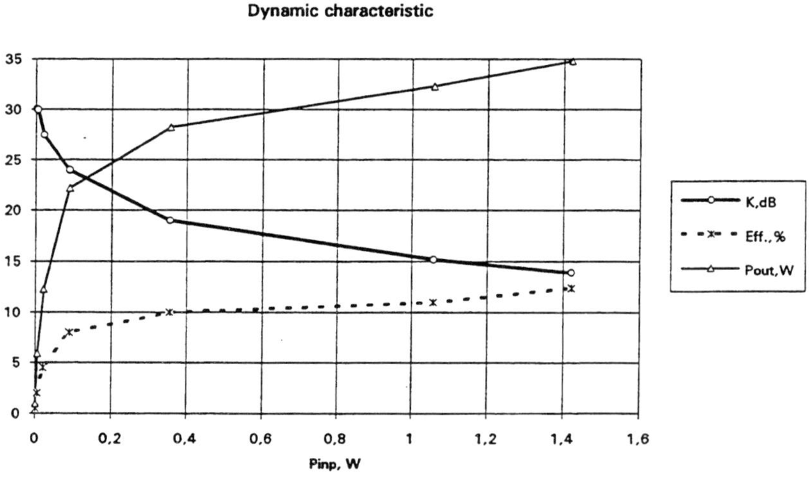

fig.10.

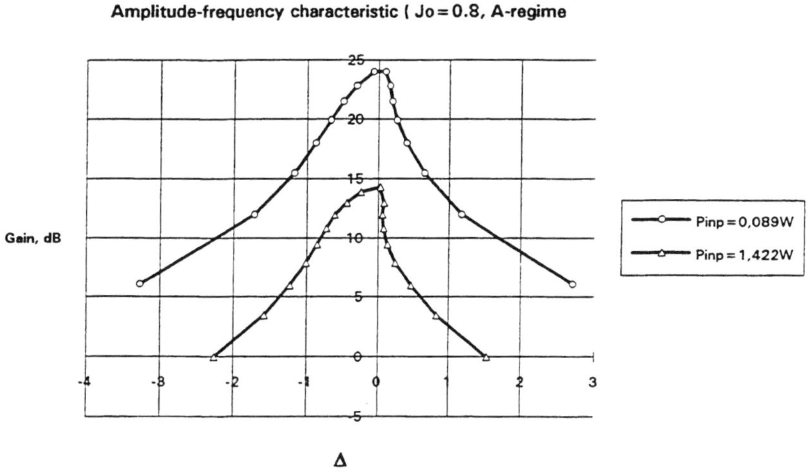

fig.11.

Numerical modeling have shown that power gain may be significantly high for wide range of input power values. And the shape of amplitude-frequency characteristic and its width remains practically without changes.

As for efficiency $\eta = F^2/G_a$ and output power $P_{out} = U_a I_{av} \eta$, the best way for their values increasing is the magnification of electron beam velocity V_0 (or phase velocity V_{ph} of synchronous surface harmonic). This conclusions follows from both, numerical experiments results (fig.12)

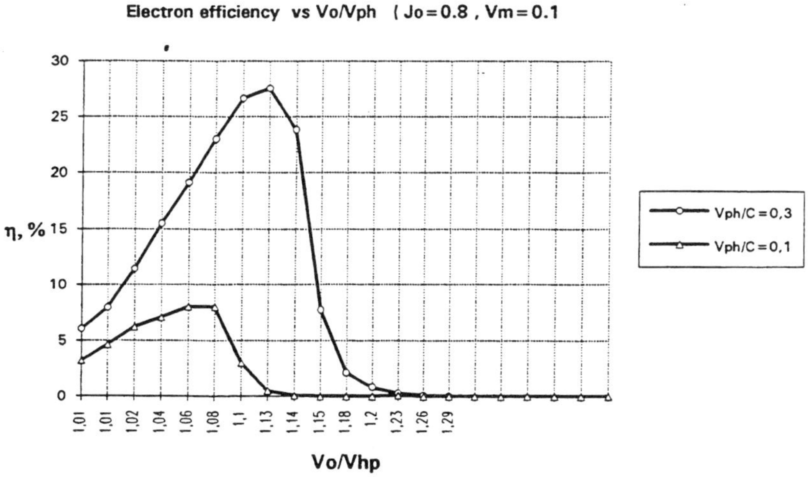

fig.12

and analytical formulas for estimations of maximum achievable efficiency η_{max}

$$\eta_{max} \approx [\, \lambda v_0 /c/R \,]*[\, 1 - 0.25\, (\lambda v_0 /c/R) \,]$$

and output power P_{max}

$$P_{max} \approx 0.65(\lambda / R)^2 * U_A^{7/2} * [1 - 0.25(\lambda v_o / R / c)].$$

where U_A is pointed in kV.

These formulas were obtained subject to the conditions that:

A) the phase velocity of slow space harmonic $v_{ph} \approx v_0 /(1 + 3(\lambda v_0 /c/R)/8)$,

B) the value of field amplitude $F = 1.9\pi^2 [\, (\gamma_o+1)\, \gamma^3_\phi / \gamma^2_o\, (v_{ph}/v_0)^2 / (\Phi_\phi w^{1/2})$),

C) the input signal power is large enough for $Uinp/U_g \geq 0.1$,

where : $R = \sqrt{(\lambda / \pi)\sqrt{d(r - d)}}$ is the radius of the open resonator field spot on the diffraction grating surface; λ - wave length of the input signal; d-the distance between the resonator mirrors is generally taken to be small compared with the curvature radius r of the upper mirror. For miniature short focus open resonators values of r stood at $3 \div 5\ \lambda$ and $d \sim 0.5\ r$, so $R \leq \lambda$..

The conditions mentioned above were obtained through the analytical investigation of relativistic electrons dynamic in resonant systems with distributed interaction [20] and they corresponds to the situation when electron beam -resonator field interaction is most effective. Figure 13 illustrate this process on the phase plane of canonical variables $\eta_e = 1 - [\gamma_e(\xi,\varphi) - 1] / [\gamma_0 - 1]$-relative variation of electron kinetic energy on its trajectory and $\Psi(\xi,\varphi) / 2 - \pi / 4$ – the variation of electron phase on the trajectory-

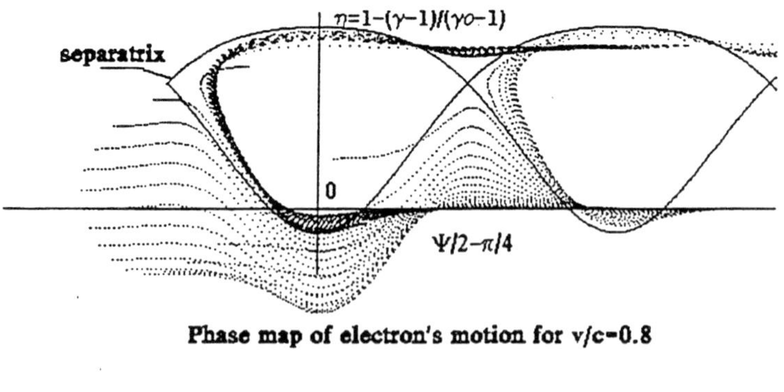

Phase map of electron's motion for v/c=0.8

fig. 13.

Example . Let us obtain the estimation of P_{max} and η_{max}, for hemispherical short focus open resonator with r=3.5 λ, d=0.48 r and λ=4.26 mm applied in experiments with miniature orotron . The choice of U_a = 2.4 kV yields the estimations for maximum CW output power and η_{max} : $P_{max} \cong 24$ W and $\eta_{max} \cong 13\%$.

V. CONCLUSION

The latest achievements in the VM technology and miniature open resonators design used for development of the Smith-Purcell devices allow to create compact high stability amplifiers with CW output power P ≈1÷100 W and efficiency η~10÷40% in the MM and SBMM ranges of electromagnetic wave spectrum. Furthermore, the **orofemitron** type construction has attached interest as synchronous generators , multipliers and mixers for the same frequency bands.

VI. REFERENCES

1. S.J. Smith and E. M. Purcell, Phys. Rev.,**92**, 4, p. 1069 (1953).

2. F.S. Rusin and G.D. Bogomolov, Pis'ma Zh. Eksp. Teor. Fiz. , **4**, p. 236 (1966).

3. V.P. Shestopalov, Diffraktsionnaya elektronika. Kharkov, 1976.

4. K. Mizuno, S. Ono and Y. Shibata, IEEE Trans. ED-20, p. 749, (1973).

5. A.F. Aleksandrov, S.Yu. Galuzo, V.I. Kanavets, Pis'ma v Zh. Tekn. Fiz., **7**, 10, p. 12 (1981).

6. D.M. Vavriv, O.A. Tretyakov and A.A. Shmat'ko, Radioteknika i electronika, **24**, 4, p.812 (1979).

7. R.P. Leavitt, D.E. Wortman and C.A. Morrison, Appl. Phys. Lett., **35**, 5, p. 363 (1979).

8. E.M. Marshall, P.M. Phillips and J.E. Walsh, IEEE Trans. PS - 16, 2, p. 199 (1988).

9. D.M. Vavriv,and O.A. Tretyakov, Radioteknika i electronika, **29**, 9, p.1798 (1984).

10. D.M.Vavriv and O.A. Tretyakov, Theory of the Resonant Amplifiers with distributed interaction of O-type, (Kiev: "Naukova Dumka", 1989).

11. D.M. Vavriv and Yu. A. Romantsov ," Current status and further progress in the Smith-Purcell amplifiers research", Proc. of 21st. European Microwave Conf., **Vol.1** ,pp.824-829 , Stuttgart, Sept., 1991.

12. V.K. Korneenkov and V.S. Miroshnichenko, Izv. Vuzov-Radioelektronika, **26**, 3, p. 93 (1983).

13. D. M. Vavriv, S. A. Sharapov, O.A. Tretyakov, Elektronnaya Tekn., Ser. 1, Elektronika SHF, 4, p. 10 (1988).

14. Special issue on vacuum microelectr. devices, IEEE Trans. on Electron. Devices, **Vol. 36, n. 11, Part II** ,1989

15. W.P. Dyke, " Field emission, a new practical electron source", IRE Trs. on Military Elect., **Vol. 4**,pp. 38-45,Jan.,1960.

16. Yu. A. Romantsov, Orofemitron - the new type of Smith-Purcelll amplifiers", Proceedings of International Conf. on Infrared and Millimeter waves , pp. 26-27, Colchester,.U.K., September 1993.

17. R.E. Neibert, P. Phillips, S. Smith and C. Spindt, IEEE Trans. on ED, **38**, 3, p. 661 (1991).

18. A.V. Nesterenko, A.I. Tsvyk and V.P. Shestopalov, Doklady of USSR Academy of Sci., **277**, 1, p. 84 (1984).

19. P.M. Morse and H. Feshbach, Metods of Teoretical Physics (N.Y.: McGraw-Hill Book Co., 1953).

20. Yu. A. Romantsov, The investigation of relativistic particles dynamics in resonant systems with distributed interaction, Preprint n.47, Institute of Radio Astronomy, Kharkov, 1990.

Low Temperature and
General Plasmas
M. Milosavljevic and Z. Petrovic

ISBN 1-56072-399-8

SPATIAL AND TIME CHARACTERISTICS OF LOW TEMPERATURE RF PLASMAS FOR MATERIALS PROCESSING

S. Djurović, J.R. Roberts and JK. Olthoff**
Institute of Physics, Trg Dositeja Obradovića 4, 21000 Novi Sad, Yugoslavia
*National Institute of Standards and Technology, Gaithersburg, MD 20899, USA

1 INTRODUCTION

Radio frequency (rf) plasmas are utilized in many applications in material processing, such as semiconductors and integrated circuits fabrication and surface modification by etching or coating. Plasma processing has replaced older techniques, such as wet chemistry, because the latter could not provide the necessary characteristics as process demands increased. A good example of this is the reduction of the feature size in semiconductors. The present critical dimension for semiconductor processing is $0.8\ \mu m$ and is anticipated to be $\leq 0.25\ \mu m$ by the year 2000. Semiconductor manufacturing needs ensure that each product wafer going through the same treatment as much as possible. This requires investigations for control plasma diagnostic development. Although there has been a rapid growth in the application of rf discharge plasma in the last decade, knowledge of processes in the rf plasma is still far from satisfactory.

Optical emission spectroscopy have been employed for characterizing the low temperature rf discharge. By detecting light from the electronic transitions of atoms and molecules it is possible to identify and monitor the different processes in the plasma without perturbing the process. Furthermore, this technique could provide i mmediate feedback for automatic adjustment of individual process parameters.

The number of studies investigating the spatially and time varying optical emission from rf plasmas at frequencies near $13.56\ MHz$ are limited. De Rosny et.al. [1] have observed the time dependent emission from a short lived state of Si in a pure SiH_4 rf plasma. Donnelly and coworkers [2] reported the effectes of frequency on the time varying optical emission from chlorine containing rf discharges. Bletzinger and De Joseph [3] investigated the temporally resolved optical emission from a nitrogen plasma as a function of position between the rf electrodes. More recently Tochikubo et. al. [4,5], Kokubo et. al. [6] and Köhler et. al. [7] reported spatially and temporally resolved optical emission data for neutral and ionic transitions that occur in argon plasmas. Tochikubo et. al. [4,5] calculated the relative net excitation rates by deconvoluting the optical emission profile, Kokubo et. al. [6] developed the diagnostic technique of the rf discharge by improving the spatiotemporally resolved opticall emission spectroscopy, while Köhler et. al. [7] performed Boltzmann calculations to derive excited state populations for comparison with the time modulated emission intensity. Petrović et. al. [8] space and time resolved emission spectroscopy applied to obtained the information on the kinetic of rf discharge. Recent modeling of low pressure rf glow discharges has begun to predict many of the time varying aspects of high frequency rf plasmas [9 – 13]. However, the presently available spatially and time resolved optical emission data is insufficient to fully validate the conclusions derived by plasma modeling, either because the data do not cover an adequate range of experimental conditions or because the experimental conditions re not adequately defined. Additionally, the experimental data presently in the literature consist only of relative emission intensities, except Tochikubo and Makabe [14], which do not allow for comparison with calculations of absolute excitation rates or absolute population densities.

The emphasis of the present experimental work is therefore to measure spatially resolved optical emission intensities and the absolute time and spatially resolved optical emission intensities from "well defined" argon plasmas over a wide range of pressures ($6.7\ to\ 133\ Pa$) and applied rf voltages ($75\ to\ 200\ V$). All experiments have been performed on a GEC RF Reference Cell [15]. This is a parallel plate discharge chamber with $102\ mm$ diameter electrodes separated by $25\ mm$. The top electrode is grounded and the bottom electrode is powered by a $13.56\ MHz$ rf power supply isolated with a blocking capacitor. Results of these measurements are presented for both an atomic transitions and an ionic transitions.

The optical emission studies have been performed in National Institute of Standards and Technology, Gaithersburg, USA, RF Plasma Laboratory.

2 EXPERIMENTAL

In the experiments presented here, the plasma source was the Gaseous Electronic Conference (GEC) Radio Frequency Reference Cell. The details of the design of the cell are given in Refs. 15 and 16. Briefly, the GEC RF Reference Cell is a paralel plate discharge chamber with 102 mm diameter electrodes separated by 25 mm. The electrodes are cylindrically sy mmetric and their surfaces are horizontal. The top electrode contains 169 holes with 380 μm diameters to provide a showerhead gas inlet. The cylindrical vacuum chamber is constructed of stainless steel and has 8 radially looking cooper gasket side ports at the chamber midplane. Two 203 mm diameter ports are fitted with 136 mm diameter quartz windows for spectroscopic observations. Two additional ports, orthogonal to these, are 152 mm diameter flanges, one of which acco mmodates a turbo molecular pump for establishing a base pressure of $< 10^{-5}$ Pa. Four 70 mm diameter ports at 45° with respect to the four larger ports are also mounted at the cell midplane. The bottom of the vacuum chamber is constructed so the pumpout of the gas is accomplished by four sy mmetrically placed 70 mm diameter ports piped into a single exhaust line to a mechanical vacuum pump.

The top electrode is grounded to the chamber on the outside of the vacuum interface. The bottom electrode is powered by a 13.56 MHz rf power supply isolated either with a 0.1 μF blocking capacitor for the time resolved measurements or with a filter box in external circuit for spatially resolved measurements. The filter box is Π network with two 1.85 μH series inductances, 100 pF input and output shunt capacitances and 150 pF middle shunt capacitance, designed as in Ref. 17.

For the data presented here, whole flow rates were 10 standard cubic centimeters per minute (sc cm). Gas pressures were 6.67, 13.33, 33.33, 66.67 and 133 Pa and the peak-to-peak applied rf voltages was 200 V.

The experimental setup for spectroscopic measurements is shown schematically in Fig. 1a and 1b. The spectroscopic apparatus consist of a 1 m focal-length monochromator with a 2400 lines/ mm grating (0.416 nm/ mm reciprocal linear dispersion). The monochromator is fitted with a quartz prism predisperser at its entrance slit to eliminate scattered light and multiple order effects in the observations. The vertical monochromator entrance and exit slits are 50 or 100 μm wide and 2 mm

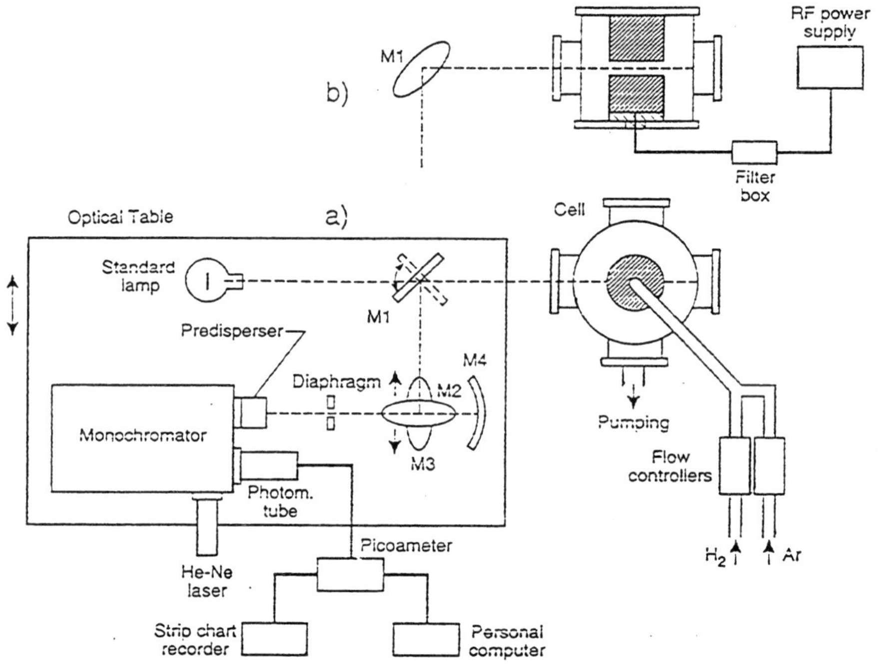

Figure 1: Schematic diagram of the GEC RF Reference Cell and experimental arrangement

high. The monochromator is also equipped with a cooled $5\,cm$ diameter end-on photomultiplier for detection of the optical emission signal. There is a retracable mirror near the monochromator exit slit so a He-Ne laser may be substituted for optical alig nment purposes. The photon flux produced in the plasma is focused on the entrance slit of the monchromator by a combination of three flat mirrors (M1, M2, M3), each $152\,mm$ in diameter and one concave mirror with $650\,mm$ focal length (M4) with the same diameter. These mirrors are arranged to act as a periscope bringing the level of the plasma emission to the same height as the monochromator entrance slit. This arrangement also rotates the image of plasma by $90°$ producing a spatial resolution of $0.5\,mm$ vertical and $5\,mm$ horizontal. By this rotation the electrode surfaces are imaged parallel to the long dimension of the entrance slit, thus permitting observations close to the electrode surfaces. Because of the periscope, vertical scanning of the plasma emission between the electrodes can be accomplished by translating mirror M2, while horizontal scanning is accomplished by translating the optical table (Fig. 1a). For absolute intensity mesurements the standard lamp was also mounted on the optic table.

3 SPATIALLY RESOLVED OPTICAL EMISSION MEASUREMENTS

For spatially resolved measurements, the photomultiplier tube is connected to a picoameter (Fig. 1a). The signal from the picoameter is fed to a strip chart recorder through an analog interface and to personal computer through IEEE-488 interface. The ArI 750.4 nm and $ArII$ 427.8 nm lines were observed in the pressure range of 6.67 to 66.67 Pa and 200 V applied rf voltage. The time averaged intensity distributions along the discharge axis for ArI 750.4 nm line are shown in Figure 2. The measurements are performed in pure argon rf discharge [18]. The intensity is not symmetric to the midpoint of the discharge axis because the top electrode is grounded and the bottom electrode

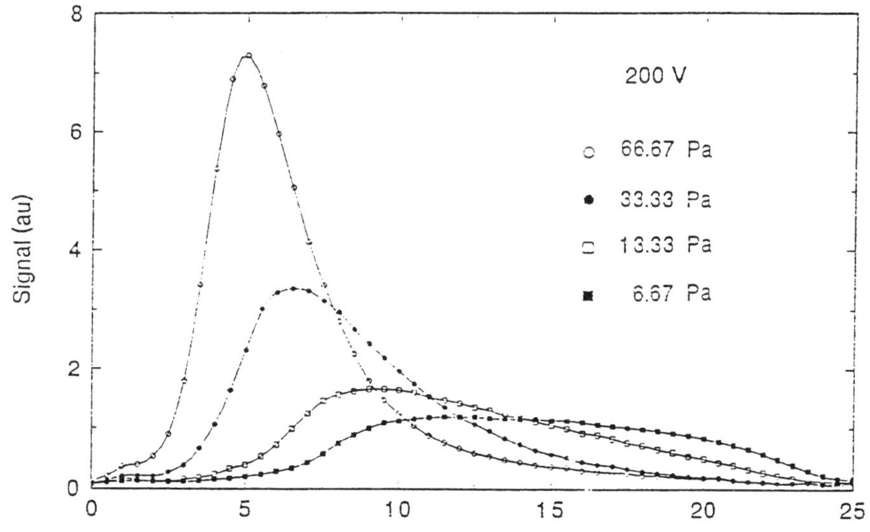

Figure 2: Spatially resolved time averaged optical emission for the Ar I 750.4 nm line. The position 0 corresponds to the surface on the powered electrode

is powered. Position of 0 mm in Fig. 2 corresponds to the bottom electrode and position of 25 mm corresponds to the top electrode.

The measurements of optical intensity distributions of $ArII$ 427.8 nm are shown in Figure 3. These measurements are taken under same con-

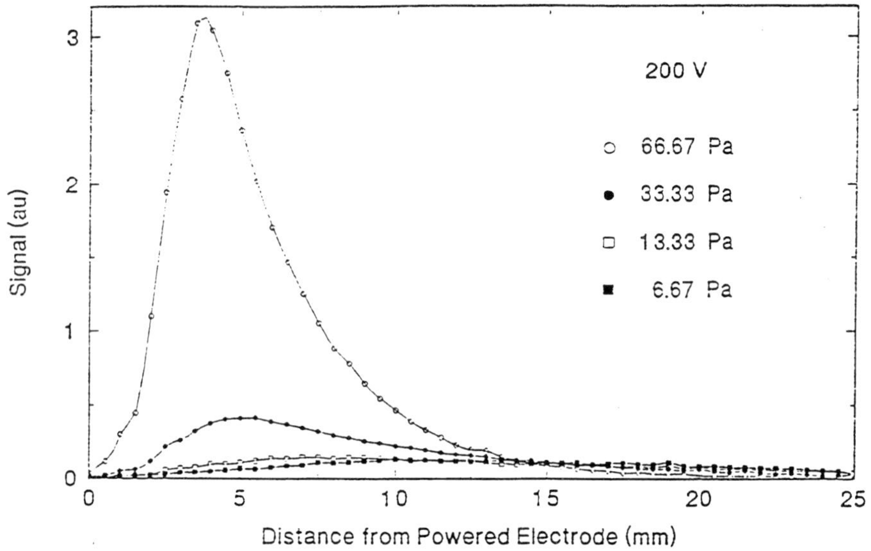

Figure 3: Spatially resolved time averaged optical emission for the Ar II 427.8 nm line. The position 0 corresponds to the surface on the powered electrode

ditions as the measurements of ArI line and show similar behaviour.

The intensity of both ArI and $ArII$ lines are highly dependent upon the discharge pressure. As the pressure is increased from 6.67 to 66.67 Pa , the most intense regions of emission shift toward the powered electrode and the formation of well defined sheaths (dark zones near the electrode surfaces) are observed. The bright band of emission is excited by electrons propagating into the discharge from the electrode [1,3]. The propagation distance is a function of the electron velocity, which in turn depends on the electron collision frequency. Changes in pressure therefore will change the distance in which the exciting electrons can propagate away from the electrodes. This effect one can see in Figs. 2 and 3.

Analysis of the spatial variation in the optical signal intensity indicates that is bright band of emission peaks, for a given pressure, closer to the powered electrode in the $ArII$ emission than for the ArI emission. This is more evident in Fig. 4 where the normalized time averaged intensity distributions of the ArI and $ArII$ lines are shown. This Figure also indicates that for all plasma conditions the $ArII$ emission is spatially broader than the ArI profile, with the $ArII$ emission extending closer to both electrodes.

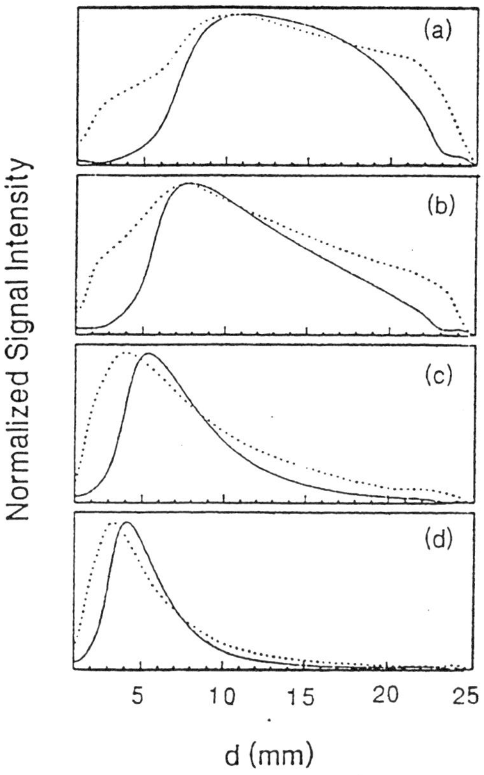

Figure 4: Normalized, time averaged optical emission spatial profiles for Ar I 750.4 nm (solid) and Ar II 427.8 nm (dashed) lines. The applied rf volrage was 200 V and the gas pressures were (a) 6.67 Pa, (b) 13.33 Pa, (c) 33.33 Pa and (d) 66.67 Pa. The position d = 0 corresponds to the surface of the powered electrode

To investigate the effect of changes in plasma parameters on the radial Ar emission profile, the horizontal distribution of neutral Ar line at 415.9 nm was observed in a pure Ar plasma as a function of pressure. The plasma was scanned at the electrode midplane in a direction parallel to the electrode surfaces by moving the optical table supporting the spectroscopic apparatus horizontally. Because of side on observations, the measured intensity of the plasma emission contains contributions from many different plasma layers. In order to separate the contributions of the different layers and to obtain real radial distribution of the plasma emission the Abel inversion procedure [19] was used. The normalized signal intensity of the plasma emission as well as the Abel inverted results are presented in Figure 5. The extent of the electrodes is also indicated for comparison. As can be seen the Abel inverted profiles show a

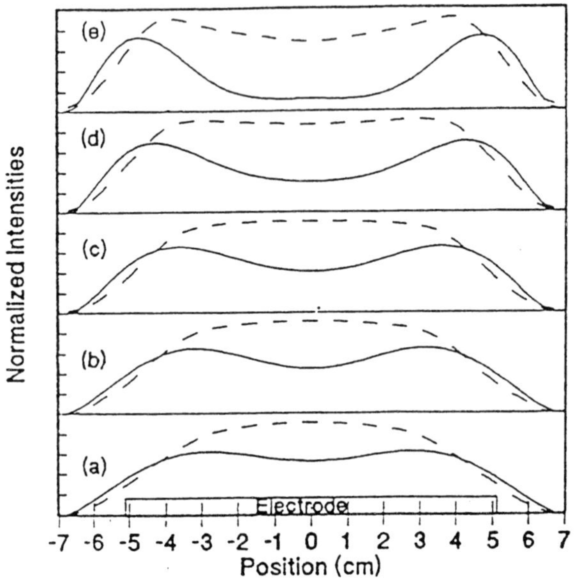

Figure 5: Abel inverted radial emission profiles of the Ar I 415.9 nm line (solid) and the normalized horizontal emission intensity (dashed) for (a) 6.67 Pa, (b) 13.33 Pa, (c) 33.33 Pa, (d) 66.67 Pa and 133.3 Pa pressures. The electrode limits are indicated

substantial difference from the horizontally scanned data as the pressure increases. Also the plasma extends outside the electrode region. These results indicate that uniformity of the plasma across the electrodes is seen to be fairly constant at low pressures, but becoms distinctly nonuniform at high pressures.

4 TIME RESOL VED OPTICAL EMISSION MEASUREMENTS

The time varying output from the photomultiplier was recorded with a system utilizing a time-to-amplitude converter (TAC) and a multichannel analyzer (MCA). A detailed schematic diagram is shown in Figure 6. The timing cycle of the TAC is iniated by a fast pulse derived by a 200 MHz discriminator from a 10 : 1 voltage probe attached to the bottom of the powered electrode. The timing cycle is stopped by a photon initiated pulse from the photomultiplier system, or by the ending of the preset TAC timing period. For the data presented here, the maximum time measured by the TAC was set to 200 ns (slightly less

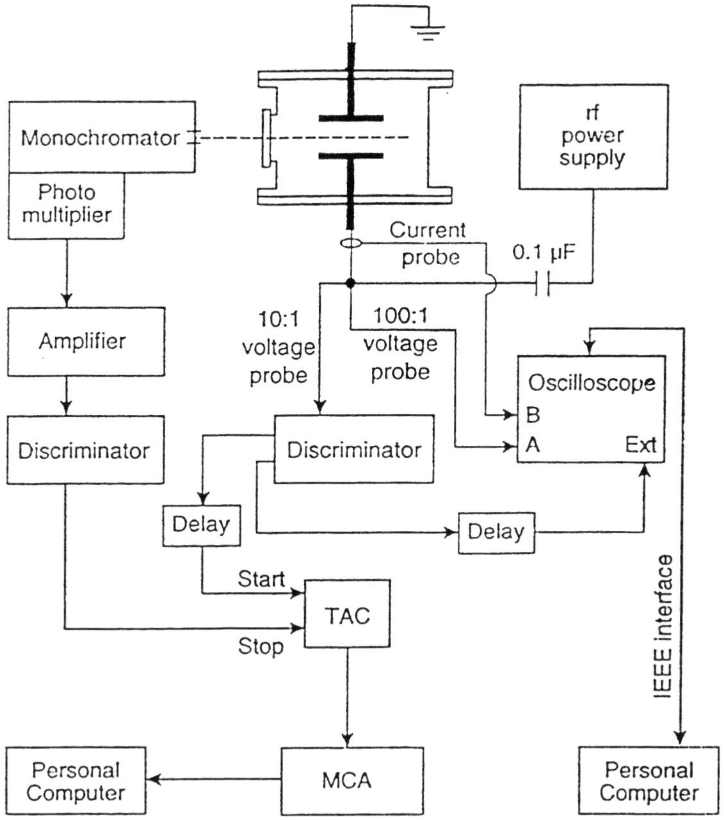

Figure 6: Schematic diagram of the timing electronics utilized to measure the temporal variation of the optical emission signal from argon plasma for comparison with the applied rf voltage waveform. The TAC is a time-to-amplitude converter, the MCA is a multichannel analyzer and Ext refers to the external trigger of the digital storage oscilloscope

than 3 rf cycles). The photon count rates were kept low ($< 10^5$ counts/s) by placing calibrated neutral density filters in the optical path for the measurement of the most intense optical signals. Because much less than one photon is detected, on the average, for each timing cycle, there is no discrimination against photons detected later in the timing cycle. The output pulses from the TAC, whose magnitudes are proportional to the time between the trigger from the voltage waveform and the detection of the photon, are sorted into channels by the MCA. This accumulated spectrum represents the time resolved optical emission signal for a single spatial location. Measurements were made at 10 – 15 different positions between the electrodes, depending upon the degree of spatial variation in the signal intensity. The MCA was set to record 256 channels providing a timing resolution of 0.78 $ns/channel$ (200 ns/256 channels). The time history of the ArI 750.4 nm and the $ArII$ 434.8 nm emission line signals were recorded with respect to the same time on the voltage waveform, therefore their relative phases may be compared. The ArI line has an atomic lifetime of approximately 20 ns while the $ArII$ line has an atomic lifetime of approximately 7 ns [20]. These atomic lifetimes are less than the 13.56 MHz period (73.3 ns) and this fact make possible the measurement of time dependent optical emission. Examples of correlations between the measured time dependent optical emission signals and the voltage and current waveforms are shown in Fig. 7 for argon plasma with peak-to-peak rf voltages of 200 V and

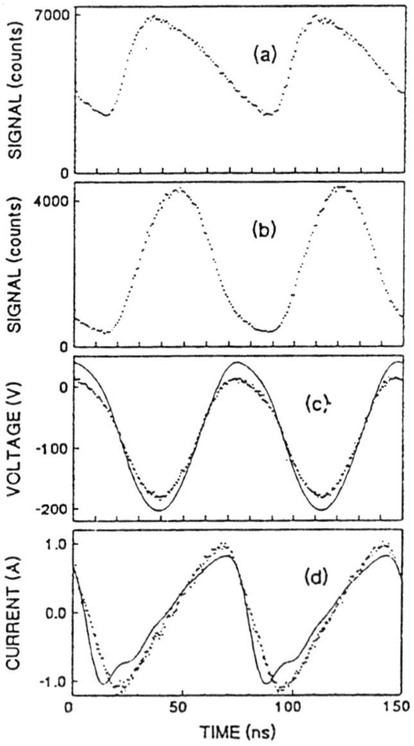

Figure 7: Measured time dependent optical emission signals and the corresponding voltage and current waveform with applied rf voltage of 200 V and pressure of 133.3 Pa. (a) non-calibrated optical emission of the Ar I 750.4 nm line, (b) non-calibrated optical emission of the Ar II 434.8 nm line, (c) (dotted) measured voltage waveform and (solid) calculated voltage waveform, (d) (dotted) measured current waveform and (solid) calculated waveform. The measured voltage and current waveforms were obtrained simultaneously with optical emission data in (a) and (b)

gas pressure of 133.3 Pa [21]. Figure 7a shows the raw ArI optical emission signal as recorded by the MCA at a point where the emission signal was a maximum (21 mm from the grounded electrode). The observation time was 3 minutes and a neutral filter with 0.6.% transmission was used to reduce the count rate to acceptable levels. The raw data for the $ArII$ line is shown in Fig. 7b for the same plasma at an observation point 22.5 mm from the grounded electrode (also the location of the maximum emission intensity). The observation time for the less intense $ArII$ line was 5 minutes and no transmission filters were used. The dotted curves in Fig. 7c and 7d show the voltage and current waveforms, acquired simultaneously by the digital oscilloscope with the data shows in Figures 7a and 7b. The solid curves represent the voltage and current waveforms at the surface of the powered electrode as calculated by the equivalent circuit model [22,23]. The presence of higher harmonics in the electrical waveforms is indicated by the non-sinusoidal character of the curves.

The optical emission profiles for the ArI and $ArII$ lines, as shown in Fig. 7 are quite different. The peaks in the time varying ArI optical emission signal are broader than in the $ArII$ profile and exhibit an asy mmetry on the right hand side of the temporal peaks. These characteristics are due to the relatively long atomic lifetime of the neutral excited state. This also contributes to the significant "background" or non-time-dependent portion of the ArI signal observed in Figure 7a. In contrast, the $ArII$ signal profile exhibit narrow temporal peaks and a large time dependent signal due to the short lifetime of the excited state as compared with rf period. Comparison of the optical emission data in fig. 7 also indicates that the ArI and $ArII$ profiles are not in phase with each other, with the ArI profile leading the $ArII$ profile. The $ArII$ time dependence is observed to be nearly 180° out of phase with the applied voltage, as was previously reported by Tochikubo et al [4]. Due to the non-sinusoidal nature of current waveform it is difficult to draw correlations between the current and the optical emission signals.

The optical emission signals as a function of observation point and time are shown for a range of plasma conditions in Fig. 8 for the ArI line (750.4 nm) and in Fig. 9 for $ArII$ line (434.8 nm). Data were taken for argon pressures of $6.67, 13.33, 66.67$ and 133.33 Pa and uncorrected applied peak-to-peak rf voltage of 200 V [21]. Figures 8 and 9 are plots of contour fits to the calibrated optical emission time profiles taken at many observation points between the electrodes.

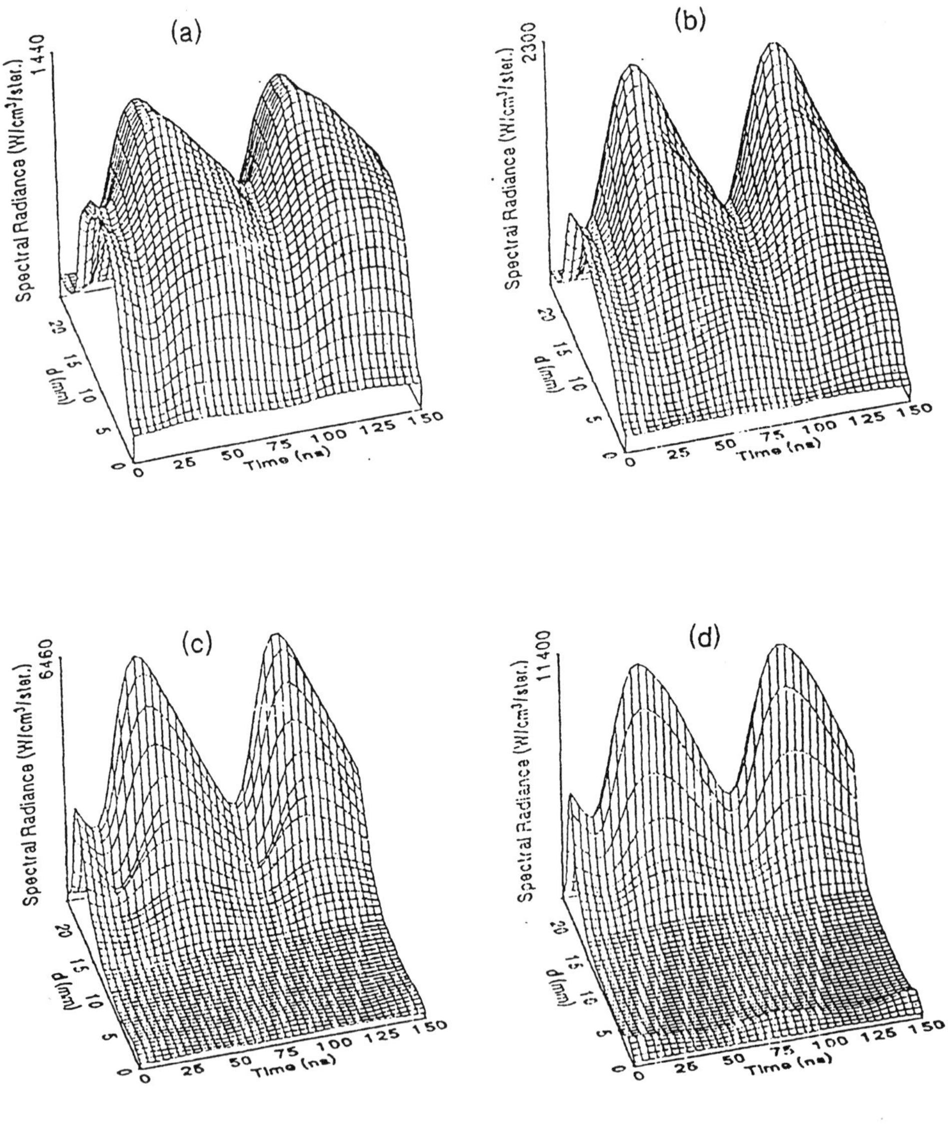

Figure 8: Optical emission measurements of the spatial profile and temporal evaluation of the Ar I 750.4 nm line. The applied rf voltage was 200 V and the gas pressures were 6.67 Pa, 13.33 Pa, 66.67 Pa and 133.3 Pa

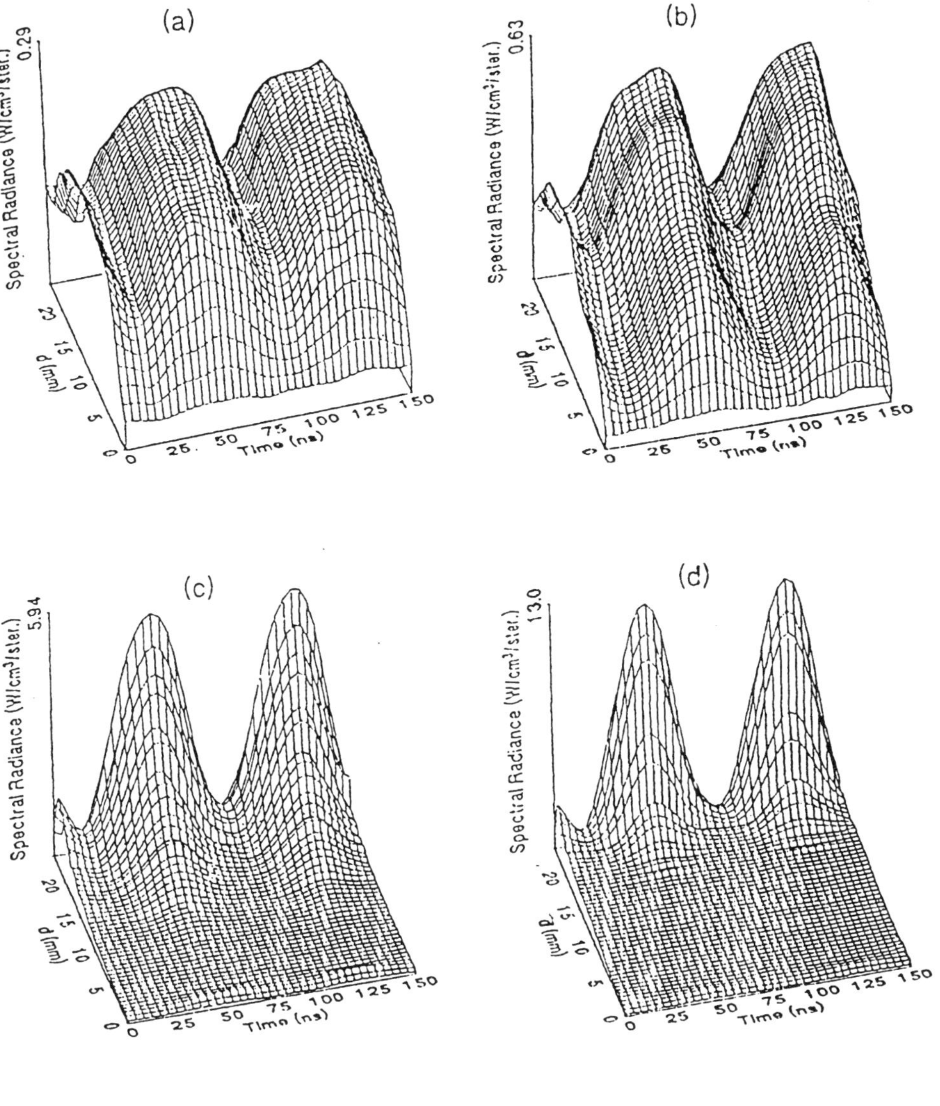

Figure 9: Optical emission measurements of the spatial profile and temporal evaluation of the Ar II 434.8 nm line. The applied rf voltage was 200 V and the gas pressures were 6.67 Pa, 13.33 Pa, 66.67 Pa and 133.3 Pa

A time range of $150\,ns$ is shown ($\sim$ 2 complete rf cycles), with each square of the grid corresponding to approximately $4ns$ on the time axis and $0.5\,mm$ on the d axis. The time $t = 0$ was chosen to correspond to the maximum of the voltage waveform. The d axis is defined as the position of the observation point as measured from the grounded electrode along an axis through the center of the rf electrodes. The vertical axis on the plots is the absolute spectral radiances measured from the discharge. The number at the top of vertical axis is the magnitude of the maximum observed signal. For the purpose of the absolute intensity measurements a tungsten ribbon filament standard lamp calibrated for spectral radiance ($\pm$ 2 %) by the Radiometric Physics Division, NIST [24] was mounted on the optics table. This lamp is used to calibrate the optics/monohromator system to obtain absolute spectral radiance measurements. The lamp was substituted for the plasma source by rotating mirror M1 (see Fig. 1a) to image the lamp filament onto the entrance slit. The calibration measurements were made with the same slit width and photomultiplier voltage as used in the plasma measurements. The intensity of the standard lamp signal was stronger than the observed spectral lines by approximately a factor of 10^3. To prevent the measurement system from saturating, metallic film neutral density filters were placed in the optical path in front of the monochromator slit for measurement of the lamp signals. To avoid both, a second order radiation from the lamp and a significant component of the signal from scattered light, bandpass interference filters were incorporated into the optical path also. The bandpass filters were used only in the standard lamp calibration procedure, and not in the plasma experiment. The metallic film neutral density filters and the bandpass filters could not be characterized using this optical monochromator system, and were therefore individually calibrated for transmission over the wavelength region of interest. The total uncertainty in the measurement of a spectral line radiance is therefore estimated to be $\pm$ 6 %.

Two complete cycles are shown in each plot (Figs. 8 and 9) to help in determining the significance of minor features in the contour plots. If a feature is not repeated in both cycles, then it is not reproducible and should not be considered to be real. This is particularly evident for the low pressure $ArII$ data shown in Fig. 9 where the signal levels were low thus resulting in noisy data.

5 DISCUSSION

As is evident from Figs. 2 – 5 and Figs. 8 and 9 the spatial and temporal profiles of the optical emission are highly dependent upon the discharge pressure. At 6.67 Pa the spatial profiles are broad with the emission essentially extending across the entire region between the electrodes. Only a small fraction of the optical emission signalexhibits time dependent behaviour at these lower pressures. At higher pressures a larger fraction of the optical signal exhibit a time dependence, which is in agreement with previous results by Seeböck and Köhler [25] for lower frequency discharges.

The bright band of emission peaks are closer to the powered electrode in the $ArII$ emission than for the ArI emission.

Formation of bright bands near the grounded electrode is observed for only a few sets of plasma conditions. For the ArI transition, a weak sheath is observed at 66.67 Pa with well defined sheath present at 133.33 Pa (see Fig. 8). If the applied voltage increases from 75 to 200 V the amplitude of the emission near the grounded electrode remains nearly constant while the emission near the powered electrode increases by a factor 2.5 . This may be related to the fact that the voltage across the sheath near the grounded electrode is nearly independent of the applied rf voltage as shown by electrical measurements [22] and ion kinetic energy measurements [26]. For the $ArII$ emission, the existance of a bright band of emission near the grounded electrode was only observed for higher pressure (Fig. 9). For these conditions a weak increase in the optical signal is detected near the grounded electrode.

The changes in the temporal and spatial distributions with applied rf voltage are less dramatic than those observed with changing pressure. As the applied voltage increases, the peak emission intensity increases, particularly at the higher pressures. Additionally, the widths of the peaks in the time dependent signal increase with increasing voltage. This may indicate that conditions are favorable for the excitation of the required transitions for emission during a grater portion of the rf cycle at higher applied voltages.

6 CONCLUSIONS

Spatially and temporally resolved optical emission profiles have been measured for argon plasmas with pressures ranging from 6.67 to

133.33 Pa and applied rf voltage of 200 V . The plasma source was the GEC RF Reference Cell [15].

Spatially resolved optical emission spectra may be used to determine sheath thicknesses and to monitor the uniformity of the plasma across the electrodes.

The magnitude of the plasma emission for two lines has been put on an absolute scale by comparison with emission from a standard lamp. This allows for a complet characterization of the optical emission signal from the rf plasma, and provides a more comprehensive data set for comparison with theoretical modelling results.

While much can be learned from optical emission spectroscopy, more research must be done to determine the aspects of the data that are of most importance for monitoring the plasma conditions relevant to etching processes.

7 REFERENCES

1. G. De Rosny, E. R. Mosburg, J. R. Abelson, G. Devaud and R. C. Kerns, J. Appl. Phys. **54,** 2272 (1983).

2. V. M. Donnelly, D. L. Fla mm and R. H. Bruce, J. Appl. Phys. **58,** 2135 (1985).

3. P. Bletzinger and C. A. De Joseph, IEEE Trans. Plasma Sci. **PS-14,** 124 (1986).

4. F. Tochikubo, T. Kokubo, S. Kakuta, A. Suguki and T. Makabe, J. Phys. D **23,** 1184 (1990).

5. F. Tochikubo, A. Suzuki, S. Kakuta, Y. Terazona and T. Makabe, J. Appl. Phys. **68,** 5532 (1990)

6. T. Kokubo, F. Tochikubo and T. Makabe, Appl. Phys. Lett. **56,** 818 (1990)

7. W. E. Köhler, R. J. Seeböck and F. Rebentrost, J. Phys. D **24,** 252 (1991).

8. Z. Petrović, F. Tochikubo, S. Kakuta and T. Makabe, J. Appl. Phys. **73,** 2163 (1993).

9. S. K. Park and D. J. Economou, J. Appl. Phys. **68,** 3904 (1990)

10. M. Surendra and D. B. Graves, IEEE Trans. Plasma Sci. **19,** 144 (1991).

11. H. W. Trombley, F. L. Terry and M. Elta, IEEE Trans. Plasma Sci. **19,** 158 (1991).

12. T. J. So mmerer, W. N. G. Hitchon, R. E. P. Harvey and J. E. Lawler, Phys. Rev. A **43,** 4452 (1991).

13. M. Meyyappan, J. Appl. Phys. **69,** 8047 (1991).

14. F. Tochikubo and T. Makabe, Meas. Sci. Technol. **2,** 1133 (1991).

15. P. J. Hargis, K. E. Greenberg, P. A. Miller, J. B. Gerardo, J. R. Torcznski, M. E. Riley, G. A. Hebner, J. R. Roberts, J. K. Olthoff, J. R. Whetstone, R. J. Van Brunt, M. A. Sobolewski, H. M. Anderson, M. Splichal, J. L. Mock, P. Bletzinger, A. Garscadden, R. A. Gottscho, G. Selwyn, M. Dalvie, J. E. Heidenreich, J. W. Butterbaugh, M. L. Brake, M. L. Passow, J. Pender, A. Lujan, M. E. Elta, D. B. Graves, H. H. Sawin, M. J. Kushner, J. T. Verdeyen, R. Horwath and T. R. Turner, Rev. Sci. Instrum. (in press).

16. J. R. Roberts, J. K. Olthoff, R. J. Van Brunt and J. R. Whetstone, Advanced Techniques for Integrated Circuit Processing, Society of Photo-Optical Instrumentation Engineers, Bellingham, USA, Vol. 1392, P. 428 (1990).

17. P. Miller, H. Anderson and M. Splichal, J. Appl. Phys. **71,** 1171 (1992).

18. S. Djurović and J. R. Roberts, Abstracts of 45th Annual Gaseous Electronics Conference, Boston USA, p. 173 (1992).

19. S. Djurović, D. Kelleher and J. R. Roberts, to be published.

20. W. L. Wiese, M. W. Smith and B. M. Miles, Atomic Transition Probabilities, Natl. Stand. Ref. Data Ser., Natl. Bur. Stand. (US) 22 Vol. II (1969).

21. S. Djurović, J. R. Roberts, M. A. Sobolewski and J. K. Olthoff, J. Res. Natl. Ins. Stand. Technol. **98,** 159 (1993).

22. M. A. Sobolewski and J. R. Whetstone, Advanced Techniques for Integrated Circuit Processing II, Society of Photo-Optical Instrumentation Engineers, Bellingham, USA, Vol. 1803, p. 309 (1993).

23. M. A. Sobolewski, J. Vac. Sci. Technol. **10,** 3550 (1992).

24. J. H. Walker, R. D. Saunders, A. T. Hattenburg, Spectral Radiance Calibrations, NBS Special Publications 250-1 (1987).

25. R. J. Seeböck and W. E. Köhler, J. Appl. Phys. **64,** 3855 (1988).

26. J. K. Olthoff, R. J. Van Brunt and S. B. Radovanov, J. Appl. Phys. **72,** 4566 (1992).

Low Temperature and
General Plasmas
M. Milosavljevic and Z. Petrovic

ISBN 1-56072-399-8

ROLE OF THE BEAM ELECTRONS IN THE HOLLOW CATHODE DISCHARGE

S. Bukvic
Faculty of Physics, University of Belgrade
P.O.B. 550, Belgrade, Yugoslavia

1. Introduction

The understanding of the hollow cathode discharge has always been a difficult task mainly because of its complex nature. Usual diagnostic techniques, such as electrostatic probe measurements, are not suitable due to the low electron density and relatively small dimensions of the hollow cathode discharge region (Burakov *et al* 1988). Determination of the electron energy distribution and electron temperature is also very difficult and the results are not reliable (van der Mulen *et al* 1980), (McNeil *et al* 1979). For the optogalvanic studies authors usually use the most simple assumption of the Maxwellian electron energy distribution (Grigorev and Eletskii 1989). This assumption, however, is questionable and definitely can not be applied to the problems concerning the hollow cathode lasers.

More acceptable seems to be an electron beam concept based on the presumption of the two separate groups of electrons: the high energy or beam electrons originating from secondary emission at the cathode surface (Warner and Person 1979),(van Veldhuizen *et al* 1983) and the thermalized bulk electrons. The latter group can be described by a quasi Maxwellian electron energy distribution (Keller and Zalewski 1980),(van Veldhuizen *et al* 1983).

In the present study we have tried to clarify the role of both: the beam electrons and also the bulk electrons in the neon hollow cathode discharge. An attention was paid to the electron excitation processes from the ground state for which only the high energy, beam electrons are responsible, and also to the electron temperature of the bulk electrons.

2. Experimental setup

The experimental setup that has been used to measure the optogalvanic and the laser induced fluorescence signals in the neon hollow cathode discharge is shown in Fig. 1.

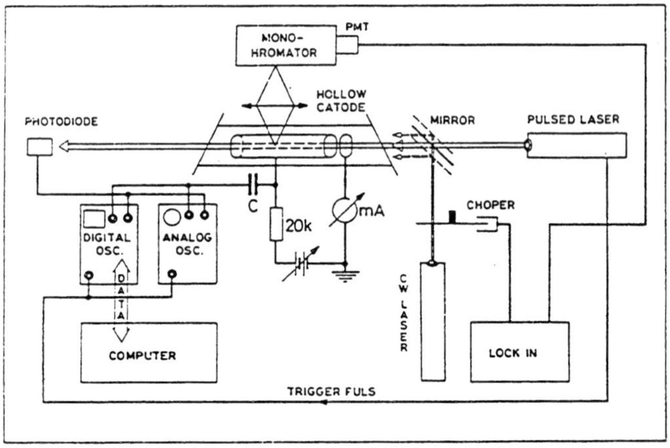

Fig. 1. Experimental Setup

See-through hollow cathode is made out of a stainless steel mesh. In order to keep the discharge inside the cathode the outer surface of the cathode was covered by the glass envelope. The inner diameter of the cathode is 0.9mm, and the length is 3cm. The anode is a metal ring separated 1mm from the cathode. The discharge tube is connected to the vacuum system, the spectroscopic grade neon was used as a working gas in the pressure range 0.5torr - 12torr. During the experiment we have kept a constant flow of the neon through the discharge tube. The discharge current was in the range 2-12mA.

Two types of lasers were used. Nitrogen laser pumped dye laser PRA LN 102 was operating on Rhodamine 6G in a repetitive mode. Duration of the pulse was of few nanoseconds, spectral bandwidth was 0.1nm. A typical peak power was in the 100kW range. The laser beam was focused into the hollow cathode along the axis (beam diameter ~ 0.2mm) and the optogalvanic signal was detected across the 20KΩ ballast resistor. A HP 54200A digital oscilloscope with 6 bit resolution was used for signal recording. To improve S/N ratio 64 successive signals have been averaged. This operation also improves resolution to 7 bit. The recorded data are transferred to the computer for further analysis.

The argon ion laser pumped Spectra Physics model 375 CW dye laser operating also on the Rhodamine 6G has been used in the fluorescence experiment. Spectral bandwidth of the laser was about 0.01nm, power in the 50mW range, and the beam diameter 0.5mm. A mirror fixed on the translation stage provides fine translation of the laser beam along the discharge radius keeping it parallel to the cathode axis.

Induced fluorescence signals were monitored in the direction orthogonal to the laser beam using a Jobin Yvon H.20 monohromator. For signal detection we used a Stanford Research model SR 510 lock-in amplifier in connection with a Stanford Research model SR 540 optical chopper. The chopper frequency was in the 1kHz range.

An analog oscilloscope has been used to monitor both, the optogalvanic signal and the photodiode response of the pulsed laser.

3. Electron temperature measurement

3.1 Theory

As mentioned before the optogalvanic effect was used for the electron temperature determination. This technique is based on the analysis of the time behavior of the optogalvanic signal recorded after the plasma has been illuminated by the short laser pulse as proposed by Veldhuizen (van Veldhuizen *et al* 1983).

Supposing that the laser is tuned to the wavelength corresponding to some emission line, the resonant absorption of the laser light will occur and this will lead to the creation of some extra ions and electrons in the illuminated area. If the laser beam illuminates plasma along the hollow cathode axis, these extra ions and electrons will propagate to the wall of the cathode due to the ambipolar diffusion (van Veldhuizen *et al* 1983), (Labat and Bukvic 1988). Change of the current in the discharge tube circuit will occur that is proportional to the flux of the ions passing through the cathode fall and reaching the wall of the cathode.

The ambipolar diffusion process of the extra ions and electrons can be described by the equation (Tihonov and Samarskii 1966)

$$\frac{\partial^2 n(r,t)}{\partial r^2} + \frac{1}{r}\frac{\partial n(r,t)}{\partial r} = \frac{1}{\mathrm{D_a}}\frac{\partial n(r,t)}{\partial t} \tag{1}$$

where n is a density of extra ions (and electrons), $\mathrm{D_a}$ is ambipolar diffusion coefficient and r is radial coordinate.

The equation (1) is true if the plasma is homogeneous along the cathode axis, if the end effects do not play a major role and if the ambipolar diffusion is a dominant loss process.

For certain initial and boundary conditions equation (1) has the analytic solution.

It was supposed that initially a distribution of the extra ions (electrons) due to the resonant absorption of the laser light is described by the function $\phi(r)$. In that case, the initial condition is:

$$n(r,0) = \phi(r) \tag{2}$$

Density of the extra ions (electrons) on the cathode wall must be zero because of the destruction processes taking place on the wall. So, the boundary condition is:

$$n(\mathrm{R},t) = 0 \tag{3}$$

where R is radius of the hollow cathode.

For the given initial and boundary conditions the diffusion equation (1) has analytic solution (Tihonov and Samarskii 1966):

$$n(r,t) = \frac{1}{R^2} \sum_{n=1}^{\infty} C_n(r) e^{-D_a (\mu_n^o / R)^2 t} \qquad (4)$$

where R is radius of the cathode, D_a is ambipolar diffusion coefficient, μ_n^o is n-th root of the Bessel J_o function, t is time and $C_n(r)$ are the coefficients depending on the $\phi(r)$. It is obvious that a particular form of the $\phi(r)$ doesn't affect the time dependence of the solution.

The solution (4) is a strict solution of equation (1) for given conditions while the solution proposed by Veldhuizen (van Veldhuizen *et al* 1983) is mathematically valid only for an infinite radius of a hollow cathode.

Attention must be paid to the following: the solution (4) is correct only if the time constant of the ionization processes in the plasma is much shorter than the time constant of the diffusion processes (van Veldhuizen *et al* 1983). In the opposite case the whole process can not be described correctly by (1).

In our experiment this condition is satisfied for currents over ~ 1mA.

As mentioned before the optogalvanic signal is proportional to the flux of ions passing through the cathode fall and reaching the cathode wall. Therefore the optogalvanic signal is given by:

$$\Delta V(t) \sim -D_a \left. \frac{\partial n}{\partial r} \right|_{r=R} = \frac{D_a}{R^2} \sum_{n=1}^{\infty} K_n e^{-(\mu_n^o)^2 (D_a/R^2) t} \qquad (5)$$

where $K_n = -\left. [\partial C_n(r)/\partial r] \right|_{r=R}$.

The series (5) converges fast enough so the one can represent $\Delta V(t)$ by the following polynomial:

$$\Delta V(t)=c\left(K_1e^{-c(\mu_1^o)^2t}+K_2e^{-c(\mu_2^o)^2t}+K_3e^{-c(\mu_3^o)^2t}+K_4e^{-c(\mu_4^o)^2t}+K_5\right) \quad (6)$$

where $c=(D_a/R^2)$.

To obtain T_e we have adjusted the coefficients $c, K_1, .., K_5$ of polynomial (6) so that the $\Delta V(t)$ function will fit the experimental data. Taking into account relation: $D_a=D_i(1+T_e/T_g)$ (where D_i is ion diffusion coefficient and T_g is gas temperature) and knowing R one can determine the electron temperature T_e.

It is important to note that the electron temperature obtained in this way has the meaning of an average energy of the low energy, thermalized electrons. This method is not sensitive to the beam electrons neither it can give us information about electron energy distribution function.

3.2 Experimental Results

For the electron temperature measurement we have used experimental setup described in Sect.2. The pulsed laser was adjusted to the wavelength 614.3nm and 609.6nm corresponding to the transitions $1s_5$-$2p_6$ and $1s_4$-$2p_4$ of neon (in Pashen notation) by using a monohromator and simultaneously monitoring the optogalvanic signal. For exact determination of the zeroth moment of the diffusion process a preview feature of the HP54200A oscilloscope has been used.

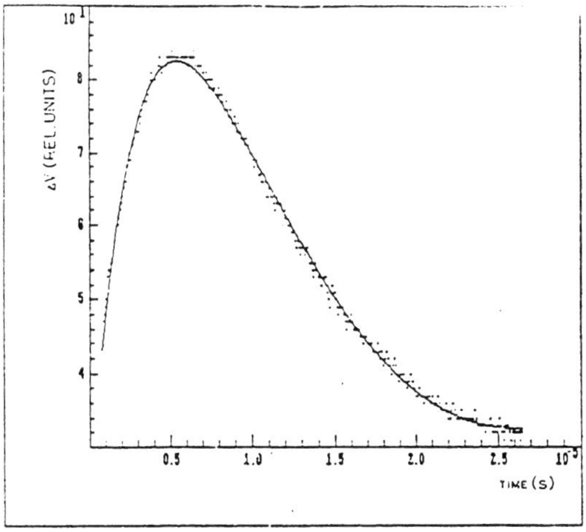

Fig.2. · Typical experimental data (dots) and best fit (solid line)

In Fig.2 we present experimental data for $\Delta V(t)$ together with the best fit to the polynomial (6) for transition $1s_5 - 2p_6$ for p=2.0torr and I=3.6mA. The polynomial (6) fits the experimental data better for higher pressures and currents. This is due to the conditions mentioned before. Namely, for low pressures and currents characteristic time of ionization processes taking place after the laser pulse is comparable with characteristic time of diffusion losses.

In Fig.3 the electron temperature obtained by this method is given as a function of the pressure and discharge current. Values for D_i and T_g are taken from (Steenhuijsen 1981) and from the experiment.

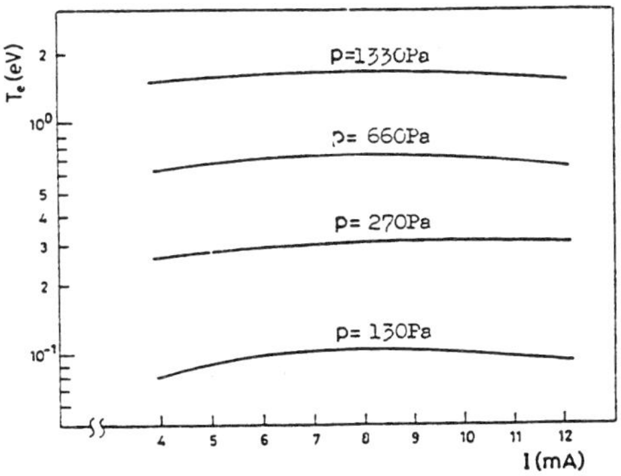

Fig.3. Temperature of the low energy electrons vs the discharge current at various gas pressure.

One can see that the electron temperature rapidly increases with the gas pressure while it is practically independent on the discharge current. For the pressures under 5 torr the electron temperature is below 0.5eV. It means that energy of thermalized electrons is too low to maintain the discharge in the neon since the energies of lowest excited states lay in the range 16.6-16.9eV. Therefore, the assumption on existence of the beam electrons originating from the cathode surface is necessary (Warner and Person 1979), (van Veldhuizen *et al* 1983).

4. Determination of relative densities of the 1s states

The experimental setup shown in Fig. 1 was used also for determination of the density of neon atoms excited to $1s_4$ and $1s_5$ states. As mentioned before the laser induced fluorescence (LIF) technique has been applied (Den-Hartog *et al* 1988). This technique has some advantages, specially when absolute measurements are not necessary. It can be successfully applied even if absorption of the laser radiation is very weak, and it is not sensitive on the effects of plasma ends.

In Fig. 4 a relative density of atoms in the metastable $1s_5$ state is shown as a function of the radial coordinate and pressure, for 7mA discharge current. During the experiment the CW dye laser has been tuned to the wavelength λ_{ex}=614.3nm and LIF has been monitored on the line λ_{obs}=630.48nm. A spatial resolution is limited by the laser beam diameter; it was in the 0.5mm range. Using the lock-in technique we have realized signal to noise ratio S/N~20.

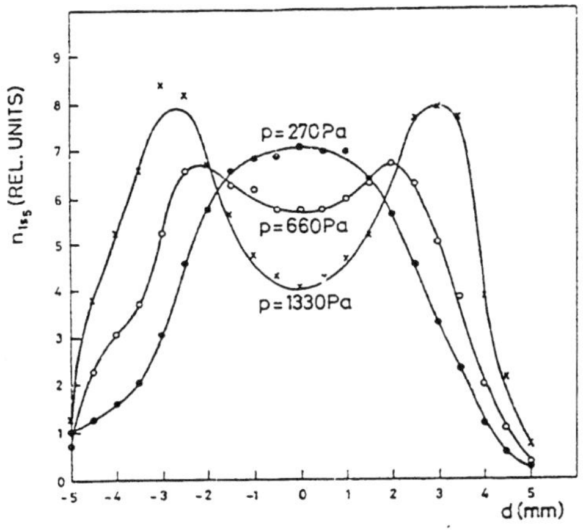

Fig. 4. Measured distribution of $1s_5$ state neon atom density along the cathode diametar at the typical values of the neon gas pressure

The radial density profile of atoms in the $1s_5$ state shows two characteristic shapes: for the low pressures ($p \leq 2$ torr) density has maximal value on the cathode axis and monotonously decreases toward the cathode wall becoming zero at the wall. For pressures above 3 torr the profile is quite different. On the cathode axis the density has minimal value and increases toward the wall reaching a maximal value at the distance 2.5-3mm from the cathode axis. Farther toward the cathode wall density decreases to the value zero at the wall. Absolute value of the minimum of the density on the cathode axis decreases with the gas pressure.

A similar shape has been obtained for atoms in the resonant $1s_4$ state of neon.

It is important to say that density of atoms excited to $1s_5$ and $1s_4$ states does not reach the maximal value in the cathode fall region of thickness about 0.5mm (Broglia *et al* 1987), (Keller *et al* 1983). The maximum of the density always appears in the negative glow, a region free of electric field in the hollow cathode discharge.

We note that similar curves were obtained in a high current neon hollow cathode discharge for density of lowest excited states using absorption technique (van Veldhuizen 1983) and in a helium hollow cathode discharge for He^+ using a mass spectroscopy method (Kuen *et al* 1981).

5. Monte Carlo Simulation

The Monte Carlo simulation was developed by taking into account only the most important processes in the hollow cathode discharge. The simulation is based on the beam concept and our intention is to explain obtained experimental facts, primarily for low gas pressures, as a result of an interaction of the high energy electrons with the gas atoms.

The discharge can be considered as being maintained purely by the beam electrons. Secondary electrons created on the cathode surface are accelerated in the cathode fall region reaching the negative glow with energies close to the cathode fall voltage (Gill and Webb 1977). The negative glow region acts as a potential well and the electrons oscillate within and loose their energy on ionization and excitation of the gas atoms (Warner and Person 1979),(Person 1965). Consequently, it will be assumed that also the creation of electron-ion pairs and lowest excited states is primarily by the beam electrons. Also, diffusion and radiative deexcitation will be considered as the only two loss processes for atoms in $1s_4$ and $1s_5$ states.

This simple picture is valid for low pressures only, for $p \geq 5$torr the role of thermalized electrons can not be neglected. Even medium and high pressure discharges are not maintained purely by the beam electrons, however, contribution of the high energy electrons can be estimated using proposed simple model.

In that case a spatial distribution of $1s_4$ and $1s_5$ atoms can be described by the time-independent diffusion equation:

$$D\left(\frac{d^2n(r)}{dr^2} + \frac{1}{r}\frac{dn(r)}{dr}\right) - \alpha n(r) + S(r) = 0 \qquad (7)$$

with boundary conditions:

$$n(r)\big|_{r=R}=0 \qquad (7a)$$

$$(\partial n/\partial r)\big|_{r=0}=0$$

where $n(r)$ is density of atoms in $1s_5$ or $1s_4$ states, α is effective probability for radiative deexcitation and $S(r)$ is a source function, i.e. the number of atoms excited in level $1s_i$ (i=4,5) per unit volume per unit time due to the beam electrons, R is radius of the hollow cathode and D is diffusion coeficient.

The Monte Carlo simulation has been applied to determine the source function $S(r)$.

5.1 Basis of the Model

By a computer simulation, using the Monte Carlo method, we have examined some of the details of electron motion in the negative glow. Instead of a complex cylindrical geometry of the hollow cathode we have used a simple linear geometry. It means that the motion of an electron has been traced in one dimensional potential well instead of two dimensional. As a consequence, our model does not take into account angular scattering of electrons, and the major features of the simulation will be determined by inelastic collisions. This simplification is usual (Tron Ngoc An *et al* 1977),(Parker and Johanson 1971) and it is based on the fact that 70% of the electrons having energies above 40eV are scattered by collisions in forward directions (Labahn and Colloway 1964),(Register *et al* 1984).

Thickness of the cathode fall is d=0.5mm (Keller *et al* 1983) and the electric field is described by the linear function of coordinate (Kawakita *et al* 1983),(Tron Ngoc An *et al* 1977).

$$E(x)=2\frac{U}{d}\left(1-\frac{x}{d}\right) \tag{8}$$

where U is cathode fall voltage and x is a distance from the cathode. In our experiment voltage across the tube was about 150V, and we have assumed that nearly whole voltage drop occured in the cathode fall region. The radius of the cathode was R=0.45cm.

In the course of simulation the following procedure has been applied: diameter of the cathode has been divided into 360 intervals of equal size (Δx) so that the probability that more than one collision occurs in any interval is negligibly small (Tron Ngoc An *et al* 1977).

The total probability of the collision is given by (Present 1958):

$$P_T=\int_0^{\Delta x} N\,\sigma_T(\varepsilon)\,\exp(-N\,\sigma_T(\varepsilon)x)dx \tag{9}$$

where N is density of the gas atoms, $\sigma_T(\varepsilon(x))$ is a total inelastic cross sections, $\varepsilon(x)$ is kinetic energy of the electron and x is coordinate. For electron motion in the cathode fall the integral (9) has been calculated numerically according to (8). For motion in the negative glow we take $P_T= 1-\exp(-\Delta x N\sigma_T(\varepsilon))$.

In each mesh point x_i the probability P_i is compared with a pseudo random number R_{1i}, uniformly distributed between 0 and 1. If $R_{1i}\le P_i$ a collision is assumed to have occurred. Four types of inelastic collisions are considered: ionization, integral excitation to all electron states except $1s_4$ and $1s_5$ and excitation to $1s_4$ and $1s_5$, therefore:

$$\sigma_{ion}+\sigma_{exc}+\sigma_4+\sigma_5=\sigma_T$$

where σ denotes appropriate cross sections.

Similarly, for probabilities we can write following expression:

$$P_{ion}/P_T+P_{exc}/P_T+P_4/P_T+P_5/P_T=p_{ion}+p_{exc}+p_4+p_5=1$$

where *p* denotes a relative probability.

Type of the process is chosen by the next random number R_2 in the following way: if $0<R_2<p_{ion}$ the ionization is assumed to have occurred; if $p_{ion}<R_2<p_{ion}+p_{exc}$ the excitation is assumed to have occurred and so on. The same random number is used for determination of the energy of an electron created if the ionization process have occurred: $E_{new}=R_2(E_{curr} - E_{ion})$. This procedure is repeated in the next mesh point for current electron and for all electrons created in ionization processes while their energies are above E_{min}=20eV. The number of avalanche electrons processed in this way is relatively small, about 4-6 for the cathode fall voltage 150V. When the energy of each electron drops below E_{min} a new one is emitted from the cathode surface with the energy randomly distributed in the range 0-7eV.

During the simulation a number of acts of ionization, integral excitation and excitation to the states $1s_4$ and $1s_5$ has been recorded for every mesh point x_i. A function N(x) obtained in this way is assumed to be proportional to the S(r) - the source function.

In order to obtain electron energy distribution function a number of electrons with appropriate energy in every mesh point has been recorded.

Approximately 30000 of initial electrons were necessary for a good statistics.

The simulation was realized on a personal computer using a short recursive algorithm. The cross sections data, Fig.6, were taken from (Register *et al* 1984). To prevent sequential correlation of the random numbers we have used a random number generator suggested in (Press *et al* 1988).

5.2. Results of Monte Carlo Simulation

5.2.1. Source functions

The source functions S(r) for $1s_5$ level and excitation obtained by the Monte Carlo simulation are shown in Fig.5.and Fig.6.

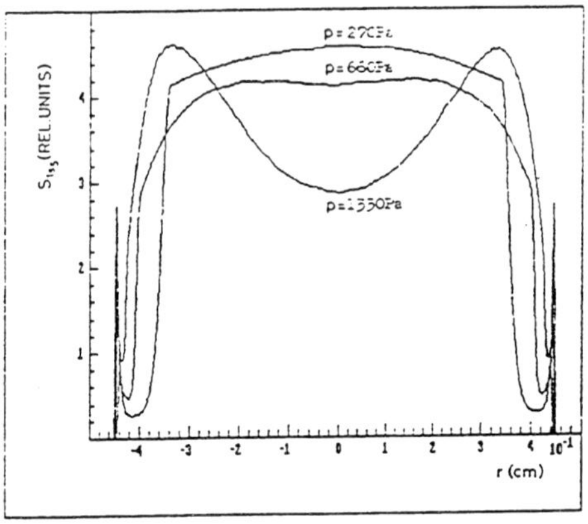

Fig. 5. Source function S(r) for $1s_5$ level neon atoms vs radial coordinate for three values of the gas pressure.

For low pressures (p≤2torr) the source functions for excitation and ionization are practically constant over whole negative glow region. Due to the small probability for inelastic collisions, the electron must pass distance 2R few times before it significantly loses its energy. Therefore, the probability for excitation or ionization is equal along the hollow cathode radius.

For higher pressures (2torr<p<10torr) the mean free path of the electrons decreases and electrons loose their energy more rapidly. Maximum of the ionization probability is at the boundary of the cathode fall - negative glow region because the electron kinetic energy is maximal at this point. The source function for ionization has for this reason a maximum at the boundary. (see Fig. 6)

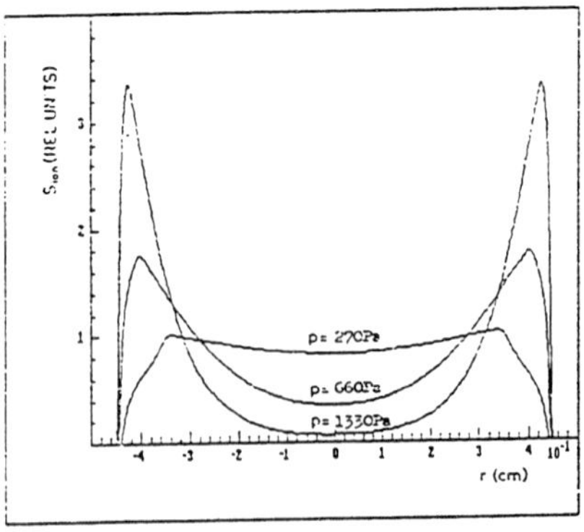

Fig. 6. Source function S(r) for ionization of neon vs. radial coordinate for three values of the gas pressure

Interesting effect appears for pressures $p \geq 10$torr. As a consequence of short free mean path the electron will undergo more than a few collisions in the cathode fall. Because of that it will reach a maximum of the kinetic energy inside the cathode fall (Tron Ngoc An *et al* 1977). The ion source function will also have a maximum at this position. (see Fig.6)

The source functions for excitation processes, (Fig.5) are slightly different because of the different shape of the excitation cross sections. Probabilities for excitation are maximal for electron energies of about 20-30eV. Therefore, the maximum of any of the excitation source function must be at the larger distance from the cathode fall - negative glow boundary, in the area where electron energy drops in the 20-30eV range. Furthermore, because of the electric field in the cathode fall region, kinetic energy of a electron is lower than in the negative glow and an additional peak on the excitation source function will appear in this region. (see Fig.5.)

In Fig.7 and Fig.8 the electron energy distribution function is given obtained by the Monte Carlo simulation for electrons with $E \geq 20$eV and for pressures $p=2.0$torr and $p=10.0$torr. One can see that for pressure $p=10.0$torr the number of electrons with initial energy of about 150eV decreases rapidly toward the axis becoming negligibly small in the middle of the cathode. For $p=2.0$torr group of the electrons with the initial energy exists on the cathode axis and we can talk about real electron beam.

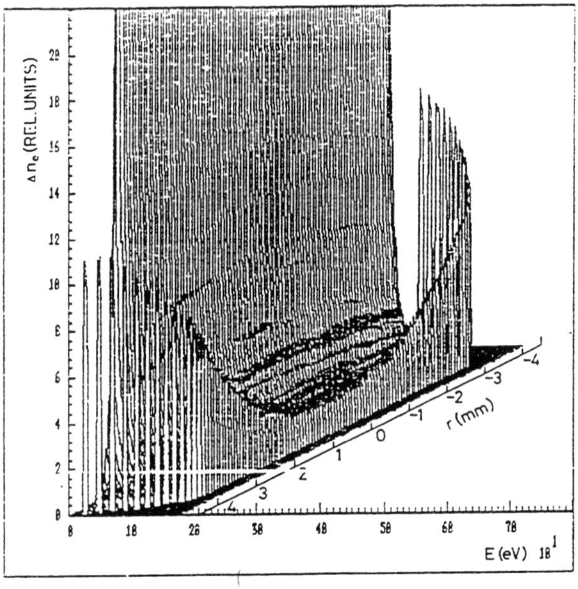

Fig. 7.

Calculated electron energy distribution function for p=270Pa

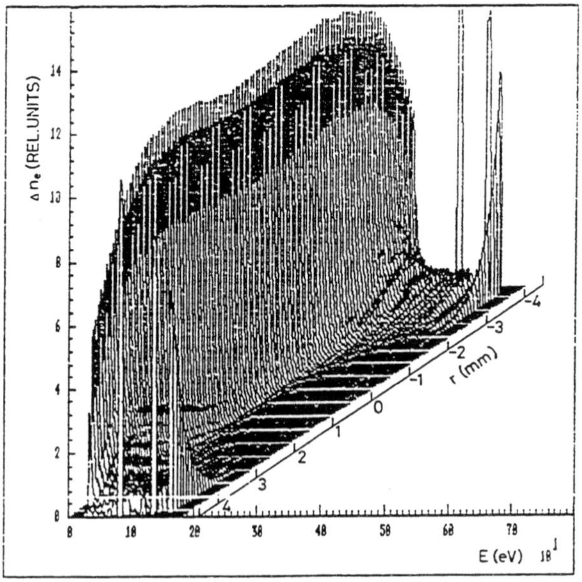

Fig. 8.

Calculated electron energy distribution function for p=1330Pa

6. Conclusions

We are now able to compare experimental data shown in Fig. 4 with a spatial distribution of atoms in $1s_5$ and $1s_4$ states predicted by the equation (7). To do this the diffusion equation (7) has been solved numerically taking for the source function S(r) values from the Monte Carlo simulation. The results are shown in Fig. 9. The equation (7) is not very sensitive on the value of α. For level $1s_5$ we estimated $\alpha=2000s^{-1}$ and for $1s_4$ $\alpha=20000s^{-1}$.

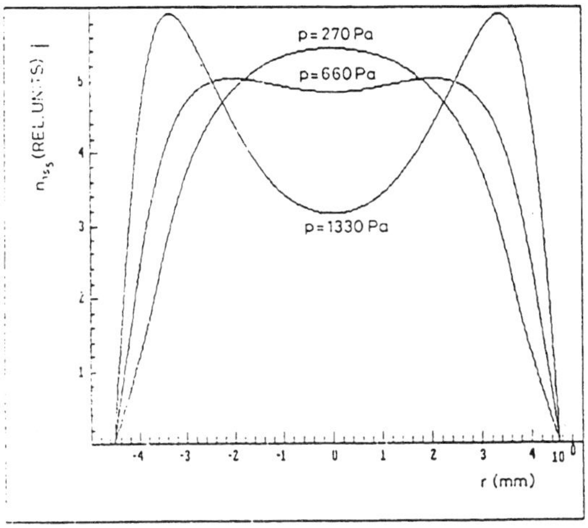

Fig. 9. Calculated density of $1s_5$ neon atoms vs radial coordinate for three values of the gas pressure

We note that results presented in Fig. 9 are obtained taking into account only contribution of the beam electrons. This is correct for low pressures, but for pressures above 5torr contribution of the thermalized electrons increases. However, characteristic pot-like shape of the radial profile of atoms excited in $1s_5$ states is due to the beam electron excitation.

Generally speaking, for low filling gas pressure, plasma in the hollow cathode is transparent for the beam electrons and discharge is maintained purely by the high energy electrons. For pressures above 10torr the electron beam does not penetrate much into the plasma, it heats just the edge of the plasma. In the central part plasma is maintained by the thermalized electrons with the energy that increases with the gas pressure. For medium pressures ($p \approx 5$torr) both of these processes are taking place in generation of the plasma.

It is important to note that for some conditions equation (7) has a simple analytical solution. Namely, if

$$\alpha \ll D/\Lambda^2 \qquad (*)$$

($\Lambda = R/2.4$) the equation (7) becomes:

$$D\left(\frac{\partial^2 u(r}{\partial r^2} + \frac{1}{r}\frac{\partial u(r)}{\partial r} \right) + S(r) = 0 \qquad (10)$$

with boundary conditions:

$$(\partial u/\partial r)|_{r=0} = 0 \quad \text{and} \quad u(r)|_{r=R} = 0$$

In our case above condition is satisfied for level $1s_5$ for pressure $p \sim 1$torr. It has been shown that for $p < 2$torr the source function for $1s_5$ is constant along the radius of the hollow cathode, e.g. $S(r) = k$ and the solution of (10) is:

$$U(r) = k(R^2 - r^2)/4D. \qquad (11)$$

Expression (11) describes a spatial distribution of metastable atoms in state $1s_5$ in a neon hollow cathode discharge for $\alpha \ll D/\Lambda^2$.

REFERENCES

Broglia M, Catoni F, Mantone A and Zampetti P
1987 *Phys. Rev. A* **36** 705

Burakov V S, Naumenkov P A and Tarasenko N V
1988 *Zh. Prikl. Spectr.* **12** 204

Den-Hartog E A, Doughty D A and Lawler J E
1988 *Phys.Rev.A* **38** 2471

Gill P and Webb C
1977 *J. Phys. B* **10** 299

Grigorev G Yu and Eletskii A V
1989 *J. Tech. Phys* **59** 90 (in Russian)

Kawakita K, Nakajama T, Adechi Y, Maeda S and Hirase C
1983 *Opt.Commun.* **48** 121

Keller R A and Zalewski E F
1980 *Appl. Optics* **19** 3301

Keller R A, Warner B E, Zalewski E F, Dyer P, Engleman Jr. and Palmer B A
1983 *J.Phys. (Paris) Colloq.* **44**, C7-23

Kuen I, Howorka F and Stori H
1981 *Phys.Rev.A* **23** 829

Labahn R W and Colloway J
1964 *Phys.Rev* **135** 1539

Labat J M and Bukvic S
1988 *J.Phys. D: Appl.Phys.* **21** 1396

McNeil J R, Collins G J and DeHoog F J
1979 *J.Appl.Phys.* **50** 6183

Parker A B and Johanson P C
1971 *Proc.R.Soc. A* **325** 511

Person K B
1965 *J. Appl. Phys.* **36** 3086

Present R D
1958 *Kinetic Theory of Gases* (McGraw-Hill Book Company)

Press W H, Flannery B P, Teukolsky S A and Vetterling W T
1988 *Numerical Recipes,* (Cambridge University Press)

Register D F, Trajmer S and Steffansen G
1984 *Phys.Rev.A,* **29** 1793

Steenhuijsen L W G 1981 *Beitr. Plasmaphys,* **21** 301

Tihonov A N and Samarskii A A
1966 *Equations of Mathematical Physics (in Russian)*
(Nauka, Moscow)

Tron Ngoc An, Marode E and Johanson P C
1977 *J.Phys.D: Appl.Phys.* **10**, 2317

van der Mulen J J, van der Sijde B and Schram D C
1980 *Phys Lett.* **79A** 51

van Veldhuizen E M
1983 *PhD Thesis* Eindhoven University of Technology

van Veldhuizen E M, de Hoog F J and Konings L U E
1983 *International Conference on Phenomena in Ionized Gases, Contrib. Papers of ICPIG '83 (Dusseldorf)* p. 406

Warner B E and Person K B
1979 *J. Appl. Phys.* **50 (9)** 5694

Low Temperature and
General Plasmas
M. Milosavljevic and Z. Petrovic

ISBN 1-56072-399-8

PREPARATION AND PROPERTIES OF DIAMOND LIKE FILMS

B. Tomčík

Institute of Physics, P. Box 57, 11001 Belgrade, Yugoslavia

Introduction

The history of carbon films is as long as that of the glow discharge techniques in hydrocarbon gases. These films often appeared as the undesirable product which was hard to remove or clean from the reactor wall made mainly of glass. The first systematic investigations of the film properties started with the works of Aisenberg and Chabot [1], followed by Holland and Ojha [2,3,4] and Weissmantel [5]. Since that time great number of experimental techniques has been employed with a common aim to produce diamond or diamond-like films with the properties as more close to diamond as possible. The most of these techniques used various kinds of glow discharges under a non-equilibrium conditions as a response to the established industrial diamond synthesis with a high pressure and high temperature procedure in the thermodynamic equilibrium.

With a large success diamond-like films can be easily deposited on various kind of substrates including metals, ceramics, semiconductors and numerous glasses. In many cases the high internal film stresses, with it connected low adhesitivity as well as an opacity in the visible region, limit the application of the films. Up to now many of the review papers about the diamond and diamond-like films have been published [6,7,8,9,10].

Diamond-like films are often marked as a-C:H films, to the analogy of a-Si:H films. The abbrevation points out the amorphous structure in which there are some hydrogen atoms incorporated- hydrogenated amorphous carbon. These films could be distinguished from other hydrocarbon compounds by means of e.g. diagram grammatom density of atoms to the atomic percentage of hydrogen being in the film [11,12].

Depending on the art of production amorphous hydrogenated carbon films contain about 20 to 60 at.% hydrogen. In the plasma sustained techniques of the film formation, the energy of the impinging species together with the substrate temperature is the main parametar determining the structure and properties of a-C:H film.

According to the mean energy of particles incorporated in the film growth, a-C:H films could be divided into two groupes: a) hard diamond-like films and b) polymer films [13]. The hard a-C:H films are usually obtained at the impinging ion energies between 100-300 eV. The resulting highly crossed structure with the prevailed C-H bonds shows the large volume stresses. Films are usually deposited at the room substrate temperature. By heating to the temperature over 300^{0}C, either during deposition or later on, the grahitisation takes place as a concequence of a hydrogen deliberation. This lead to the lost of desirable mechanical properties - high microhardness and good abrasion resistance against many foreign influences.

Polymeric films are usually obtained at the ion impinging energies less than 50 eV. The structure is poorly crossed and consists mainly from the long C chains with a lot of CH_3 radicals. The high optical gap, low mechanical hardness and small internal stress are the common features of polymer films.

The new class of carbon films-polycrystalline diamond film is of the most interest nowdays. The working definition of a diamond film is such a coating that a) shows a crystalline morphology on electron microscopy, b) has a single phase crystal structure revealable either by X-ray or electron diffraction and c) posses a Raman spectar typical of a crystalline diamond structure.

There are many experimental techniques applicable to the diamond film synthesis. But still there are difficulties about: a) film deposition on different substrates, b) too high film surface roughness as a result of the microcrystal growth independently from each other, c) small deposition rates and d) the need of a high substrate temperature.

The concentration of CH_3 methyl radicals is the most critical parametar for the proper film growth. It is believed that the hydrogen reacts with graphite disabling the creation of sp^2 bonds and favouring the tetrahedral combination among the free carbon atoms.

Experimental techniques

Among the numerous carbon film deposition techniques it is possible to make classification according to the relevant parameters involved: carbon precursor, working pressure, ion energy, substrate temperature etc. Initial carbon can be used as a solid target to be sputtered or evaporated. For the sputtering of a target an argon ion beam at energies around 1 keV is mainly used. It is possible to enhance a film thickness, transparency and crystallinity by the application of an additional ion beam oriented to the growing film at lower energy, e.g. around 500 eV [14]. The scheme of the used ion beam techniques is shown in Fig. 1a.

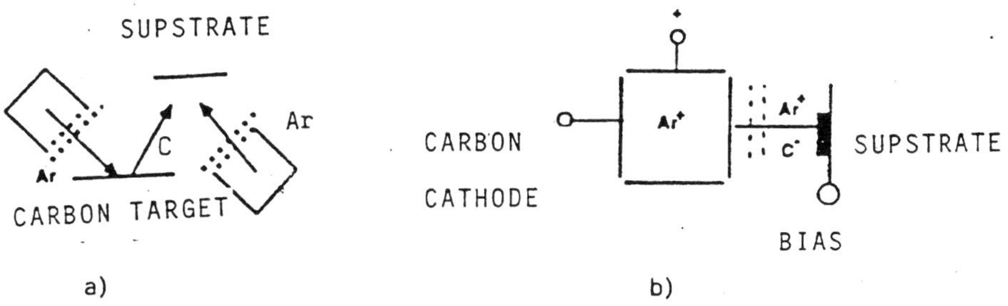

Figure 1. Carbon film deposition from the solid carbon target:
a) sputtering in ion beam arrangement
b) dc or rf excitation with the extraction electrodes-grids

The erosion of the solid target can be achieved in a dc arc discharge or by sputtering in rf discharge at lower pressures, as shown in Fig. 1b. The energy of the ion beam can be changed in a broad range. In the flux of deposited species beside the carbon atoms there is some amount of argon ions being partially incorporated in the film. In a vacuum arc the clusters of carbon atoms were observed leading to the unwanted film properties. Various hydrocarbons can also serve as the carbon precursors. The increase of deposition rate was observed with the use of higher orders alkanes C_nH_{2n+2}. Remarkable high film growth rate is achieved in benzene C_6H_6. With the higher molecular weight of the gas used increases its ionization probability. Therefore, the larger number of carbon atoms per ion contributes the higher film growth rate.

Various kinds of glow discharges are used for the gas decomposition into chemically active species such are ions and radicals. It is found in a-C:H films that the use of different kinds of hydrocarbons do not have a predominant influence on the resulting film properties. With a good reproducibility a-C:H films can be deposited in a broad pressure range, from 10^{-1} to 10^{-4} mbar and from dc to the microwave excitation. In low pressure plasma (10^{-3} - 10^{-4} mbar) the energy of ion species can be directly measured. For the estimation of the ion impinging energy in the pressure range 10^{-1} -

10^{-3} mbar, a dc self-bias potential developed on the rf coupled electrode or the voltage amplitude of the power generator is monitored.

During film deposition at higher pressures the non-elastic collisions between plasma ions and neutrals take place in a dark region. The symmetric charge exchange between ions and neutrals of the same type in Ar-discharge is observed as the main lost mechanism of the ion energy in a near electrode region. Additionally, numerous ion-molecular reactions in hydrocarbons are responsible for the ion energy lost in the electrode vicinity.

The self-bias potential is used as the main macroscopic parameter for the comparison of the film qualities in rf discharges. One empirical formula based on the energy measurement of ions extracted through the small hole on a cathode relates the mean ion energy E, gas pressure p and self-bias potential V_b as

$$E = V_b/p^m \ , \ 0 \leq m \leq 1 \qquad (1)$$

If the mean free path of a particle is λ and L - the thickness of a plasma boundary region, the two boundary conditions may be defined:

1) collisionless acceleration $\lambda > L$, $E \sim V_b$, $m=0$
2) collisional domination $\lambda < L$, $E \sim V_b/p$, $m=1$

For the estimation of the actual ion energy the values of the self-bias potential V_b on the coupled electrode and of the gas pressure must be known.

The higher deposition rate in rf discharges at 13.6 MHz in comparison to the low-frequency excitation at 50 KHz, as well as a need to deposit a-C:H films on isolators, resulted in a wide-spreaded use of rf glow-discharges.

The "in situ" functional investigation of the film quality versus incident carbon ion energy and its dosage has been performed with mass separation of the carbon beam through the laser dissociative ionization of CO and CO_2 gas [15,16].

Under certain conditions and elevated substrate temperatures, e.g. over 600^{0}C, the monocrystalline diamond films can be deposited. The methane-hydrogen gas mixtures containing only couple of at.% C, or a mixture with oxygen containing organic molecules that may even have a nitrogen, can lead to the successful diamond film synthesis [17]. In spite of the numerous applied experimental techniques, diamond films still posses many imperfections that limit the range of their potential applications nowdays.

Recently, it was shown that the C-H-O atomic gas phase composition, independently of the deposition technique used and the type of the gases introduced, is a key factor determining a type of the film to be obtained: diamond film, diamond-like film or carbon film at all [10]. The model predicts the existance of a relatively large diamond synthesis domain around the so called CO line in a triangular atomic gas phase diagram. Each side of the triangle represents the atom fraction of the gas phase of one of the binary systems C-H, H-O and O-C.

Film structure and composition

The basic structure of a-C:H films is a random network of carbon atoms in tetrahedral coordinations with nearest neighbour atoms. The deviations of the bond lenghts and angles from the tetrahedral diamond lattice arrangement are small even in the cases when the distribution of the first and higher neighbours is strongly disturbed. Based on the measurement of the radial distribution function of diffracted electrons and on the theoretical calculations of the profile of radial distribution function of diffracted electrons in diamond crystals of various sizes, it was found that a-C:H films posses short range diamond structure with the crystallites sizes not larger than 0.6 nm [18,19]. According to the IR apsorption spectra analysis and taking into account the dimensions of a diamond lattice (0.15 nm) and graphite (0.14 nm), it may be concluded that the basic tetrahedral carbon network is modificated with C=C double bonds (sp^2 configuration) and the hydrogen incorporated. In average, one carbon atom is connected to 0.5 hydrogen atoms through the CH and CH_2 configurations.

Heating the film to the temperatures over 350^{0}C or exposing it to argon ion bombardment with energies over 1000 eV leads to the transformation of the film structure through hydrogen deliberation i.e. conversion of sp^3 bonds into sp^2 bonds [20]. Beside carbon and hydrogen as the main film constituents, there are always some trace elements presented: oxygen, nitrogen or argon. Oxygen originates mainly afrom the ocluded water vapor on the chamber walls and its subsequent dissociation upon ignition of the glow discharge. Nitrogen is present as a residual gas in a chamber or as a constituent of the gas mixture used.

From the practical point of view of great importance is a knowledge of the surface and in-depth film composition as well as of the nature of the film interface to the substrate.

For the measurement of the integral hydrogen content in a film the chemical burning method of a film deposited on a platin substrate is most widespreaded. Effusion measurements, if applicable, can also give the value of a total hydrogen content-bounded and intersticially not bounded. Depth distribution of a total hydrogen content can be revealed using one of the nuclear analytical methods e.g. by means of ^{15}N-nuclear resonant reaction [11].

$$^{15}N + {}^{1}H \rightarrow {}^{12}C + {}^{4}He + \gamma \qquad (2)$$

High energy nitrogen ions fall onto the film, penetreting it till the certain depth where they still posses an energy of 6.4 MeV, neccessarry for the above resonant reaction. It is possible to measure the concentration of hydrogen at different film depths with changing the ion incident energy. The hydrogen concentration is determined by measuring the intensity of the accompained 4.43 MeV γ-radiation.

The total concentration of only bounded hydrogen can be determined using the integrated intensity of C-H absorption bands in the wavenumber region 3000-2840 cm^{-1} [11,22].

The compositional in-depth analysis of diamond-like films is usually performed using one of the electron or mass spectroscopy methods: Auger electron spectroscopy AES, photoelectron spectroscopy XPS or secondary neutral mass spectroscopy SNMS. In Fig.2a is shown one electron beam SNMS depth profile of diamond-like film deposited on Si substrate [22].

The homogenous carbon depth profile, under the assumption of the stable film deposition conditions, must be accompained with uniform hydrogen distribution along a film depth. If there is a substantiall amount of oxygen incorporated in the film than a dark brown film colour becomes yellowish. A depth scale corresponds to the erosin rate measured on a film using 5-keV Ar^+ ions at incident angle of 70^0.

Nature of the film interface to the silicon substrate can be determined e.g. using the chemical shifts of Si or C peaks in the energy spectra of analyzing electrons as shown in Fig. 2b. After the mechanical peeling of a film from the substrate, i.e., prior the Ar ion bombardment, a considerable amount of oxidic and carbidic contribution is revealed. With sputter removal of atomic layers at first disappears the oxide contribution and afterwards a carbide one. The width of the interface is of the order of a couple of nanometers.

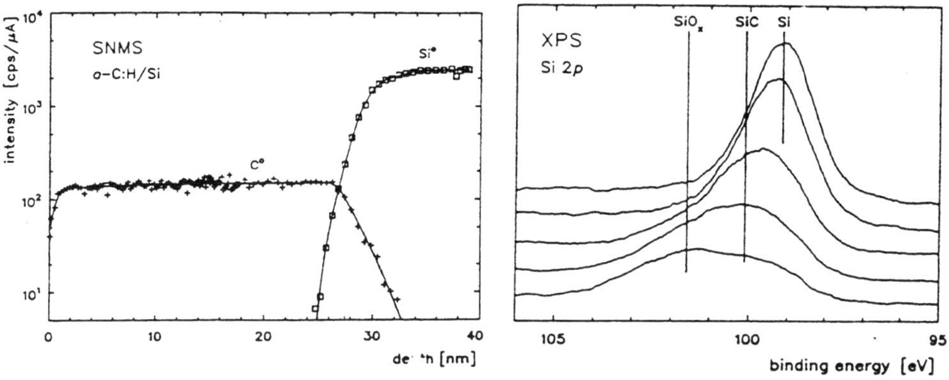

a) b)

Figure 2 a) Electron beam SNMS depth profile of the main constituents of an a-C:H film close to the Si substrate
b) XPS Si 2p lines during depth profiling of the interface after mechanical removal of a layer. Depth increases from bottom to top curves [22].

The adhesion of the film is influenced by the overlap of the interface to the surrounding layers of the substrate and the film itself and to the respective lattice matching.

Mechanical and electro-optical film properties

Internal stress

Desired mechanical and electro-optical film properties play an important role in the selection of necessary deposition parameters and deposition techniques. Mechanical properties as the film stress, adhesion, microhardness, friction coefficient, stability against foreign influences are in many cases of vital importance for the application of a film as a biocompatible coating, infra-red transparent and protective coating etc.

The total internal stresses σ_{tot} are inevitable in a film during film growth and are the result of: thermal σ_{therm} and volume σ_{vol} stress, stress in the interface film-substrate σ_{int} and external stress σ_{ext}

$$\sigma_{tot} = \sigma_{therm} + \sigma_{vol} + \sigma_{int} + \sigma_{ext} \quad (3)$$

Thermal stress is present when a thermal expansion coefficient of a film is different from that of a substrate or if the film is deposited at temperature different from that at which the stress is evaluated. With diamond-like films deposited at room temperature this contribution may be neglected.

Volume stress originates on the structural film disorder as well from the side of impurities as from the atoms of a film with energy positions outside of the potential minimum. Different bond lenghts and distorted bond angles between the atoms of a film during its growth contribute also to the volume stresses.

Stresses in the interface are caused from different lattice parameters of a film and a substrate. There is also an influence from the foreign atoms-impurities short after initiating the discharge. The best matching between film and substrate is achieved on nickel underlayer.

It is possible to change the intensity of volume stresses by varying the deposition parameters: substrate temperature, deposition rate, incident angle of a beam, residual gas pressure, energy and composition of a beam. Incorporation of inert or reactive gases can lead to the complete change of a stress i.e. from compressive strains into the tensile strains.

The total stress in a film can be determined from a curvature of a thin strip caused by the deposited film, using a formula [23]

$$\sigma = \frac{4\,E\,D^2\,\delta}{3\,L^2\,(1-\nu)\,T} \quad (4)$$

where E is Young's modul of a substrate, D is thickness of a strip, δ is sagitta at the center of a strip, L is the lenght of a strip, ν is the Poasson's substrate coefficient and T is the film thickness.

An increase of energy of deposited ions leads to decrease of total stresses in a-C:H film produced in methane at room substrate temperature. Typical a-C:H films posses large compressive stresses of the orders of magnitude up to 10^{10} Pa[12]. These internal stresses can rise with increase of a film thickness and very often lead to the spontanous delamination of a film from the substrate.

Developing of sp^3 C-H and C-C structure at the expence of C=C bonds results in a volume enlargment of a film and in the increase of the bond lenghts between atoms (from 0.134 nm to 0.154 nm)[20]. The enlarged volume of a film is always followed by the compressive stress.

Adhesion

The adhesion of diamond-like films is of the most practical importance. The best adhesion is achieved on substrates that form carbides in the interface e.g. Si, Fe, Ti and Mo. Good adhesion is also obtained on germanium and quartz, but a bad one is observed on Au, W and GaAs [22].

Worthmentioning are the improvements of a-C:H film adhesion on a substrate previously covered with a thin metal film. The procedure can be simplified with the substrate pretreatment through the e.g. argon sputtering and subsequently metal film sputter deposition. Gradually, the concentration of hydrocarbons prevails in the mixture with argon and on the substrate the a-C:H film can grow over carbon-metal, metal, metal-substrate and substrate layers.

Various film deposition techniques were used for the improvement of the film abrasion resistance. Good results were achieved by incorporating into the film a small percentage of W and Ta atoms [24].

For the quantitative determination of the film adhesion the "scratch test" method can be applied. The film adhesion σ_A (N/mm^2) can be determined knowing a critical force at which the specially formed diamond indenter provokes the film delamination from the substrate. The ahesion can be calculated using a suggested formula [24]

$$\sigma_A = k\left(\frac{L_c\, H}{\pi\, R^2}\right)^{1/2} \qquad (5)$$

where L_c is the critical force (N), H is the substrate hardness (N/mm^2), R is a top radius of the diamond head (mm) and k is a coefficient to be experimentally determined.

A lot of parameters have an influence on a value of a critical force L_c: internal volume stresses of a film, friction coefficient between diamond indenter head and a film, film thickness e.t.c. In a temperature range between 50⁰C and 800⁰C the largest adhesion was achieved on substrates held at 300⁰C. At this temperature a mechanism of hydrogen deliberation in a molecular form was observed by means of effusion measurements. Hydrogen molecules leave the film through the void structures. Therefore this leads to the better interconnection between substrate atoms and the growing film as well as between the atoms within a film.

Friction and microhardness

The friction coefficient of diamond-like films is small in comparison to other films in use. In the literature published values are in the range 0.01 - 0.28 [25,26]. The microhardness of a-C:H films can be related to the extente of presence of tetrahedral coordination between carbon atoms and carbon-hydrogen atoms. The film microhardness can be larger for the factor 2 by increasing the hydrogen content in a film from 30 to 35 at.%. According to Angus et al.[12] the hardness of a film passes through the maximum at 100 eV ion deposition energy. Depending on the deposition parameters the most common reported values of the Knoop microhradness were in the range 1245-1976 HK. Diamond posses the hardness of 7000 HK. Because of the film elastic properties the real microhardness is somewhat lower.
For many applications the elastic behavior of a diamond-like protective film is a very desirable property.

Electrical conductivity

Four-point probe resistivity measurements reveal very pronounced dependance of the film resistivity versus ion deposition energy. With increasing an ion energy the film resistivity decreases for many orders of magnitude as a result of the smaller hydrogen content. The free carbon dangling bonds in a disordered carbon network are then less saturated with hydrogen atoms. The widespreading of π orbitals throughout the whole volume of a film is a result of the increased orbital overlapping of the carbon valence electrons i.e. the transformation of sp^3 structure into sp^2 and sp^1 [27].

The electrical resistivity measurements of a-C:H film in various directions show no signs of electrical anisotropy. The electrical conductivity can be increased with increasing the substrate temperature or by doping a film with e.g. diborane, phosphine or nitrogen [28]. In the temperature range 50-370^{0}C the electrical conductivity σ does not obey a simple temperature dependence $\sigma(T) = \sigma_0 \exp(-\epsilon_o/kT)$ over the complete range. This points out the presence of a complex conduction mechanisms typical for amorphous semiconductors. The films posses a high dielectric breakdown voltage of about 1×10^6 V/cm and a high dielectric constant of about 8 [1].

Optical properties

Diamond-like films produced by low pressure ion beam deposition are optically transparent and films with the thickness up to several hunderd nanometers can be water clear or pale yellow [1,29]. Infra-red absorption in the 1-10 micrometers wavelenght range is very low, except where C-H vibrations are excited or with the films deposited at higher deposition chamber pressure.

The optical transparency suggests an absence of free electrons that could interact with incident electromagnetic radiation implying that the films should also be insulating. The index of refraction is around 2 in the visible range. The absorption of diamond-like films can be described by relation

$$(\alpha\, h\nu)^{1/2} = B(E_g - h\nu) \qquad (6)$$

where $h\nu$ is the photon energy, α - the absorption coefficient and B is a constant. Values of the optical gap E_g are between 0.8 and 1.8 eV. With an increase of the ion energy the sharp decrease of the optical gap was observed. Since graphite exibits metallic conductivity without band gap, while diamond is semicondacting with a band gap of 5.45 eV, once more is confirmed that the desired film properties could be achieved by proper selecting the deposition parameters: mainly ion energy and substrate temperature.

Applications

Many of the possible applications arise from the unique properties of diamond-like films: transparency, electrical insulation, extraodinary chemical resistance to most acids, bases and solvents, barrier properties, smootness, hardness, low coefficient of friction, wear resistance, good adhesion to many substrates and deposition on low temperature substrates.

The films already found application as a protective, scratch and abrasion resistant coating on plastic lenses with a high transmission in the visible and infra-red region. Another proposed optical application is as a transparent protective coating for optical disks. Films can serve as a hermetic barrier to extend the disks archive lifetime.

Also, optical sensors to be immersed in corosive enviroment could be covered with thin optically transparent diamond-like films.

As a hermetic seal (moisture barrier) these films may find a wide application as a protective cover of optical fibers. Because of the high atomic density of a film lattice, water vapor can not reach the fiber surface and provoke the stress corrosion of it.

As a blood and tissue compatible coating diamond-like films were used as a coverage of other artificial organs and biological implants. In spite of a remarkably good coverage if the plastic substrates have structural defects larger than 100 nm the film of the same thickness can not completely cover that defect and a danger of a blood clot formation can not be excluded.

As a wear protection coating these films already demonstrated the ability to protect the Winchester disks from head crashes.

Ion beam deposited diamond-like films can exibit a thermal conductivity greater than 700W/mK [30]. This combination of high thermal conductivity and high electrical resistivity may promote the film as a good dielectric layer for high temperature devices.

References

[1] Aisenberg S. and Chabot R., *J. Appl. Phys.* 42(1971)p.2953
[2] Holland L. and Ojha S.M., *Thin Solid Films* 38 L17(1976)
[3] Holland L. and Ojha S.M., *Thin Solid Films* 48 L15(1978)
[4] Holland L. and Ojha S.M., *Thin Solid Films* 58 107(1979)
[5] Weissmantel C., Schürer C., Fröhlich R., Grau P. and Lehmann H., *Thin Solid Films* 61 L5(1979)
[6] Despandey C.V. and Bunshah R.F., *J. Vac. Sci. Technol.* A7(3)(1989)p.2295
[7] Messier R., Badzian A.R., Badzian T., Spear K.E., Bachmann P. and Roy R., *Thin Solid Films* 153(1987)p.1
[8] Weissmantel C., in *Thin Films from Free Atoms and Particles*, Kenneth Klabunde editor, Academic Press, NY,1985,p.158
[9] Green D.C., Mc Kenzie D.R. and Lukins P.B., *Materials Science Forum* Vols. 52 & 53 (1989)p.103

[10] Bachmann P.K., Leers D. and Lydith H., *Diamond and Related Materials*, 1(1991)p.1
[11] Angus J.C., Stultzt J.E., Schiller P.J., Mc-Donald J., Mitrich M. and Domitz S, *Thin Solid Films* 118 (1984) 311
[12] Angus J.C., Koidl P. and Domitz S.in CRC "*Plasma Deposited Thin Films*" ed. Mort J. and Jansen F., Cleveland 1986
[13] Dischler B., Sah R.E., Koidl P., Fluhr W. and Wokaun A., *Proc. 7th Int. Symp. on Plasma Chemistry*, Eindhoven (1985)
[14] Fujimori S., Kasai T. and Inamura T., *Thin Solid Films* 92 (1982) 71
[15] Miyazawa T., Misawa S., Yoshida S. and Gonda S., *J. Appl. Phys.* 55 (1) 1983 188
[16] Kasi R.S., Kang H. and Rabalais J.W., *J. Chem. Phys.* 88(9)1988 5914
[17] Simpson M., *New Scientist*, 10 March 1988,p.50
[18] Craig S. and Harding G.L., *Thin Solid Films* 97(1982)p.345
[19] Moss S.C. and Graczyk J.F., *Phys.Rev.Lett.*, 23(1969)p.1167
[20] Couderic P. and Catherine Y.,*Thin Solid Films* 146(1987)p.93
[21] Dischler B., *E-MRS Symp.Amorph. Hydrogenated Carbon Films*, Les Edition de Physique, Les Ulis, Strasbour 1987
[22] Sander P., Altebockwinkel M., Storm W., Wiedmann L. and Benninghoven A., *J.Vac. Sci. Technol.* B 7(3)1989 p.517
[23] Brenner A. and Senderoff S., *J. Res. Natl. Bur. Stand.*, 42(1949)105
[24] Haefer A.R., *Oberflächen-und Dünnschicht-Technologie, Teil 1 Beschichtungen von Oberflächen*, Springer-Verlag 1987
[25] Marinković Z. and Roy R., *Carbon* 14, 329(1976)
[26] Okada K., *J. Vac. Sci. Technol.*A7(2)(1989)p.132
[27] Gambino R.J. and Thompson J.A., *Solid State Communication*, 34,15(1980)
[28] Jones D.I. and Stewart A.D., *Philosophical Magazine B*, Vol.46, No 5(1982)p.423
[29] Spencer E.G., Schmidt P.H., Joy D.C. and Sansalone F.J., *Appl.Phys. Lett.* 29, (1976) 118
[30] Yoshida M., Ogawa K.,Tsuji H., Ishikawa J. and Takagi T., *Proc. 11th Symp. on ISIAT87*, Tokyo (1987)

Low Temperature and
General Plasmas
M. Milosavljevic and Z. Petrovic

ISBN 1-56072-399-8

NEUTRON EMISSION DETECTION IN PLASMA FOCUS AND CAPILLARY FUSION EXPERIMENTS

A. Zarić and D. Šević,
Institute of Physics, Beograd
11080 Zemun, Pregrevica 118, Yugoslavia

1. INTRODUCTION

In order to study the mechanisms and dynamics of fusion reactions in simple capacitor bank discharges into deuthered media, plasma focus as well as capillary fusion experiments was prepared.

Investigation of the properties of a plasma focus and capillary fusion facilities sets as a central task the problem of neutron detection. For this purpose two types of liquid scintillator neutron detectors was developed.

Neutron yield were measured by a large gadolinium loaded liquid scintillator (NE343) detector. This was a large spherical tank containing the liquid scintillator viewed by 12 photomultipliers and placed at 8 m from the plasma focus. Time and angular distribution of emitted neutrons were looked upon by a smaller liquid scintillation detector (proton recoil) were the NE213 scintillator was selected on behalf of its superior time response characteristics.

The plasma focus chamber is of the Mather type and consists of two brass coaxial electrodes (see Fig.1) separated by a glass insulator sleeve at one end, where the breakdown of a gas discharge at an initial density of about 10^{17} cm^{-3} takes place. The chamber has been designed for the currents up to 1 MA (10^{10} $neutrons/pulse$).

Up to now the experimental preparations for capillary fusion are completed, and the low inductance capacitor bank ($\approx 50nH$) used for PF experiments was adapted. The capillary device chamber is made of steel and represent a modified version of that of the Uppsala group (see Fig.2). The main parts of capillary device chamber are the capillary supports whose distance can be set from 1.5cm up to 3cm, the optical windows for diagnostics, the switch (triggered spark gap) and the nitrogen inlet valve. The chamber is tested for high pressure (up to 20atm) necessary to avoid discharge along outer wall of capillary.

In this report detailed description of neutron detectors and preliminary results of plasma focus and capillary fusion experiments are presented.

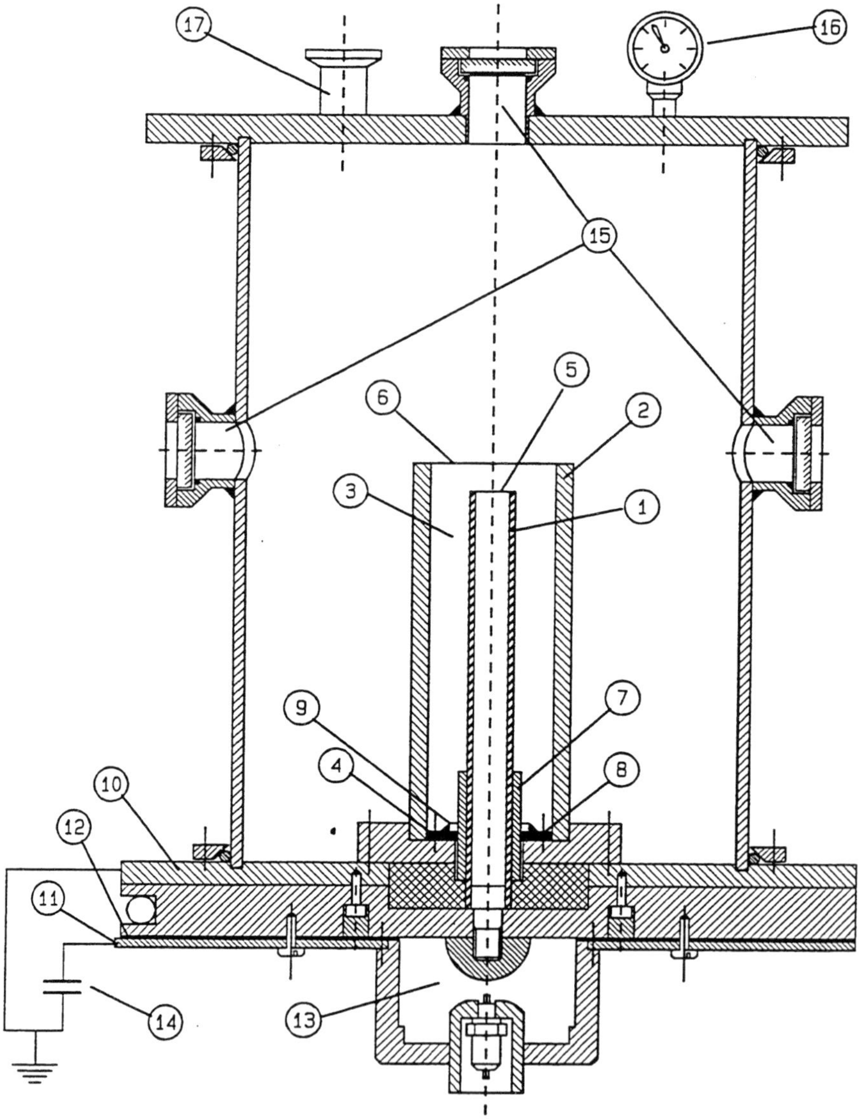

Figure 1: Shematic sectional view of the plasma focus device: 1 - inner electrode, 2 - outer electrode, 3 - interelectrode gap, 4 - breech wall, (5 and 6) - muzzle ends, 7 - insulator sleeve, 8 - field distortion element, 9 - cylindrical brass knife edge, (10 and 11) - breech brass plates, 12 - insulator layer, 13 - switch, 14 - power supply, 15 - optical window flanges, 16 - pressure control and 17 - vacuum pump flange; proportion: 1 : 3

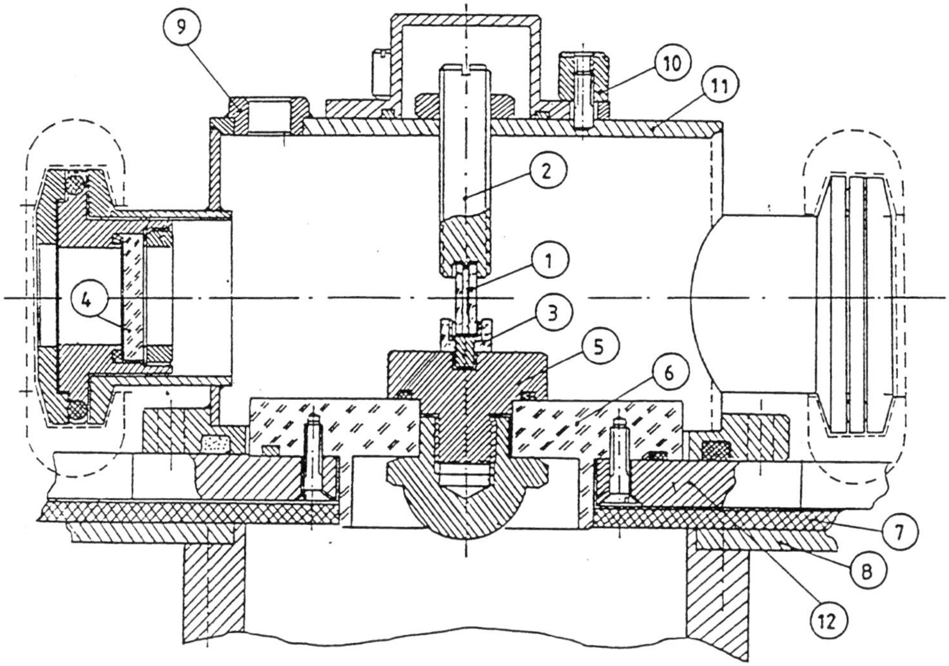

Figure 2: Shematic sectional view of the capillary fusion focus device: 1 - capillary, 2,3 - capillary support 4 - quartc window, 5 - electrode, 6 - insulator G-10, 7 - insulator, 8 - brass plate, 9 - gas inlet, 10 - safety valve, 11 - chamber 12 - stainless steel plate,;

2. SELECTION OF NEUTRON DETECTORS

Since the neutrons interact only with nuclei of target atoms, cross sections for interactions and, hence, detection efficiencies tend to be much lower than for charged particles or photons. In our plasma focus and capillary fusion facilities experiments many properties of emitted neutrons are of interest as are flux (magnitudes) angular distributions and time distributions. For this purposes we have selected neutron detectors basically satisfying the imposed requirements.

Measurements of total neutron yields are fundamental in plasma focus experiments. Thus it is necessary to select the neutron detector with maximal detection efficiency. Since thermal neutrons have the highest interaction cross sections many types of detectors are developed for this region of energy. Neutron detector types based on a large loaded liquid scintillation media have been widely used in experiments requiring high detection efficiency. The principle of this type of the counter is neutrons slowing down in a large volume of a hydrogenous liquid and absorption by loaded element [5]. In our case emitted neutrons are monoenergetic (D-D reaction), so, the measurement of the energy distribution is not necessary. So, for flux measurement we have selected a large volume liquid scintillation detector with gadolinium loaded scintillator, whose description will be given in following sections.

3. Calculation of NE343 based neutron detector efficiency using Monte Carlo method

The efficiency of a large volume liquid scintillator neutron detector strongly depends on shape and dimension of scintillator tank. In order to check detector properties before construction, simulation of detector performances, especially detection efficiency, were done using the DENIS computer program [6]. The program is organized to follow neutron detection process step by step.

Neutron are moderated by the hydrogenous material. After thermalisation and diffusion neutrons are captured by the loaded element (gadolinium $\sigma_{aps} = 3.6 \cdot 10^4$ barn). Gadolinium deexitacion by tree gamma rays emission (9 MeV total energy) and their interaction with scintillator are the second step in neutron detection. Finally, light collection and detection are considered. Consequently the total neutron efficiency of the detector is product of captured neutron efficiency, gamma rays detection efficiency and light detection efficiency. Input variables in the program are the diameter of the sphere, the initial neutron energy and the discrimination threshold of output photomultipliers signal. Capture time distribution and neutron detection efficiency as a function of discriminations threshold are output results.

Figure 3 represents calculated efficiency of neutron detection for a sphere of 1 meter in diameter as a function of the initial neutron energy. Such a detector satisfies entirely our demands about total neutron yields detection in plasma focus experiments, since the expected detection efficiency for the neutrons from D-D reaction (2.45MeV) is about 90% for source of neutrons placed in the center of the sphere. In the case of plasma focus experiments the position of the source of neutrons (plasma focus chamber) must be placed at distance of a few meters from detector (shock hazards, EMI) and the overall detection efficiency decreases about three order of magnitude.

Results of simulation for fission neutron energy spectra are used for comparison with results of calibration made by spontaneous fission californium neutron source (Fig.4 and Fig.5).

4. Construction and calibration of large liquid scintillator neutron detector

The main body of the detection system is a spherical tank of 1 meter in diameter, made of stainless steel, filled with liquid scintillating solution (Fig.6). Twelve photomultipliers (Hamamatsu R1512, 5 inches in diameter) are mounted on quartz glass viewing windows. For light collection (reflection) improvement, inner walls of the sphere are electrochemically polished. The scintillating solution was freed from oxygen with argon gas.

The photomultiplier pulses are summed in two branches (six photomultipliers in each) and via preamplifier, amplifier and discriminator transmitted to the coincidence unit. When coincident pulses are observed from both branches, signals are accepted like a neutron events and counted.

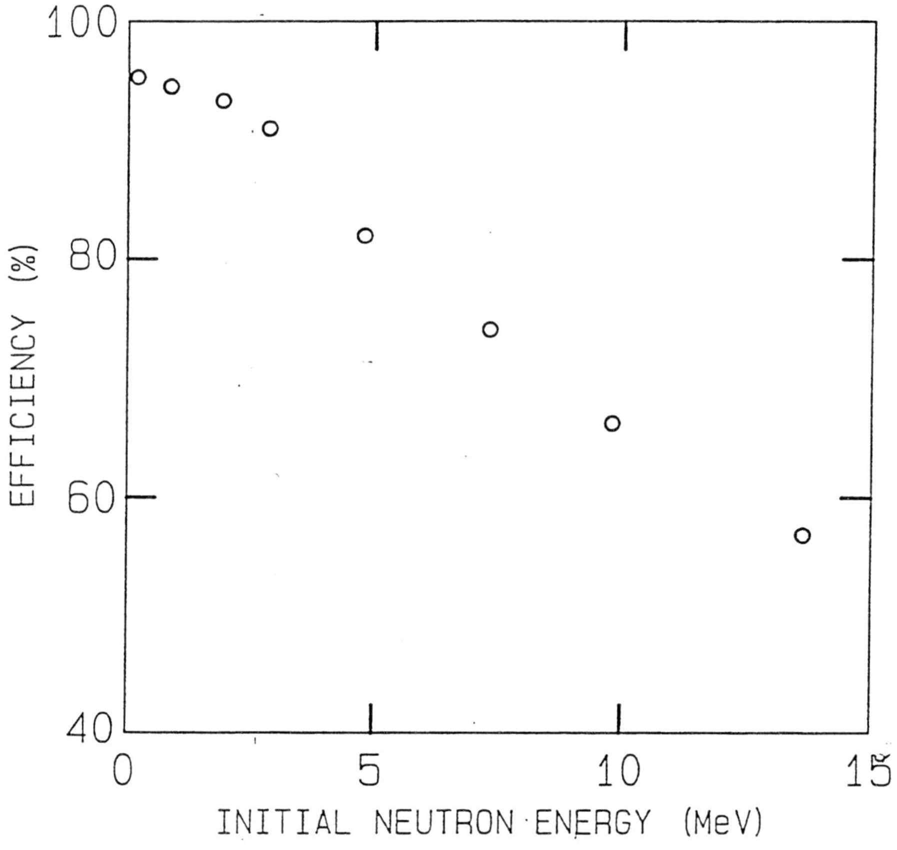

Figure 3: Efficiency of large volume gadolinium loaded liquid scintillator detector calculated using Monte Carlo method as a function of the initial neutron energy

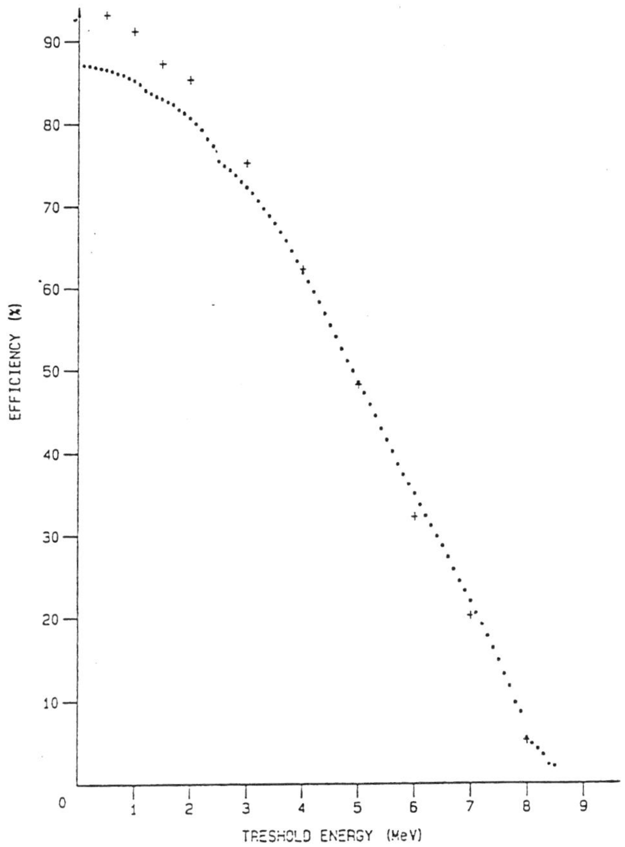

Figure 5: Neutron detection efficiency for large volume gadolinium loaded liquid scintillator detector as a function of the threshold gamma ray energy: o - calculated values and + - measured values

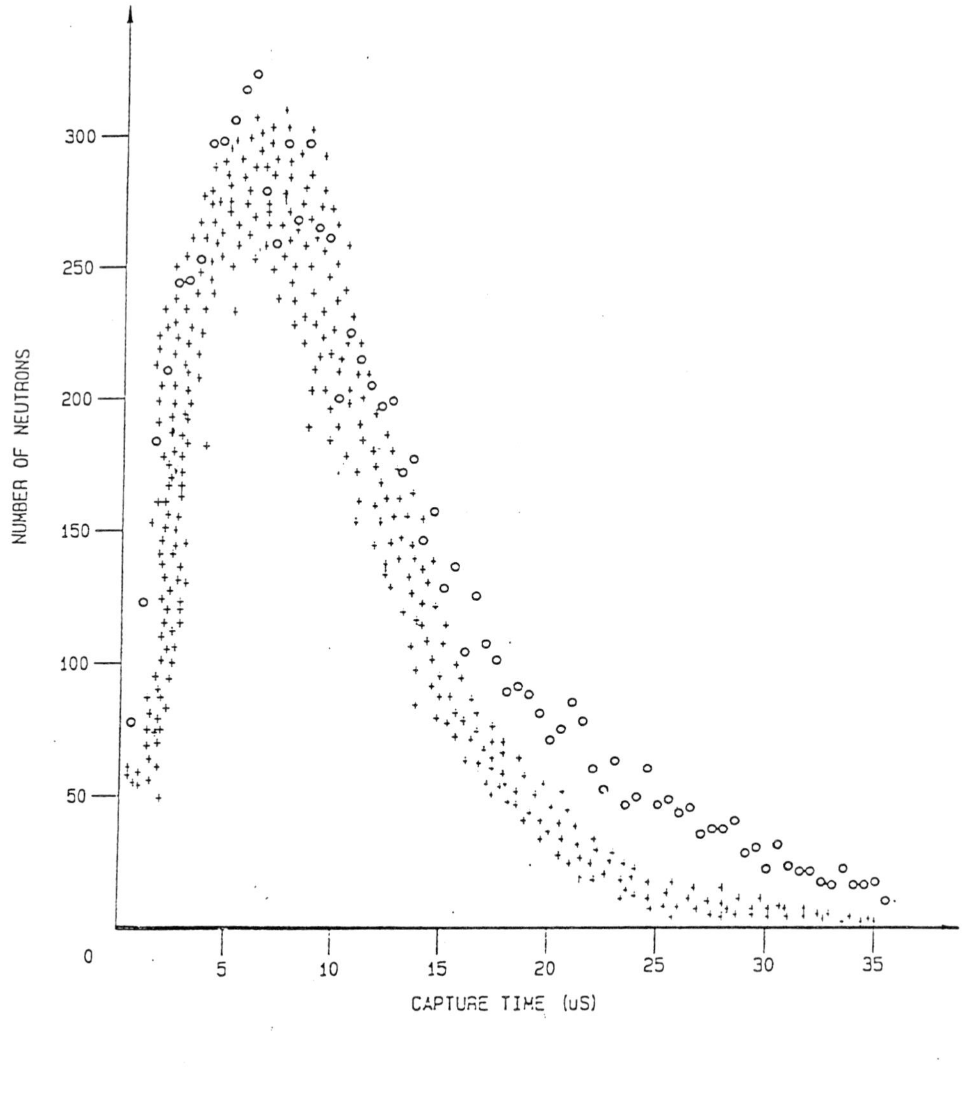

Figure 4: Neutron capture time distribution: o - calculated values and + - measured values

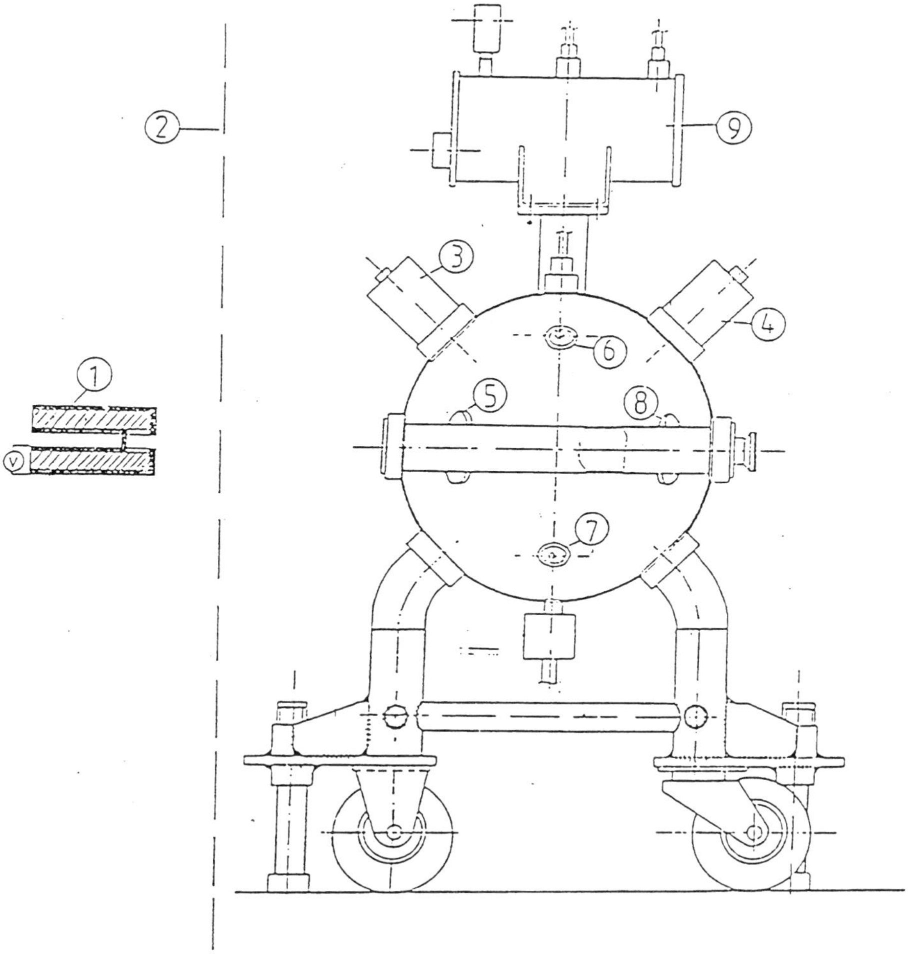

Figure 6: The large volume gadolinium loaded liquid scintillator detector: 1 - plasma focus chamber, 2 - electromagnetic shield, (3 - 8) - photomultipliers and 9 - expansion tank

Since at least two photomultipliers must give a signal, the effects of random noise from photomultipliers are minimized. Block diagram of the electronic circuit is given in Fig.7.

The capture time distributions of simultaneously emitted neutrons is broad and has a peak at about 10 μs after their production (see Fig.4). So, about 90% of all emitted neutrons (for neutron energy fission spectra) are captured within a 35 μs time interval. This fact is very important for plasma focus experiments because of high EM radiation emission due to the discharge of the capacitor bank simultaneously with neutrons.

The detection system was calibrated by a californium-252 spontaneous fission neutron source placed at the center of the spherical detector. The average neutron energy of the source of about 2.5 MeV corresponds to those of the plasma focus device. The source was placed very near to the a surface barrier counter and pulses from fission fragments were used as triggers for spontaneous fission prompt neutrons. The fission rate of the source was about one per second and neutron counting was initiated after 1μs delay to last fission. The 1μs delay was introduced to get rid of all prompt/gamma signals from the source, the 35μs gate duration was sufficient to thermalize all prompt neutron produced per fission. Since the detector is very sensitive to cosmic and laboratory background, measurements without any source were also made. The efficiency of detector is then obtained using formula $\epsilon = n_m - n_b/n_e$,

where is: $n_e = 3.72$, n_m is mean value of measured number of neutrons, n_b is background activity and n_e is the average number of neutrons emitted from Cf-252 per fission.

Detector efficiency is measured for a different gamma energy discrimination thresholds (Fig.5).

For imposed discrimination threshold corresponding to 2 MeV of gamma rays energy the detector efficiency is about 83% (see Fig.8). The neutron capture time distribution is also measured and compared with calculated values (see Fig.4). Agreement between measured and calculated values is satisfactory. So, our detection system is able to detect within a broad time interval (about 40 μs) neutrons emitted simultaneously. This is very important in the case of the plasma focus experiments since discharge produced EMI can be avoided by a few μs delay of a neutron counting start.

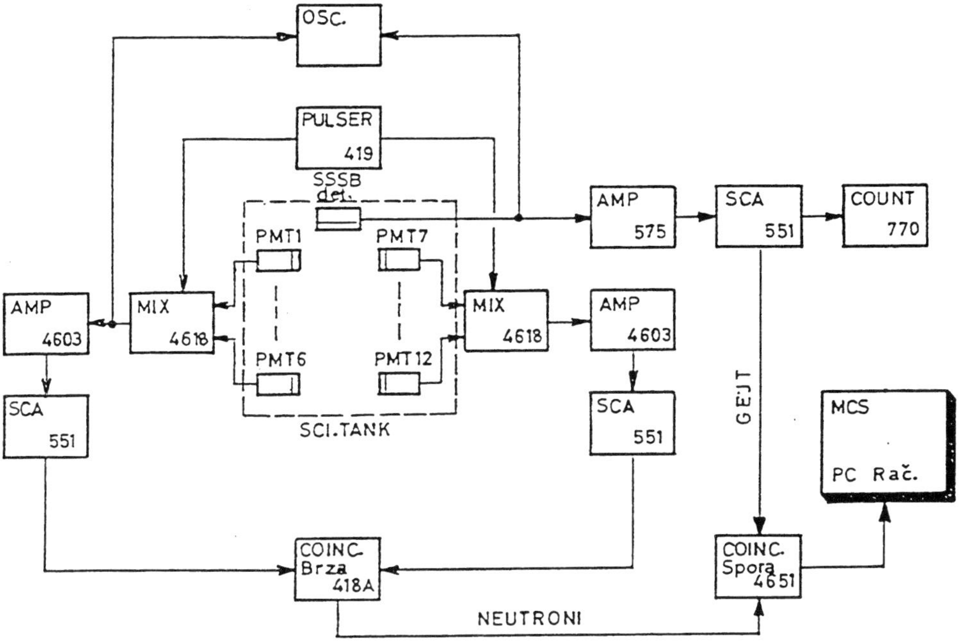

Figure 7: Block diagram of the electronic circuit for the large volume liquid scintillator detection system

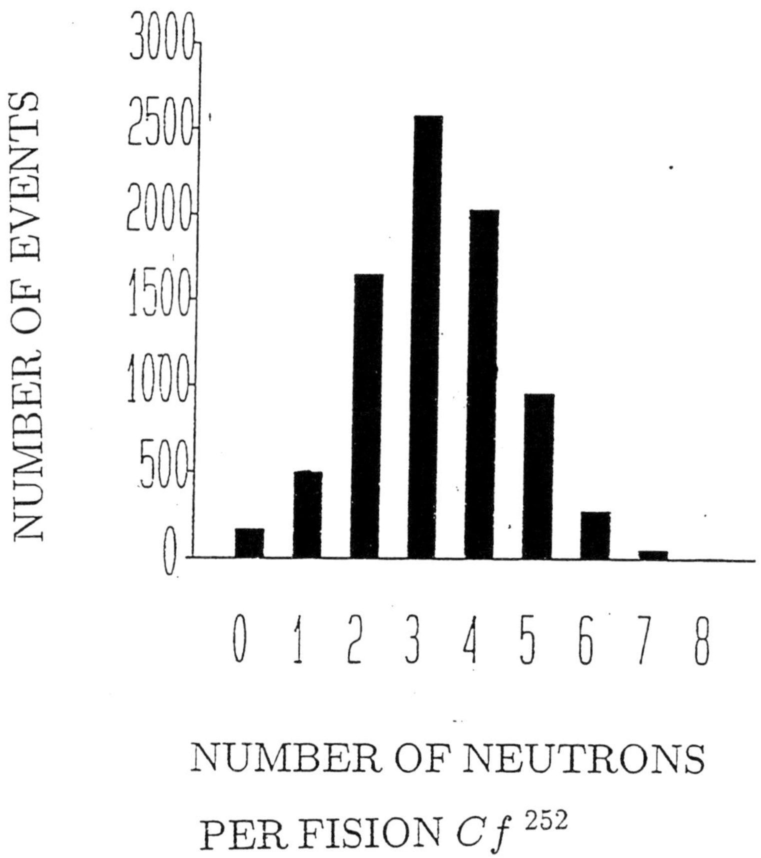

Figure 8: Neutron multiplicity distribution for Cf-252 source; average value is 3.48 and corresponding efficiency is 90%

5. Detectors for time and angular neutron distribution measurement

Because of its good time response properties NE213 liquid scintillator detector was chosen for time and angular distribution measurements. Scintillator cell is built with sizes 25 cm in diameter and height, and with a glass window 15 cm in diameter. The photomultiplier Hamamatsu R1215 is used for light detection. The possibility of neutron-gamma discrimination is also regarded because of soft X rays associated to neutrons from plasma focus.

6. Data acquisition and analysis system

Data acquisition system consist of nuclear electronic equipment (neutron detectors), probes for high voltage and current measurements, and various diagnostics for plasma characteristics. Data analysis system is comprised of PC computer and software support. Main interface between the data analysis system and the data acquisition system are digital storage oscilloscopes Tektronix 2440 (500 MSamples/s) and 2430A (150 MSamples/s). This is most convenient and accurate way of taking data from neutron detectors. Oscilloscopes also enable all voltage and current measurements. Data are transferred from the oscilloscopes to the personal computer and are thus available for numerical analysis.

Although the whole setup of experiment is properly grounded and good shielded, our nuclear electronic equipment is very sensitive to the electromagnetic interferences (EMI) generated by the discharge of the capacitor bank in the plasma focus chamber. Because of that, signals from photomultiplier tubes are mixed and recorded on the aforementioned digital oscilloscopes, which are excellently protected from EMI. Shape of signals are easy to analyze on computer, so possible errors because of EMI are reduced.

Initial idea to overcome problems connected to EMI disturbances to neutron detector measurements was to measure the EMI signal separately and to subtract it from the neutron detector signal corrupted with EMI.

There are two ways to measure EMI:

a) simultaneously, with circuitry similar to circuitry of neutron detector, on another channel of oscilloscope, and

b) successively, on the same circuitry, without neutron production (for example, plasma focus chamber filled with argon instead of deutherium).

However, results of simple subtraction of EMI signal in time domain are not promising. Main cause for this is the fact that it is not possible to obtain equal phases for the EMI signal and EMI corruption of neutron detector signal from the obvious reasons in case a), and in the case b) because of time sampling of detector and EMI signals, even if the EMI signals are reproducible, which is not exactly true.

Because they are of different origin, detector signals and EMI signals have quite different characteristics. Even a simple Fourier analysis reveals that fact. This is the real basis for the separation of detector signals from EMI. An examples of reducing EMI from the NE213 based detector shown in Fig.9 are obtained simply removing the Fourier coefficients characteristic for EMI. More sophisticated methods in the field of Digital Signal Processing are left for future work.

Preliminary results concerning neutron yields

Neutron yields were measured using the large tank liquid scintillator(NE343) which was placed at 8m distance from the plasma focus and capillary fusion chambers. So, the threshold of the detector is 10^3 neutrons per pulse. Typical results obtained for plasma focus at constant pressure of 1.3 mbar are represented in table I.

TABLE I

Energy [kJ]	2.0	2.2	3.2	4.4	5.8	9.0
Voltage [kV]	9.2	10	12	14	16	20
J_{max} [kA]	200	210	240	290	350	450
Yield [n/pulse]	$< 10^3$	8×10^5	1×10^5	2×10^6	2×10^5	$< 10^3$

At higher pressure the neutron yields are also higher and so, at 4.5 mbar the maximal neutron yield is about 10^9 neutrons per pulse(see figure

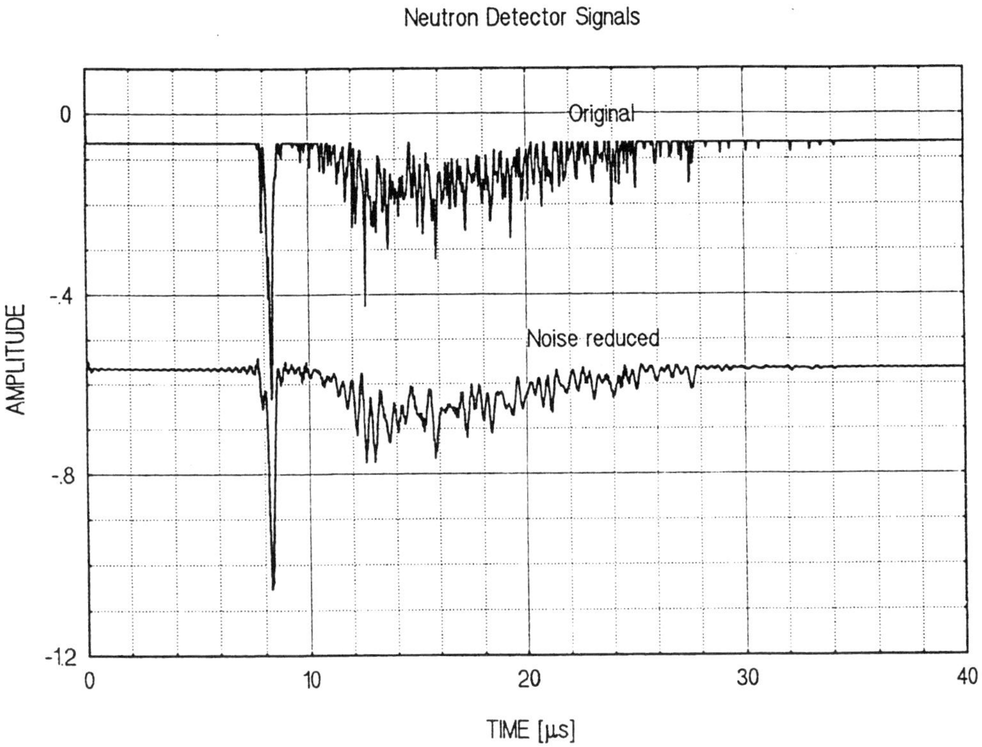

Figure 9: EMI reduction in the case of NE213 based detector

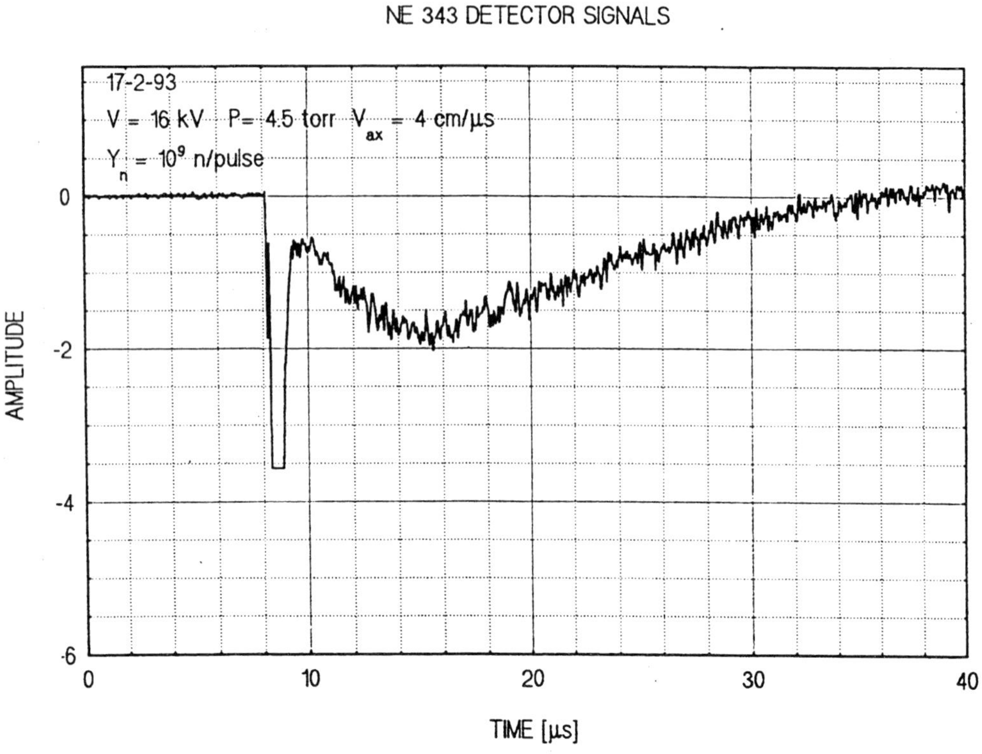

Figure 10: Signal from NE343 based neutron detector for plasma focus

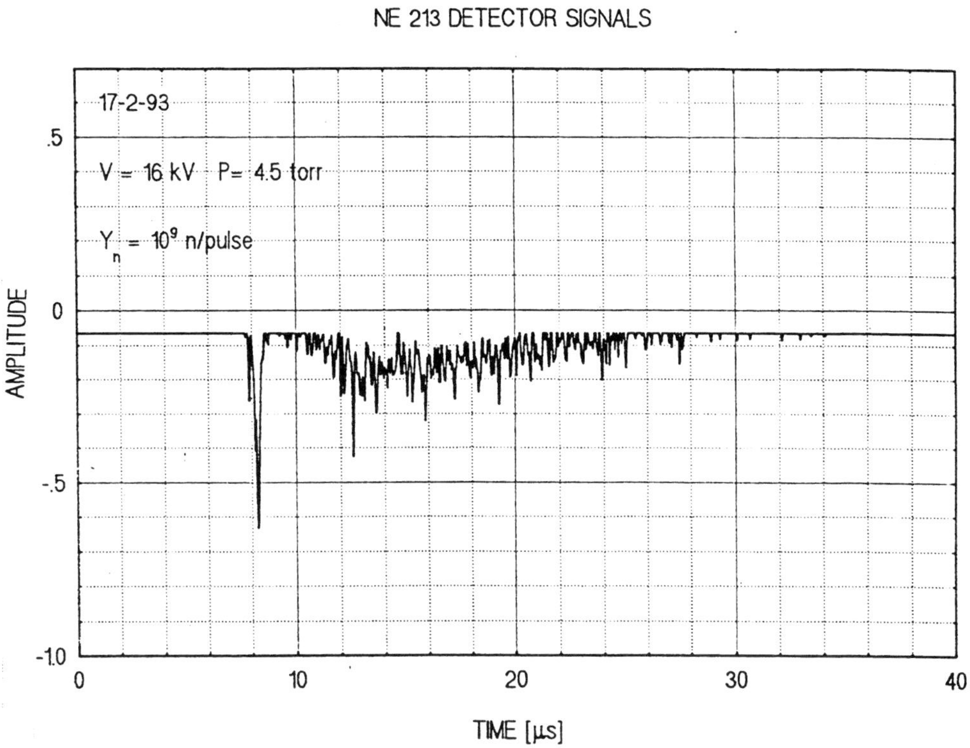

Figure 11: Signal from NE213 based neutron detector for plasma focus

10). In the case of neutron yield higher than 10^6 neutrons the detector was calibrated using a Am-Be neutron source($10^6 n/s$).

Figure 11 presents time evolution of process of neutron production in plasma focus obtained with the NE213 based liquid scintillator detector. The result is in agreement with the prediction that neutron emission starts about 9 microseconds after the capacitors discharge which duration is $\approx$ 50ns.

Up to now neutrons from capillary fusion were not detected, probably because of high detection threshold of 10^3 neutrons per pulse and further experiments demand better neutron detection efficiencies.

Conclusion

The large volume liquid scintillator neutron detector(NE 343) satisfies all needs concerning detection of the total neutron yield in the plasma focus experiments. The problem of improvement of the detection efficiency for capillary fusion experiments will be resolved simply by replacement of neutron detector at smaller distance from capillary fusion facility.

The time evolution of neutron emission is regarded using NE213 based neutron detector and obtained results are in agreement with previous results and prediction.

So, the study of neutron production in the low input energy fusion equipments is possible to continue in our institute.

References

[1] ITER WORKSHOP ON RADIATION EFFECTS ON DIAGNOSTIC COMPONENTS, by F. Engelmann et al., St. Petersburg, October 14-17, 1991

[2] Proceedings of International Panel on 14 MeV Intense Neutron Sources Based on Accelerators for Fusion Material Study, Tokyo, January 14-16, 1991

[3] Proceedings of the Japan - U.S. Workshop P-119 on 14 MeV Neutron Source for Material R&D Based on Plasma Devices, June 7-10, 1988

[4] V. Nardi et al., IEEE Trans. Plas. Sci., Vol. 16, p. 368 - 373 and p. 374 - 378, 1988

[5] B. C. Diven et al., Phys. Rev., Vol. 120, p.556, 1960

[6] J. J. Poitou et al. Nucl. Instr. Mat., Vol. 114, p.113, 1974

GENERAL PLASMA

Low Temperature and
General Plasmas
M. Milosavljevic and Z. Petrovic

ISBN 1-56072-399-8

POINT VORTEX DESCRIPTION OF DRIFT WAVE VORTICES: DYNAMICS, SELF-ORGANIZATION AND CHAOS

Mitsuo Kono and Kentaro Yabuki[†]
Faculty of Policy Studies,
Chuo University, Hachioji, Tokyo, 192-03, Japan
[†] Interdisciplinary Graduate School of Engineering Sciences,
Kyushu University, Kasuga, Fukuoka, 816, Japan

Abstract

The dynamical behavior of drift wave vortices is studied based on the modulated point vortex model associated with the Hasegawa-Mima equation and is shown to be consistent with the results obtained by solving the Hasegawa-Mima equation numerically. In a bounded region vortices exhibit chaotic behavior when the number is small and are self-organized to emerge into larger vortices when the number is fairly large. The boundary effect can be taken into account with an aid of the circle theorem which replaces the boundary by the image system of the vortices inside the circle and ensures that the potential for the vortices is zero on the boundary. A self-consistent theory describing vortex-wave interaction is also formulated by extending the modulated point vortex model so as for the energy and enstrophy to be conserved. Dynamical behaviors of vortices under the interaction with waves are numerically studied.

1. Introdution

Vortical motions are characterized by the term of $[\ \psi, \nabla^2\psi\ ]$ where ψ is a stream function and [,] denotes the Poisson blacket. In magnetized plasmas this term exists commonly, implying vortices are universal nonlinear excitations which play an important role in dynamics and transport. In fact dynamics of drift vortices is thought to be one of the key issues to get a full understanding of anomalous transport in a magnetically confined plasma as well as the nature of turbulence. Numerical works[(1)~(5)] based on the Hasegawa-Mima equation[(6)] revealed dynamical properties of the drift vortices which are found to be nicely recovered by the modulated point vortex model[(7)~(9)].

An advantage of introducing point vortices is to convert a nonlinear partial differential equation into a system of ordinary differential equations which are easier to solve. A crucial difference of the Hasegawa-Mima equation from Euler's equation is the existence of the drift term giving rise to dispersive waves, which requires that the vorticity attached to each point vortex is no longer constant but varies in space and time. This is in contrast with the point vortex description of

Euler's equation characterized by constant strength vorticities. This modulated point vortex model was first introduced by Zabusky-McWilliams[7] based on the fact that the Hasegawa-Mima equation conserves the vorticity along the trajectory. They studied the configurations of the vortices corresponding to a stationary solution of the Hasegawa-Mima equation and the stability of the configurations. Then Kono and Horton[8] and Hobson[9] studied dynamical properties of the vortices based on the modulated point vortex model and showed excellent agreements with the results of numerical simulations for the Hasegawa-Mima equation. Chaotic motion is observed for the system of a small number of point vortices confined in a bounded region. The modulated point vortex model is also able to describe the self-organization which can be compared with the numerical simulations that the collitional drift wave instability evolves through a turbulent state into an organized state which is nothing but the stationary solution of the Hasegawa-Mima equation [4],[5]. The number of the point vortices seems to be crutial to determine which is dominant, chaotic motion or self-organized motion.

It should be noted that Reznik[10] and Gryanik[11] also developed point vortex models for stationary atomospheric vortices. However the modulated point vortex model has a definite advantage to describe the dynamical behavior of the vortices.

In spite of the fact that the modulated point vortex model successfully describes many results of the numerical simulations, there is one problem that the energy and enstrophy are not conserved in the modulated point vortex medel while they are originally conserved for the Hasegawa-Mima equation. Drift wave turbulence is certainly have two main components, coherent structures on one side and background field on the other, whose interactions are not taken into account in the modulated point vortex model, resulting in the lack of the conservation laws. As a matter of fact the numerical simulations have shown that a vortex-pair emits wake fields though small when it changes its propagation direction. Reznik[12] formulated the vortex-wave interaction based on the model he developed previously. However his theory is too much involved and hard to treat. We rather extend the modulated point vortex model so as to include the vortex-wave interaction.

In Sec. 2 we summarize vortex dynamics based on the modulated point vortex. In Sec. 3. the chaotic behaviors and self-organized motions are numerically demonstrated. In Sec. 4 we extend the modulated point vortex model to include the vortex-wave interactions and show that the energy and enstrophy are conserved. In Sec. 5 the self-consistent dynamics of the vortex-wave system is numerically studied. Discussions are given in the last section.

2. Modulated point vortex dynamics

The modulated point vortex model for the Hasegawa-Mima equation in a slab geometry

$$\frac{\partial \Pi}{\partial t} + [\psi, \Pi] + v_* \frac{\partial \psi}{\partial y} = 0, \tag{1}$$

where [,] denotes the Poisson bracket and $\Pi = \psi - \nabla^2 \psi$, was introduced by Zabusky and McWilliams[7] first and then later Kono and Horton[9] and Hobson[10] in the following form:

$$\frac{d\mathbf{r}_\alpha}{dt} = \hat{\mathbf{z}} \times \nabla \psi(\mathbf{r}_\alpha), \tag{2}$$

$$\psi(\mathbf{r}) = \frac{1}{2\pi}\sum_{\alpha}(\kappa_{\alpha 0} + v_* x_\alpha)K_0(|\mathbf{r} - \mathbf{r}_\alpha|), \tag{3}$$

where K_0 is the modified Bessel function of the first kind. We also assume v_* positive constant and $\kappa_{\alpha 0}$ constant.

The Hamiltonian is given by

$$H = \sum_{\alpha}\psi(\mathbf{r}_\alpha), \tag{4}$$

from which we have

$$\frac{dx_\alpha}{dt} = -\frac{\partial H}{\partial y_\alpha} = \frac{1}{2\pi}\sum_{\beta}(\kappa_{\beta 0} + v_* x_\beta)\frac{y_\alpha - y_\beta}{r_{\alpha\beta}}K_1(r_{\alpha\beta}), \tag{5}$$

$$\frac{dy_\alpha}{dt} = \frac{\partial H}{\partial x_\alpha} = -\frac{1}{2\pi}\sum_{\beta}(\kappa_{\beta 0} + v_* x_\beta)\frac{x_\alpha - x_\beta}{r_{\alpha\beta}}K_1(r_{\alpha\beta}), \tag{6}$$

where $r_{\alpha\beta} = |\mathbf{r}_\alpha - \mathbf{r}_\beta|$. Since H is translationally invariant along the y-axis, the translational momentum in x is conserved:

$$P = \sum_{\alpha}(\kappa_{\alpha 0} + v_* x_\alpha)^2. \tag{7}$$

For two vortices in addition to Eq. (7) the relative distance of the vortices is a constant of motion,

$$r_{12}^2 = (x_1 - x_2)^2 + (y_1 - y_2)^2 = r_0^2, \tag{8}$$

which gives the solution as

$$cos\theta = \gamma\frac{1 - \beta^2 sn^2(\omega t, k)}{1 - \alpha^2 sn^2(\omega t, k)}, \tag{9}$$

where $x_1 - x_2 = r_0 cos\theta$, $A = [\sqrt{2P} + (\kappa_{20} - \kappa_{10})]/v_* r_0$, and $B = [\sqrt{2P} - (\kappa_{20} - \kappa_{10})]/v_* r_0$. In Eq.(9) for $v_* r_0 + |\kappa_{10} - \kappa_{20}| > \sqrt{2P} > |v_* r_0 - |\kappa_{10} - \kappa_{20}||$ we have

$$\alpha^2 = \frac{1+B}{A+B}, \quad \beta^2 = A\alpha^2, \quad \gamma = -sgn(\kappa_{10} - \kappa_{20})$$

$$\omega = \sqrt{2(A+B)}\frac{v_* K_1(r_0)}{4\pi}, \quad k = \sqrt{\frac{(1+A)(1+B)}{2(A+B)}}, \tag{10}$$

and for $\sqrt{2P} < |v_* r_0 - |\kappa_{10} - \kappa_{20}||$ we have similar expressions for $\alpha, \beta, \gamma, \omega$ and k .

Equations for the center of gravity are given by

$$\frac{1}{2}(x_1 + x_2) = -\frac{\pi r_0}{v_* K_1(r_0)}\frac{d\theta}{dt} - \frac{\kappa_{10} + \kappa_{20}}{2v_*},$$
$$\frac{1}{2}\frac{d}{dt}(y_1 + y_2) = \frac{K_1(r_0)}{4\pi}\left[\kappa_{10} - \kappa_{20} + v_* r_o cos\theta\right] cos\theta,$$

whose explicit solutions are readily obtained by using Eq. (9).

A vortex-pair $(\kappa_{10} + \kappa_{20} = 0)$ propagates in the $y-$direction without oscillation in the orbit for $\theta_0 (= \theta(t = 0)) = n\pi$ (n : integer) and with oscillation for $\theta_0 \neq n\pi$ as is shown in Fig. 1, which has been numerically observed by Makino, Kamimura, and Taniuchi[1] based on the Hasegawa-Mima equation. In fact the frequency given by Eq. (10) shows the same type of v_*-dependence as those observed in numerical simulations by Makino et al. Recently Nycander and Isichenko[13] derived the equation for the center of gravity of a vortex-pair from the moment equations of the Hasegawa-Mima equation and obtained the frequency of the trajectory which is also well fitted to the results by Makino et al. A Non-propagating solutions shown in Fig. 1 are realized for such initial angles of the symmetry axis of the vortex-pair to the $x-$axis that the time-averaged velocity of the vortex-pair is zero.

Two like-signed vortices $(\kappa_{10} = \kappa_{20})$ are mutually trapped, rotating around the center of gravity. This mutual trapping leads to coarse-graining of the correlations over directions and may be considered a mechanism behind the fusion of vortices in the sense that a group of point vortices positioned sufficiently near one another act at large distances as a single vortex with the sum intensity, $\kappa_T \cong \sum_\alpha \kappa_\alpha$. A coalescence of like-signed vortices and a long-lived monopole numerically observed by Horton[14] may be interpreted by this mutual trapping process. The inverse cascade[4],[5] of the energy associated by the conservation of enstrophy is also regarded in the vortex representation as a trapping as a kind of snowballing process.

Collision processes of two vortex-pairs are shown to recover those observed numerically[1],[3]-[4] for elastic cases with zero (Fig.2(a)) and non-zero impact parameters (Fig.2(b)-(f)). Since our point vortex model does not take into account effects of interaction with the wake fields, the inelastic collisions with an emission of wake fields observed by McWilliams and Zabusky[2] cannot be described by the vortex field component in Eq. (3) alone. However, the position dependence of the vorticity provides a variety of complicated behavior including an exchange scattering and a boomerang scattering , indicating that our point vortex system is likely to become turbulent when many vortices are involved. However the potential structure constructed from Eq. (3) is quite orderly: mutually trapped vortices behave as a single vortex though dynamics of constituent point vortices is very complicated as is shown in Fig.3 which corresponds to Fig.2(e). Therefore the complication of the dynamics of the point vortices is rather analogous to complicated behaviors of constituent partices in an ordinary gas or fluid dynamics and averaged properties may be of primary importance although the dynamical properties of the point vortex system are academically interesting since the chaotic behavior may be characterized by intermittent structures-clusters of vortices, in which local order is a preferred state because of the short-range interaction force between point vortices. The range of the interaction $\rho_s = c(m_i T_e)^{1/2}/eB$ is given by the parallel electron motion shielding the charge separation in the Euler vortex.

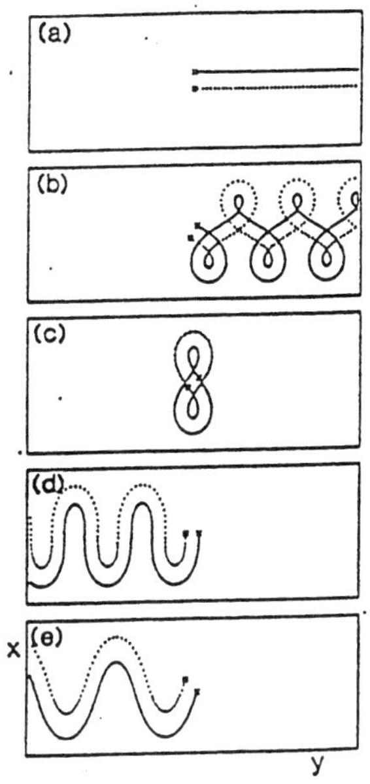

Fig.1. Trajectories of an opposite-signed vortex pair of $\kappa/v_* = 2.0$ and $r_0 = 0.4$: solid lines and dotted lines indicate a positive and negative vortex, respectively. Angles between the symmetry axis and x axis are (a) $\theta_0 = 0$, (b) $\pi/6$, (c) 0.8976, (d) $\pi/2$, and (e) $2\pi/3$.

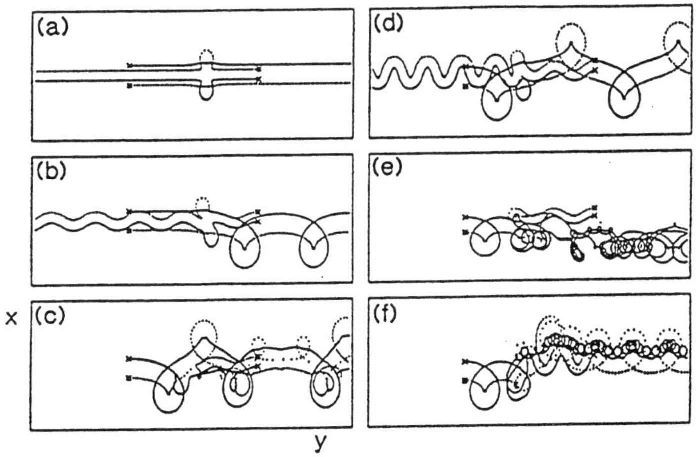

Fig.2. Head-on collision between two opposite-signed vortex pairs of $\kappa_1/v_* = -1.0$, $\kappa_2/v_* = 1.0$, and $\kappa_3/v_* = 2.0$, $\kappa_4/v_* = -2.0$. The initial positions are: $x_1 = x_{10} + \Delta x_1$, $x_2 = -x_{10} + \Delta x_1$, $x_3 = x_{20} + \Delta x_2$, $x_4 = -x_{20} + \Delta x_2$, $x_{10} = 0.25$, $x_{20} = 0.50$, $y_1 = y_2 = -y_3 = -y_4 = 4.0$ with $(\Delta x_1, \Delta x_2)$: (a) (0.0,0.0), (b) (0.15,0.0), (c) (0.15,0.15), (d) (0.25,0.25), (e) (0.25, 0.50) and (f) (0.50,0.50).

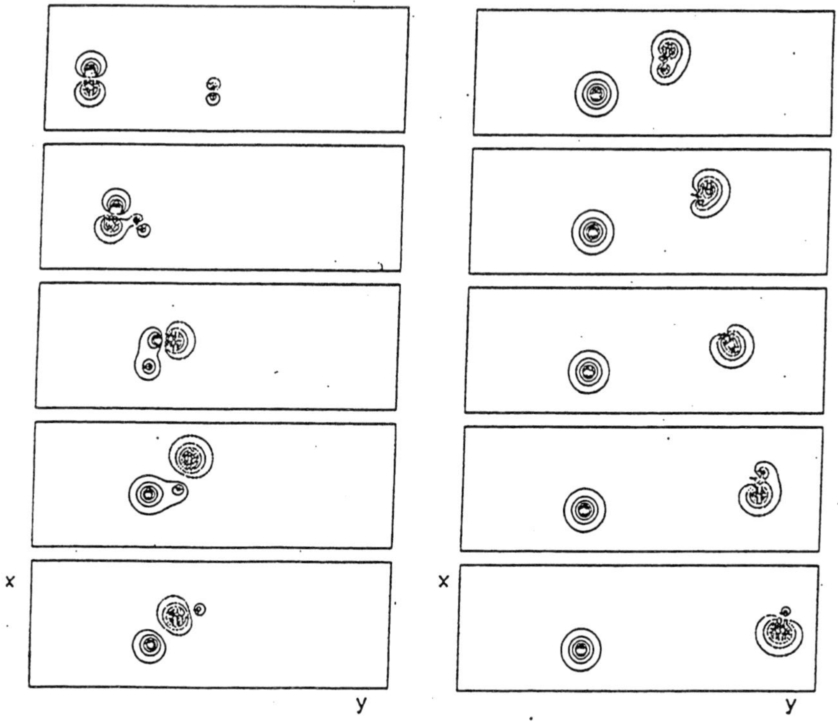

Fig.3. Contours of stream function obtained from Eq.(3) using the point vortex trajectory in Fig.2(e).

Fig.4. (shown on the next page) (a) Redistributions of vortices to form an ordered states in the course of time evolution with $\kappa/\nu_* = 100$, (b) Redistribution of vortices to form an ordered states in the course of time evolution with $\kappa/\nu_* = -100$, (c) Time evolution of a cyclone ($\kappa/\nu_* = 10$), (d) Time evolution of an anticyclone ($\kappa/\nu_* = -10$), (e) Coalescence of the liked-signed clusters: cyclone-cyclone ($\kappa_1/\nu_* = \kappa_2/\nu_* = 100$), (f) Coalescence of the liked-signed clusters: anticyclone-anticyclone ($\kappa_1/\nu_* = \kappa_2/\nu_* = -100$), (g) Interacting clusters with relatively small $|\kappa_1/\nu_*|$: cyclone-cyclone ($\kappa_1/\nu_* = \kappa_2/\nu_* = 10$), (h) Interacting clusters with relatively small $|\kappa_1/\nu_*|$: cyclone-anticyclone ($\kappa_1/\nu_* = -\kappa_2/\nu_* = 10$).

For a cylindrical plasma ($\mathbf{v}_* = (v_T^2/\Omega)\hat{\mathbf{z}} \times \nabla ln n_0$), we have

$$\frac{d\mathbf{r}_\alpha}{dt} = \hat{\mathbf{z}} \times \nabla\psi(\mathbf{r}_\alpha), \tag{11}$$

$$\psi(\mathbf{r}) = \frac{1}{2\pi}\sum_\alpha(\kappa_{\alpha 0} - \hat{v}_* r_\alpha^2)K_0(|\mathbf{r} - \mathbf{r}_\alpha|), \tag{12}$$

where the background density is assumed to be a Gaussian. A vortex-pair revolves around the center of the cylinder while the constituent vortices rotate around each other, that is, the motion is characterized by two frequencies, implying the motion is mostly quasi-periodic. A single cluster consisting of many vortices with the same sign of the vorticity ($\kappa_{\alpha 0} = \kappa$) turns round the center of the cylinder clockwise, while the vortices rotate around the center of mass clockwise for $\kappa > 0$ and counterclockwise for $\kappa < 0$. When $|\kappa| > v_*(r_\alpha^2 - r_\alpha^2(t=0))$, the vortices redistribute themselves to lattice points to form a circle (Fig.4(a) and (b)). When $\kappa(>0)$ decreases so as for some outer vortices in the cluster not to satisfy the above condition, those vortices rotate in the direction opposite to that of the core of the cluster and are left behind (Fig.4(c)). Thus cyclons ($\kappa > 0$) are not long lived while anticyclons ($\kappa < 0$) are long-lived as is shown in Fig.4(d). This situation is just similar to the observations by Antipov et al [15] . Two clusters coalesce into one when κ s' are the same sign and become large (Fig.4(e) and (f)). However they are almost independent on each other and their motion is a superposition of that of each cluster when κ is small compared with the relative distance between the clusters (Fig.4(g) and (h)).

3. Chaotic dynamics and self-organized motion in a bounded region

In this section we study the dynamical properties of point vortices in a bounded region. The equation of motion for point vortices is rewritten as

$$\frac{dz_\alpha}{dt} = \frac{1}{2\pi i}\sum_{\beta\neq\alpha}(\kappa_\beta + v_*\frac{z_\beta + \bar{z}_\beta}{2})\frac{z_\alpha - z_\beta}{|z_\alpha - z_\beta|}K_1(|z_\alpha - z_\beta|), \tag{13}$$

where z_α is a complex variable related to the coordinate (x_α, y_α) of the center of a point vortex z_α through $z_\alpha = x_\alpha + iy_\alpha$ and $\bar{z}_\alpha$ is the complex conjugate to z_α. If, for simplicity, we take a circle with a radius R as a bounded region, the circle theorem[16] gives the potential for the vortices inside the circle as

$$\psi(z) = \frac{i}{2\pi}\sum_\beta(\kappa_\beta + v_*\frac{z_\beta + \bar{z}_\beta}{2})\Big\{K_0(|z - z_\beta|) - K_0(|\frac{R^2}{z} - \bar{z}_\beta|)\Big\}. \tag{14}$$

On the circle $(\bar{z} = R^2/z)$, $\psi(z)$ is zero, implying the boundary is a stream line or an equipotential line. In addition, Eq.(14) has no singularity at $z = 0$. Then the equation describing dynamics of vortices in the circle is given by

$$\frac{dz_\alpha}{dt} = \frac{1}{2\pi i}\sum_{\beta\neq\alpha}(\kappa_\beta + v_*\frac{z_\beta + \bar{z}_\beta}{2})\frac{z_\alpha - z_\beta}{|z_\alpha - z_\beta|}K_1(|z_\alpha - z_\beta|)$$
$$+ \frac{1}{2\pi i}\sum_\beta(\kappa_\beta + v_*\frac{z_\beta + \bar{z}_\beta}{2})\frac{R^2}{\bar{z}_\alpha^2}\frac{R^2/z_\alpha - \bar{z}_\beta}{|R^2/z_\alpha - \bar{z}_\beta|}K_1(|R^2/z_\alpha - \bar{z}_\beta|). \tag{15}$$

For the case of one vortex in the circle, Eq.(15) is integrable. If z is put as $\rho e^{i\theta}$, Eq.(15) becomes to

$$\frac{d\rho}{dt} = 0$$
$$\frac{d\theta}{dt} = \frac{-R^2}{2\pi\rho^3}(\kappa + v_*\rho\cos\theta)K_1(R^2/\rho - \rho), \tag{16}$$

whose solution is given by

$$\tan\frac{\theta}{2} = -\frac{A\tan[\omega(t-t_0)] - \tan(\theta_0/2)}{1+(1/A)\tan[\omega(t-t_0)]\tan(\theta_0/2)}, \tag{17}$$
$$A = \sqrt{\frac{\kappa + v_*\rho}{\kappa - v_*\rho}},$$
$$\omega = \frac{R^2}{4\pi\rho^3}K_1(R^2/\rho - \rho)\sqrt{\kappa^2 - (v_*\rho)^2}.$$

Note that ω and A should be replaced by $i\omega$ and iA for $(v_*\rho)^2 > \kappa^2$. The trajectory is a circle for $(v_*\rho)^2 < \kappa^2$ and is an arc connecting the initial position $\rho e^{i\theta_0}$ with one of the fixed points $\rho e^{\pm i\theta_*}$ where $\theta_* = \cos^{-1}(-\kappa/v_*\rho)$ for the opposite case.

The dynamics of vortices is sensitive to their initial configuration and the magnitude of the value of v_*. First we consider the case of a vortex-pair with vorticities of the same magnitude and opposite sign $(\kappa_1 = -\kappa_2 = \kappa > 0)$ for which the fixed points are obtained by putting $z_\alpha = \rho_\alpha e^{i\theta_\alpha}$ for the two cases of the configurations symmetric with respect to the y-axis (the diamagnetic direction) $(z_1 = -\bar{z}_2;\ \rho_1 = \rho_2 = \rho$ and $\theta_1 + \theta_2 = \pi)$ and the origin $(z_1 = -z_2;\ \rho_1 = \rho_2 = \rho$ and $\theta_1 - \theta_2 = \pi)$. For the first case the fixed point is given by $(\rho_1, \theta_1) = (\rho_*, 0)$ where ρ_* is determined by

$$K_1(2\rho_*) + (R/\rho_*)^2[K_1(R^2/\rho_* + \rho_*) - K_1(R^2/\rho_* - \rho_*)] = 0. \tag{18}$$

For $R = 1$, ρ_* is equal to 0.4629. The stability of the fixed point can be analyzed by linearizing Eq.(15) around the fixed point with respect to perturbations $\delta\rho_1$ and $\delta\theta_1 \propto e^{i\omega t}$. The eigen value ω is expressed as

$$\omega = \frac{\alpha}{2\pi}\kappa K_1(2\rho_*), \tag{19}$$

where α is a positive constant given by

$$\alpha^2 = \frac{2}{R^2 - \rho_*^2}\Big\{1 + \frac{R^2 - \rho_*^2}{\rho_*}\frac{K_0(2\rho_*)}{K_1(2\rho_*)} + \frac{R^2(R^4 - \rho_*^4)}{2\rho_*^5}\frac{K_0(R^2/\rho_* - \rho_*)}{K_1(2\rho_*)}$$
$$+ O\Big(\frac{K_1(R^2/\rho_* + \rho_*)}{K_1(2\rho_*)}\Big)\Big\}. \tag{20}$$

Equation (19) indicates that the fixed point is stable. This result is applied to the case that the initial positions are largely deviated from the fixed point as long as two vortices are initially located symmetrically with respect to the y axis. Each of them follows a closed trajectory centered at the fixed point $\rho = \rho_*$, independent of the magnitude of v_*. The vortices propagate in a pair up to the boundary, and then change their partners with their own mirror images. New pairs move along the boundary of the circle to trace semicircles. Then they again exchange their partners with the original ones and repeat the previous trajectories. The periodic orbit suggests the existence of an integral for this case although we have not found it yet.

On the other hand for the latter case the fixed points are homogeneously distributed on the circle of the radius ρ_*, that is, θ_1 remains arbitrary. The stability analysis provides the same result as before. However, since the fixed points are distributed densely on the circle, the orbit with an initial position slightly deviated from one of the fixed points drifts back and forth along the circle of $\rho = \rho_*$, rotating around the fixed points. This drift on the circle can be seen from the equation for θ_1 :

$$\frac{d\theta_1}{dt} = -\frac{\kappa_2}{2\pi}K_1(2\rho_*)\alpha^2\delta\rho + \frac{v_*}{2\pi}K_1(2\rho_*)\delta\theta \cdot \sin\theta_1, \tag{21}$$

where $\delta\rho$ and $\delta\theta(= \delta(\theta_1 - \theta_2))$ are given by linearized equations as

$$\begin{aligned} \delta\rho &= \delta\rho_0 \cos(\omega t), \\ \delta\theta &= 2\alpha\delta\rho_0 \sin(\omega t), \end{aligned} \tag{22}$$

where $\delta\rho_0$ is an initial deviation from ρ_*. Differentiating Eq.(21) with respect to t and taking time average of the resultant equation, we obtain

$$\frac{d^2}{dt^2}\xi = -\Omega^2 \sin\ \xi, \tag{23}$$

where $\xi = 2 < \theta_1 > +\pi$ and $\Omega = (v_*/2\pi)K_1(2\rho_*)\sqrt{< \delta\theta^2 >}$. In deriving Eq.(23), we have used $\Omega/\omega = 2v_*\delta\rho_0/\kappa << 1$. Equation (23) is nothing but a pendelum equation and its solution is given by

$$\sin\frac{\xi}{2} = k\ sn(\Omega t + K(k), k), \tag{24}$$

where $k = \sin\frac{\xi_0}{2}$ and $K(k)$ is the complete elliptic integral. From Eq.(24), the maximum oscillation angle is given by

$$< \theta_1 >_{max} = \frac{1}{2}[\sin^{-1} k - \pi] = \theta_1(t = 0), \tag{25}$$

which is confirmed by the numerical results. The above result suggests that the orbit is stable as long as the condition $\Omega/\omega << 1$ is satisfied.

When the initial position of the vortex-pair is deviated from one of the fixed points, the trajectory is periodic for $v_* = 0$, quasiperiodic for small v_* and chaotic for finite v_* (Fig.5). For small values of v_*, the trajectory is basically a semicircle (Fig.5(a)) and drifts back and forth to cover the region given by Eq.(25)(Fig.5(b)). As v_* increases, the length of the drift on the circle centered at ρ_* becomes

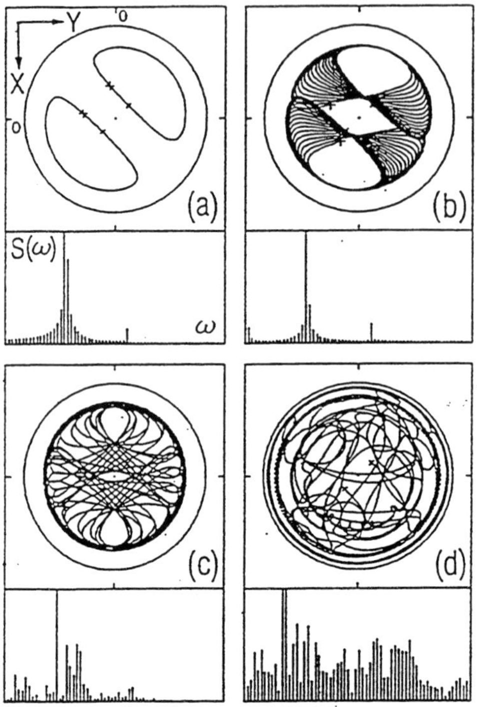

Fig.5. The trajectories of a vortex pair ($\kappa_1 = -\kappa_2 = 4.0$) located initially symmetric with respect to the origin ($x_1 = -x_2 = \sqrt{3}/10$, $y_1 = -y_2 = 0.1$): (a) $\nu_* = 0$, (b) $\nu_* = 0.005$, (c) $\nu_* = 0.5$ and (d) $\nu_* = 2.5$.

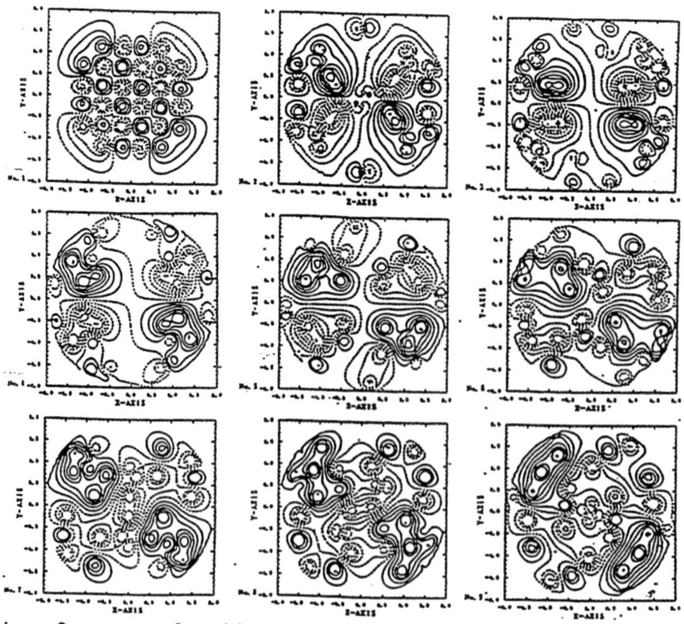

Fig.6. Evolution of a group of positive and negative vortices initially alternately located at lattice points in a circle ($|\kappa/\nu_*| = 100$).

large and at the same time its magnitude changes cycle by cycle, leading to mixing which is an indication of onset of chaos in the trajectories (Fig.5(d)). The onset of chaos is also confirmed by a dramatic change in frequency spectrum of the time sequence of the orbit, that is, the change from a line spectrum to a broad spectrum. By examining the frequency spectrum, we find that the critical value of v_* for the onset of chaos is approximately 1.75 for the particular parameters used in Fig.5.

A system of three vortices with different sign of vorticities initially located at the vertices of an equilateral triangle exhibits chaotic motion independent of the magnitude of v_*. For vortices on the vertices of an equilateral square, the motion is periodic for $v_* = 0$ and chaotic for $v_* \neq 0$ (Fig.9). When point vortices of like sign are initially located at the vertices of an equilateral polygon, then they rotate together for $v_* = 0$ while they follow quasiperiodic orbits for $v_* \neq 0$.

As is shown above vortices in a bounded region exhibit complicated behaviors similar to a billiard problem. It is the diamagnetic drift that has a tendency to drive the system into a chaotic motion.

Similar numerical simulations(17) have been done based on two dimensional Euler vortices whose interaction potential is a logarithm. When the bounded region is taken as a circle, a system with two vortices is complete integrable and does not exhibit chaotic motion(18). One of the bounded regions for a vortex-pair to show chaotic motion for the Euler vortices is shown to be a semi-circle(19). The reason we have chaotic motion of a vortex-pair in a circle boundary is that our vorticity depends on the vortex position because of the diamagnetic drift which breaks the rotational symmetry and causes mixing.

One interesting problem is to examine whether a coherent structure (an aggregation of like sign vortices) is formed when many point vortices are initially distributed in a bounded region. When the vortices with positive and negative vorticities are initially distributed alternately on square lattice points, the vortices with the same sign have a tendency to emerge each other to form larger vortices as is shown in Fig.6. This might be a kind of self-organization observed in the numerical simulations (4),(5).

4. Self-consistent description of vortex-wave dynamics

There are two conserved quantities for the Hasegawa-Mima equation; one is the energy

$$E = \int d\mathbf{r}\left[\psi^2 + (\nabla\psi)^2\right] = \int d\mathbf{r}\psi\Pi, \tag{26}$$

and the other is the enstrophy

$$W = \int d\mathbf{r}\left[(\nabla\psi)^2 + (\nabla^2\psi)^2\right] = -\int d\mathbf{r}(\nabla^2\psi)\Pi. \tag{27}$$

In the point vortex description the energy and enstrophy are expressed as

$$E = \frac{1}{2\pi}\sum_{\alpha\neq\beta}(\kappa_{\alpha 0} + v_* x_\alpha)(\kappa_{\beta 0} + v_* x_\beta)K_0(r_{\alpha\beta}), \tag{28}$$

$$W = \frac{1}{8\pi}\sum_{\alpha\neq\beta}(\kappa_{\alpha 0} + v_* x_\alpha)(\kappa_{\beta 0} + v_* x_\beta)[K_0(r_{\alpha\beta}) + K_2(r_{\alpha\beta})], \tag{29}$$

where the self-energy has been subtracted since it diverges. Equations (28) and (29) are not conserved in general since

$$\frac{dE}{dt} = -\frac{1}{2}\sum_{\alpha} v_* \frac{\partial}{\partial y_\alpha}\psi^2(\mathbf{r}_\alpha), \quad \frac{dW}{dt} = -\frac{1}{2}\sum_{\alpha} v_* \frac{\partial}{\partial y_\alpha}[\nabla_\alpha \psi(\mathbf{r}_\alpha)]^2.$$

The descrepancy comes from the fact that Eqs. (26) and (27) take account of contributions from both vortices and waves, while Eqs. (28) and (29) are based on vortices only. Therefore for the vortex-pairs corresponding to the stationary solutions of the Hasegawa-Mima equation, that is, for those propagating in straight in the y direction, Eqs. (28) and (29) are certainly constants of motion since the wave are never excited. In general, however, changes in energy and enstrophy of the vortex system in the course of time evolution may occur whenever waves are involved in the fundamental processes of the vortices such as emission or absorption of waves by the vortices, which is the subject in this section.

For the Hasegawa-Mima equation we split the potential and vorticity into a vortex part and a continuous field part as

$$\psi = \psi_v + \psi_w, \quad \Pi = \Pi_v + \Pi_w, \tag{30}$$

where subscripts v and w denote the vortex part and wave part, respectively. Since linear eigen-frequency of the drift wave is well separated from the turn-over frequency of the vortex, the above separation is always possible. Following Ref.8, we introduce vortices as

$$\Pi_v = (1 - \nabla^2)\psi_v = \sum_{\alpha} \kappa_\alpha(t) V_\alpha(\mathbf{r} - \mathbf{r}_\alpha(t)), \tag{31}$$

where $V_\alpha(\mathbf{r} - \mathbf{r}_\alpha(t))$ is a localized function centered at $\mathbf{r} = \mathbf{r}_\alpha(t)$ and $\mathbf{r}_\alpha(t)$ is determined by the characteristics of Eq.(1)

$$\frac{d\mathbf{r}_\alpha(t)}{dt} = \mathbf{z} \times \nabla(\psi_v + \psi_w)\,|_{\mathbf{r}=\mathbf{r}_\alpha(t)}. \tag{32}$$

Substituting Eq.(30) into Eq.(1), we obtain

$$\sum_{\alpha} \frac{d\kappa_\alpha}{dt} V_\alpha(\mathbf{r} - \mathbf{r}_\alpha) - \sum_{\alpha} \kappa_\alpha \left\{ \frac{d\mathbf{r}_\alpha}{dt} - \mathbf{z} \times \frac{\partial \psi(\mathbf{r}, t)}{\partial \mathbf{r}} \right\} \cdot \frac{\partial}{\partial \mathbf{r}} V_\alpha(\mathbf{r} - \mathbf{r}_\alpha)$$
$$+ \frac{\partial \Pi_w}{\partial t} + [\psi,\ \Pi_w] + v_* \frac{\partial \psi}{\partial y} = 0. \tag{33}$$

Since $V_\alpha(\mathbf{r} - \mathbf{r}_\alpha)$ is a function localized around $\mathbf{r} = \mathbf{r}_\alpha(t)$, we may replace $\mathbf{r}$ appeared in the coefficients of $V_\alpha(\mathbf{r} - \mathbf{r}_\alpha)$ by $\mathbf{r}_\alpha\ (t)$. Then the second term of the left-hand side of Eq.(33) vanishes according to Eq.(32). The remaining is

$$\frac{\partial \Pi_w}{\partial t} + [\psi,\ \Pi_w] + v_* \frac{\partial \psi}{\partial y} + \sum_{\alpha} \frac{d\kappa_\alpha}{dt} V_\alpha(\mathbf{r} - \mathbf{r}_\alpha) = 0. \tag{34}$$

Multiplying Eq.(21) by $V_\beta(\mathbf{r}-\mathbf{r}_\beta(t))$ and integrating with respect to $\mathbf{r}$, we have

$$\frac{d}{dt}(\kappa_\beta - v_* x_\beta + \Pi_w(\mathbf{r}_\beta, t)) = 0, \tag{35}$$

which gives for a constant diamagnetic drift velocity v_*

$$\kappa_\alpha(t) = \kappa_{\alpha 0} + v_* x_\alpha(t) - \Pi_w(\mathbf{r}_\alpha(t),\ t). \tag{36}$$

This shows that the vorticity changes under the influence of the wave fields, which is the reaction of emission or absorption of the wave fields by the vortex.

The vortex part of the potential is obtained through Eq.(31) as

$$\psi_v(\mathbf{r},t) = \int d\mathbf{r}' G(\mathbf{r}-\mathbf{r}')\Pi_v(\mathbf{r}') = \sum_\alpha \psi_\alpha(\mathbf{r}-\mathbf{r}_\alpha), \tag{37}$$

where the Green funcction $G(\mathbf{r})$ is difined by

$$(1-\nabla^2)G(\mathbf{r}) = \delta(\mathbf{r}), \tag{38}$$

whose solution is given by

$$G(\mathbf{r}) = \frac{1}{2\pi} K_0(|\mathbf{r}|) \tag{39}$$

Now we examine conserved quantities for our system. In our present description the energy and enstrophy are carried by both vortices and waves, while in the previous description they are carried by only vortices. First we examine the energy conservation. The time derivative of E is given as

$$\begin{aligned}\frac{dE}{dt} &= 2\int d\mathbf{r}\psi\frac{\partial \Pi_w}{\partial t} + 2\sum_\alpha \frac{d\kappa_\alpha}{dt}\psi(\mathbf{r}_\alpha,\ t) - 2\sum_\alpha \kappa_\alpha \int d\mathbf{r}\psi(\mathbf{r},\ t)\frac{d\mathbf{r}_\alpha}{dt}\cdot\frac{\partial}{\partial \mathbf{r}}V_\alpha(\mathbf{r}-\mathbf{r}_\alpha)\\ &= 2\int d\mathbf{r}\psi\frac{\partial \Pi_w}{\partial t} + 2\sum_\alpha \frac{d\kappa_\alpha}{dt}\psi(\mathbf{r}_\alpha,\ t) + 2\sum_\alpha [\psi(\mathbf{r}_\alpha,\ t),\ \psi(\mathbf{r}_\alpha,\ t)]\\ &= 2\int d\mathbf{r}\psi\frac{\partial \Pi_w}{\partial t} + 2\sum_\alpha \frac{d\kappa_\alpha}{dt}\psi(\mathbf{r}_\alpha,\ t).\end{aligned} \tag{40}$$

On the other hand, multiplying Eq.(34) by ψ and integrating over the space gives

$$\int d\mathbf{r}\psi\left\{\frac{\partial \Pi_w}{\partial t} + [\psi,\ \Pi_w]\right\} + \sum_\alpha \frac{d\kappa_\alpha}{dt}\psi(\mathbf{r}_\alpha,\ t) = 0. \tag{41}$$

Noting an identy

$$\int d\mathbf{r}\varphi[\psi,\ \phi] = \int d\mathbf{r}\phi[\varphi,\ \psi], \tag{42}$$

which holds for a periodic boundary condition, Eq.(40) reduces to

$$\int d\mathbf{r}\psi\frac{\partial \Pi_w}{\partial t} + \sum_{\alpha}\frac{d\kappa_\alpha}{dt}\psi(\mathbf{r}_\alpha,\ t) = 0. \tag{43}$$

Thus the energy is conserved.

In a similar way the enstrophy is also conserved. Equation (27) is rewritten as

$$W = \int d\mathbf{r}\Pi^2 - E, \tag{44}$$

from which we have

$$\frac{dW}{dt} = 2\int d\mathbf{r}\Pi\frac{\partial \Pi_w}{\partial t} + 2\sum_{\alpha}\frac{d\kappa_\alpha}{dt}\Pi(\mathbf{r}_\alpha,\ t) + 2\sum_{\alpha}\kappa_\alpha[\psi(\mathbf{r}_\alpha,\ t),\ \Pi(\mathbf{r}_\alpha,\ t)]. \tag{45}$$

Multiplying Eq.(34) by Π and integrating the resultant equation over $\mathbf{r}$, we have

$$\int d\mathbf{r}\left\{\Pi\frac{\partial \Pi_w}{\partial t} + \Pi[\psi,\ \Pi_w] + v_*\Pi\frac{\partial \psi}{\partial y}\right\} + \sum_{\alpha}\frac{d\kappa_\alpha}{dt}\Pi(\mathbf{r}_\alpha,\ t) = 0. \tag{46}$$

Noting that

$$\int d\mathbf{r}\Pi\frac{\partial \psi}{\partial y} = \frac{1}{2}\int d\mathbf{r}\frac{\partial}{\partial y}\left\{\psi^2 + (\nabla\ \psi)^2\right\} = 0, \tag{47}$$

and

$$\begin{aligned}\int d\mathbf{r}\Pi[\psi,\ \Pi_w] &= \int d\mathbf{r}\left\{\Pi[\psi,\ \Pi] - \Pi[\psi,\ \Pi_v]\right\}\\ &= \int d\mathbf{r}\Pi_v[\psi,\ \Pi]\\ &= \sum_{\alpha}\kappa_\alpha[\psi(\mathbf{r}_\alpha,\ t), \Pi(\mathbf{r},\ t)].\end{aligned} \tag{48}$$

Eq.(45) reduces to

$$\int d\mathbf{r}\Pi\frac{\partial \Pi_w}{\partial t} + \sum_{\alpha}\left\{\frac{d\kappa_\alpha}{dt}\Pi(\mathbf{r}_\alpha,\ t) + \kappa_\alpha[\psi(\mathbf{r}_\alpha,\ t),\ \Pi(\mathbf{r}_\alpha,\ t)]\right\} = 0, \tag{49}$$

which implies that

$$\frac{dW}{dt} = 0. \tag{50}$$

5. Self-consistent dynamics of vortex-wave system

In this section we study the self-consistent dynamics of vortex-wave system governed by Eqs.(32),(34),(36) and (37). It is noted that Eq.(34) is free from singularity since the motion associated with the vortex cores is subtracted because of Eq.(36), That is , Eq.(32) is symbolically rewritten with the aid of Eq.(36) as

$$\left\{\frac{\partial \Pi_w}{\partial t} + [\psi_w, \Pi_w] + v_* \frac{\partial \psi_w}{\partial y} + [\psi_v, \Pi_w - v_* x]\right\}_{\bar{D}} = 0, \tag{51}$$

where $\bar{D}$ implies that the vortex trajectory is subtracted from the whole space. Here we need some approximations to Eq.(51). First we may omit the wave nonlinearity $[\psi_w, \Pi_w]$, since this vector nonlinearity produces vortical motions which are already subtracted as vortices and therefore the wave is assumed small. Because of the same reason, we may approximate the last term of Eq.(51) as

$$[\psi_v, \Pi_w - v_* x] \simeq v_* \frac{\partial \psi_v}{\partial y}. \tag{52}$$

Furthermore the vortex part of the potential is approximated as

$$\begin{aligned}
\psi_v(\mathbf{r},\, t) &= \sum_\alpha \int d\mathbf{r}' G(\mathbf{r} - \mathbf{r}') V_\alpha(\mathbf{r} - \mathbf{r}') \\
&\simeq \sum_\alpha \int d\mathbf{r}' [G(\mathbf{r} - \mathbf{r}_\alpha) - \mathbf{r}'\mathbf{r}' : \nabla\nabla G(\mathbf{r} - \mathbf{r}_\alpha)] V_\alpha(\mathbf{r}') \\
&\simeq \sum_\alpha \frac{\kappa_\alpha}{2\pi} K_0(|\mathbf{r} - \mathbf{r}_\alpha|),
\end{aligned} \tag{53}$$

Thus our basic system of equations for the vortex-wave interaction are

$$\frac{d\mathbf{r}_\alpha}{dt} = \mathbf{z} \times \nabla \psi(\mathbf{r}_\alpha,\, t), \tag{54}$$

$$\frac{\partial \Pi_w}{\partial t} + v_* \frac{\partial \psi_w}{\partial y} + v_* \frac{\partial \psi_v}{\partial y} = 0, \tag{55}$$

$$\psi(\mathbf{r},\, t) = \psi_v(\mathbf{r},\, t) + \psi_w(\mathbf{r},\, t). \tag{56}$$

Numerical simulations are performed for various initial configulations of the vortices. In solving Eq.(55) the contributions from the vortex core are subtracted. When a small wave exists initially, a monopole is driven to propagate in contrast with the simple modulated vortex model in which a monopole cannot propagate since there is no field acting on it. A vortex-pair whose symmetry axis initially parallel to the diamagnetic drift propagates straight without emitting wake fields. When it is tilted initially, it emits wake fields behind it, and the wake affects the motion of the vortex-pair depending on the difference between the average velocity of the vortex-pair and the drift velocity. Vortex-pairs with small initial tilted angles paropagate faster than the drift waves and are not affected appreciably (Fig.6). However the others are fairly affected as if they are under an external field.

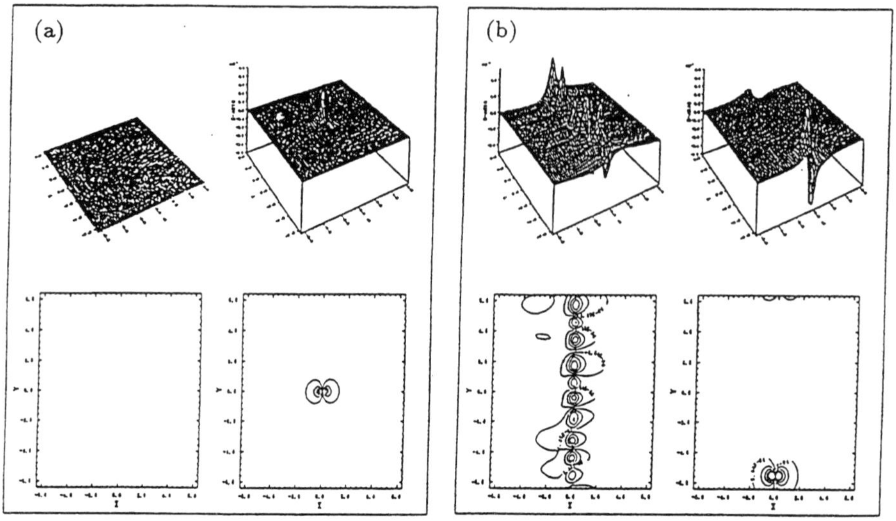

Fig.7. Propagation of a vortex pair at (a) the initial moment and (b) the time when the vortex pair reaches the boundary. In each box the left two are for wakes and the right two for both the vortex pair and the wakes.

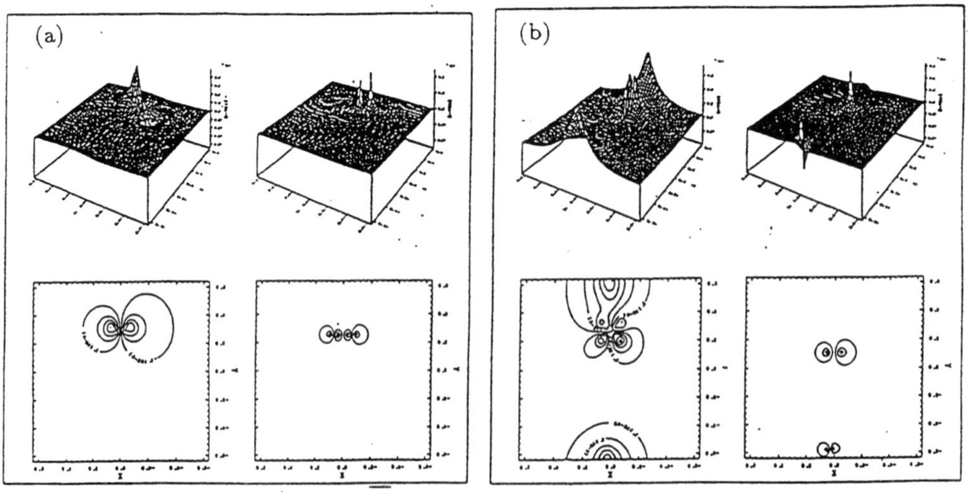

Fig.8. Head-on collision of two vortex pairs at (a) the instance of collision and (b) some time after the collision.

Those travelling in the direction opposite to drift waves are certainly not affected. Because of the same reason collision processes with zero impact parameter are not affected greatly. Figure 7 shows one of the head-on collision which can be compared with the result by McWilliams and Zabusky [2]. Those with non-zero impact parameters are inevitably under the influence of the wake fields.

6. Discussion

In this paper we have studied the dynamical behaviors of drift wave vortices based on the modulated point vortex model associated with the Hasegawa-Mima equation and shown them to be consistent with the results obtained by solving the Hasegawa-Mima equation numerically. In a bounded region vortices exhibit chaotic behavior when the number is small and are self-organized to emerge into larger vortices when the number is fairly large. A self-consistent theory describing vortex-wave interaction is also formulated by extending the modulated point vortex model so as for the energy and enstrophy to be conserved. Dynamical behaviors of vortices under the interaction with waves are numerically studied.

References

(1) M. Makino, T. Kamimura and T.Taniuchi: J. Phys. Soc.Jpn. **50**, 980 (1981)
(2) J. C. McWilliams and N. J. Zabusky : Gesphys. Astrophys. Dyn. **19**, 207 (1982)
(3) J. D. Meiss and W. Horton: Phys. Fluids **26**, 990 (1983)
(4) M. Kono and E. Miyashita: Phys. Fluids **31**, 326 (1988)
(5) V. Na ulin, K. H. Spatschek, A. Hasegawa: Phys. Fluids **B** (1992)
(6) A. Hasegawa and K. Mima: Phys. Rev. Lett. **39**, 205 (1977)
(7) N. J. Zabusky and J. C. McWilliams: Phys. Fluids **25**, 2175 (1982)
(8) M. Kono and W. Horton: Phys. Fluids **3**, 3255 (1991)
(9) D. D. Hobson: Phys. Fluids **A3**, 3027 (1991)
(10) G. M. Reznik: Oceanology **26**, 119 (1986)
(11) V. M. Gryanik: Oceanology **26**, 126 (1986)
(12) G. M. Reznik: J. Fluid Mech. **240**, 405 (1992)
(13) J. Nycander and M. B. Isichenko: Phys. Fluids **B2**, 2042 (1990)
(14) W. Horton: Phys. Fluids **B1**, 524 (1989)
(15) S. V. Antipov, M. V. Nezlin, V. K. Rodionov, A. Yu. Rylov, E. N. Snezhkin, A. S. Trubnikov and A. V. Khutoretskii: Sov. J. Plasma Phys. **14**, 648 (1988)
(16) L.M.Milne–Thomson, **Theoretical Hydrodynamics** (Macmillan, London, 1964) p. 154
(17) J.C.Hardin and J.P.Mason, Phys. Fluids **27**, 1583 (1984)
(18) Y. Kimura, Y. Kusumoto and H. Hashimoto, J. Phys. Soc. Jpn. **53**, 2988 (1984).
(19) Y. Kimura and H.Hashimoto, J. Phys. Soc. Jpn. **55**, 5 (1986).

Low Temperature and
General Plasmas
M. Milosavljevic and Z. Petrovic

ISBN 1-56072-399-8

MAGNETOTAIL CURRENT SHEET

Lev M. Zelenyi

**Space Research Institute, 117810 Moscow,
Profsouznaya 84/32, Russia**

Earth's geomagnetic tail represent a specific magnetic trap with highly stretched field lines, which result in a very weak amplitude of field in equatorial plane.

As a result conservation of usual adiabatic invariants breaks for magnetotail ions (but still holds for magnetospheric electrons). Nevertheless the analytical description of ion dynamics based on so called QUASIADIABATIC invariants could be developed. Theory explains the acceleration and heating of ions in magnetotail, formation of fast ion beams at the edges of plasma sheet and structuring of plasma distribution in a phase space. Properties of ion dynamics result in a self-sharpening of cross-tail current profile up to the value c/ω_{pi}, formation of specific nongyrotropic pressure distribution and anisotropies.

Results of analytical studies are compared with numerical modeling of ion dynamics in realistic models of magnetotail.

Low Temperature and
General Plasmas
M. Milosavljevic and Z. Petrovic

ISBN 1-56072-399-8

ON THE EXCITATION OF ION-CYCLOTRON INSTABILITY IN WEAKLY IONIZED PLASMAS PLACED IN NON-PARALLEL ELECTRIC AND STATIONARY MAGNETIC FIELDS

B.S. Milić
Faculty of Physics, Belgrade, Yugoslavia

1. INTRODUCTION

Quasi–perpendicular (QP) ion–cyclotron (IC) modes have received much attention in the last three decades or so, as they are frequently encountered both in laboratory (e.g. Q-machines [1–3], fusion [4], gas discharges [5]) and in nature (ionosphere, radiation belts, stellar atmospheres [6–8]). It is well–known that these modes are weakly damped [9,10] and that they can be destabilized, in the form of waves of growing amplitude, by a variety of physical factors. Among these factors, the presence of an electric current (i.e. of a relative electron-to-ion drift velocity) flowing along the lines of force of the external magnetization field [3,10] or perpendicular to them [11] seems to be most frequently encountered. If the magnitude of the said current density (or, alternatively, of the relative drift velocity) exceeds some threshold value, depending generally on the plasma parameters and on the geometry of the experimental situation, the onset of the spontaneously arising IC instability can be observed, much in accordance with the theoretical predictions, the earliest of which date back to some thirty years ago.

A very convenient way of ensuring the presence of electron and ion drifts is by placing the magnetized plasma in an external d.c. electric field. Cases where the two fields, $\boldsymbol{B}_0$ and $\boldsymbol{E}_0$, are either parallel or perpendicular to each other (and where ensuing drifts of the two plasma constituents are, also, respectively, parallel or perpendicular to $\boldsymbol{B}_0$) have been studied to a considerable extent (e.g. [3,11]), However, the situations in which the two fields are neither parallel nor perpendicular to each other can also be encountered (e.g. magnetically insulated diodes [12], circumterrestrial plasma [7,8]). These situations are, thus, of considerable interest as well. and their theoretical analysis in conjunction with the IC waves and instabilities is the topic of the present lecture. Attention is focused on weakly ionized plasmas, and the theoretical model of infinite and uniform plasma is adopted for the sake of simplicity. Kinetic equations with BGK model collision integrals describing the elastic encounters between the charged particles and the neutrals are taken. The ions are assumed to be Maxwellian, and the component of their drift velocity along the magnetic lines of force is assumed to be negligible with respect to the corresponding electronic quantity. Quasi–perpendicular electrostatic ion–cyclotron (QPESIC) modes and instabilities were already studied under these assumptions not long ago [13], and the results of that study will be shortly recapitulated in this lecture for the sake of comparison. The main emphasis, however, is on the results recently arrived at for the quasi–perpendicular electromagnetic ion–cyclotron (QPEMIC) modes and instabilities. These results show that if the electron drift (or, alternatively, the current density) is gradually increased, the QPEMIC instability sets in sooner (i.e. with smaller threshold electron drift) than the corresponding (i.e. belonging to the same IC harmonic) QPESIC instability, quite in analogy with the earlier theoretical predictions pertaining to the situation where $\boldsymbol{B}_0$ and $\boldsymbol{E}_0$ were assumed to be parallel [14]. Moreover, the QPEMIC instability is very sensitive to variations of the gas pressure in the discharge tube, which is not observed in the QPESIC case. The roles of the angle γ between $\boldsymbol{B}_0$ and $\boldsymbol{E}_0$, the non-isothermality parameter $\tau = T_e/T_i$, and the magnetization parameter ω_{Bi}/ν_i (ν_i being the ion collision frequency) in determining the threshold drifts for the QPEMIC instabilities in the vicinity of various IC harmonics will be also elucidated to some extent.

2. BASIC RELATIONS

It is general knowledge [9,10,15] that the IC modes are most conveniently studied in the so–called $\boldsymbol{B}_0$-system (Oz-axis directed along $\boldsymbol{B}_0$, the wave vector $\boldsymbol{k}$ lying in xOz-plane), and this will be adopted here. In the case of non–parallel fields studied presently, several privileged planes thus emerge, the ones formed by the directions of, respectively, $\boldsymbol{B}_0$ and $\boldsymbol{E}_0$, and $\boldsymbol{B}_0$ and $\boldsymbol{k}$, being the most conspicuous ones (the wave vector $\boldsymbol{k}$ is almost at right angles to $\boldsymbol{B}_0$, as the QP modes are being studied). The angle χ between these two planes varies from 0^0 to 360^0 (Fig. 1); it is easily seen that $\chi = 0^0$ (or $\chi = 360^0$) and $\chi = 180^0$ correspond to the directions of wave propagation lying in the $(\boldsymbol{B}_0,\boldsymbol{E}_0)$–plane and forming, respectively, acute and obtuse angles with $\boldsymbol{E}_0$. It can be also inferred from Fig. 1 that (E_0,γ,χ) are actually the spherical coordinates of $\boldsymbol{E}_0$ in the coordinate system chosen for the study (E_0 denoting the magnitude of $\boldsymbol{E}_0$).

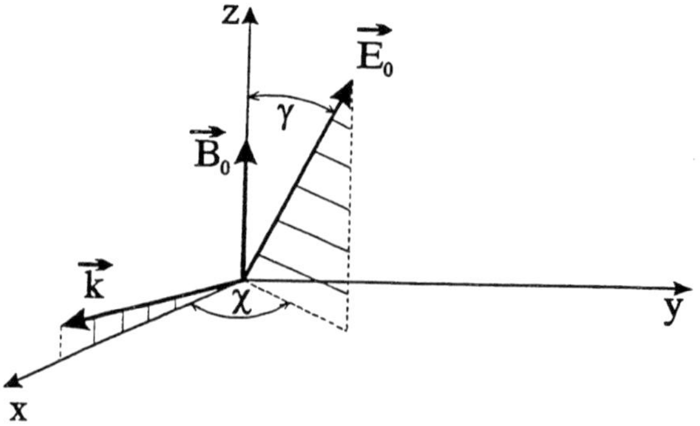

Figure 1.

Kinetic equations with BGK model collision integral in the weakly ionized plasma are of the well-known form:

$$\frac{\partial f_\alpha}{\partial t} + \boldsymbol{v}\cdot\frac{\partial f_\alpha}{\partial \boldsymbol{r}} + \frac{e_\alpha}{m_\alpha}(\boldsymbol{E}+\boldsymbol{v}\times\boldsymbol{B})\cdot\frac{\partial f_\alpha}{\partial \boldsymbol{v}} = -\nu_\alpha (f_\alpha - n_\alpha \Phi_\alpha), \quad (\alpha = i, e), \tag{1}$$

with

$$\Phi_\alpha = \left(\frac{m_\alpha}{2\pi \varkappa T_\alpha}\right)^{3/2} \exp\left(-\frac{m_\alpha v^2}{2\varkappa T_\alpha}\right), \quad (\alpha = i, e). \tag{2}$$

Here, ν_α are effective (constant) collision frequencies for elastic encounters of the charged particles ($\alpha = i, e$) with neutrals, $\varkappa$ is the Boltzmann constant, T_α are the temperatures (they are related to the thermal velocities, which will be introduced below, by $v_{T_\alpha} = \sqrt{\varkappa T_\alpha/m_\alpha}$). If (1) is written out for the steady state in the uniform plasma, the following equations are obtained for the determination of the drifts $\boldsymbol{u}_\alpha$:

$$e_\alpha (\boldsymbol{E}_0 + \boldsymbol{u}_\alpha \times \boldsymbol{B}_0) = m_\alpha \nu_\alpha \boldsymbol{u}_\alpha, \quad (\alpha = i, e). \tag{3}$$

The solutions are of the well-known form:

$$\boldsymbol{u}_\alpha = \frac{(e_\alpha/m_\alpha\nu_\alpha)}{1+(\omega_{B\alpha}/\nu_\alpha)^2}\left[\boldsymbol{E}_0 + \frac{\omega_{B\alpha}}{\nu_\alpha}\frac{\boldsymbol{E}_0\times\boldsymbol{B}_0}{B_0} + \left(\frac{\omega_{B\alpha}}{\nu_\alpha}\right)^2 \frac{(\boldsymbol{E}_0\cdot\boldsymbol{B}_0)\boldsymbol{B}_0}{B_0^2}\right], \quad (\alpha = i, e) \tag{4}$$

Here, $\omega_{B\alpha} = e_\alpha B_0/m_\alpha$ are the gyrofrequencies (including the charge sign for different senses of gyration of ions and electrons). The above drift velocities can be decomposed into components respectively parallel and perpendicular to $\boldsymbol{B}_0$: $\boldsymbol{u}_\alpha = \boldsymbol{u}_{\alpha\parallel} + \boldsymbol{u}_{\alpha\perp}$. Applying the same decomposition to $\boldsymbol{E}_0$, one obtains from (4):

$$\boldsymbol{u}_{\alpha\parallel} = \frac{e_\alpha \boldsymbol{E}_{0\parallel}}{m_\alpha\nu_\alpha}, \quad \boldsymbol{u}_{\alpha\perp} = \frac{1}{1+(\nu_\alpha/\omega_{B\alpha})^2}\left[\frac{\boldsymbol{E}_0\times\boldsymbol{B}_0}{B_0^2} + \frac{\nu_\alpha}{\omega_{B\alpha}}\frac{\boldsymbol{E}_{0\perp}}{B_0}\right], \quad (\alpha = i, e). \tag{5a,b}$$

It can be seen from the above relations that the directions of $\boldsymbol{u}_{i\perp}$ and $\boldsymbol{u}_{i\parallel}$ generally do not coincide. Consequently, two drift planes, ($\boldsymbol{B}_0$, $\boldsymbol{u}_i$) and ($\boldsymbol{B}_0$, $\boldsymbol{u}_e$), also appear in the situation considered. However, if the plasma is sufficiently magnetized (say, if $\omega_{Bi}/\nu_i \geq 5$), the above expressions for $\boldsymbol{u}_{\alpha\perp}$ can be simplified by treating the ratios $\nu_\alpha/\omega_{B\alpha}$ as small parameters. It is worth pointing out in this context, that $\nu_e/|\omega_{Be}| \ll \nu_i/\omega_{Bi}$, so that if ν_i/ω_{Bi} is a small parameter owing to strong magnetization, the same certainly holds for $\nu_e/|\omega_{Be}|$; indeed, one has

$$\frac{\nu_e}{\nu_i}\frac{\omega_{Bi}}{|\omega_{Be}|} \approx z_i \frac{\sigma_{en}}{\sigma_{in}} \tau^{1/2} \left(\frac{m_e}{m_i}\right)^{1/2}, \tag{6}$$

if one adopts the billiard-ball model for the collisions (i.e. if one puts $\nu_\alpha = n_n \sigma_{\alpha n} v_{T\alpha}$, with σ_{en} and σ_{in} being the effective cross sections for the collisions taken into account, and n_n standing for the number density of the neutrals). With such magnetizations, (5b) becomes:

$$\boldsymbol{u}_{\alpha\perp} = \frac{\boldsymbol{E}_0\times\boldsymbol{B}_0}{B_0^2}, \quad (\alpha = i, e), \tag{5c}$$

and the meaning of this result is immediately understood: in sufficiently magnetized plasma both perpendicular drifts reduce to the ($\boldsymbol{E}\times\boldsymbol{B}$)-drift. Both drift planes merge then together, the common drift plane assumes a perpendicular position with respect to the ($\boldsymbol{B_0}$,$\boldsymbol{E_0}$)–plane, and forms the angle $\chi - 90^0$ with the ($\boldsymbol{B_0}$,$\boldsymbol{k}$)–plane. This will be the situation discussed further on.

Fig. 2 shows the position of the drift plane relative to the privileged planes given in Fig. 1. It should be borne in mind that, according to (5a), the ratio $|\boldsymbol{u}_{i\|}|/|\boldsymbol{u}_{e\|}|$ is equal to the left-hand-side in (6) (and, consequently, very small), and that the angles formed by the directions of the drift and $\boldsymbol{B_0}$ are given by:

$$\tan\gamma_\alpha = \frac{\nu_\alpha}{\omega_{B\alpha}}\tan\gamma \tag{7}$$

(including the charge sign in $\omega_{B\alpha}$). In view of (6), one concludes that the angle γ_e is below 1^0 or above 179^0 (depending on whether γ is obtuse or acute) if γ is not too close to 90^0 (say, $|90^0 - \gamma| \geq 30^0$). The angle γ_i is much larger (see Fig. 2). This means that the effects caused by the presence of $\boldsymbol{u}_{i\|}$ and $\boldsymbol{u}_{e\perp}$ are negligible in comparison with those arising from $\boldsymbol{u}_{e\|}$ and $\boldsymbol{u}_{i\perp}$, respectively. On the other hand, (5c) also implies that no net perpendicular current flows in the plasma, despite the presence of the perpendicular drifts.

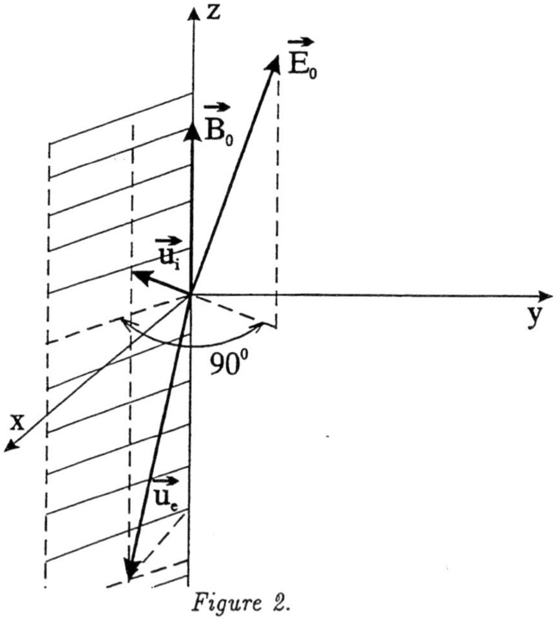

Figure 2.

The solutions of the steady–state kinetic equations obtained from (1) in the case of uniform plasma coincide, up to the first–order small quantities, with the well–known shifted Maxwellians. With the drifts determined by (5a)–(5c), these distribution functions become anisotropic in the velocity space, and if the facts given in the comment of eq. (7) are borne in mind, one approximately obtains [13]:

$$f_i^0 = \frac{n_i^0}{(2\pi)^{3/2} v_{Ti}^3}\left[1 - \frac{v_\perp u_\perp}{v_{Ti}^2}\sin(\varphi - \chi)\right]\exp\left(-\frac{v^2}{2v_{Ti}^2}\right), \tag{8a}$$

$$f_e^0 = \frac{n_i^0}{(2\pi)^{3/2} v_{Te}^3} \exp\left(-\frac{v_\perp^2}{2v_{Te}^2}\right) \cdot \exp\left[-\frac{(v_z - u_z)^2}{2v_{Te}^2}\right], \tag{8b}$$

with $v_\perp$, φ and v_z denoting the cylindrical coordinates in the velocity space.

In the frames of the linear theory, eq. (1) yields the following relation for the velocity distribution function perturbation:

$$\frac{\partial\, \delta f_\alpha}{\partial t} + \boldsymbol{v} \cdot \frac{\partial\, \delta f_\alpha}{\partial \boldsymbol{r}} + \frac{e_\alpha}{m_\alpha}(\delta \boldsymbol{E} + \boldsymbol{v} \times \delta \boldsymbol{B}) \cdot \frac{\partial f_\alpha^0}{\partial \boldsymbol{v}} + \\ + \frac{e_\alpha}{m_\alpha}(\boldsymbol{E}_0 + \boldsymbol{v} \times \boldsymbol{B}_0) \cdot \frac{\partial\, \delta f_\alpha}{\partial \boldsymbol{v}} = -\nu_\alpha (\delta f_\alpha - \delta n_\alpha \Phi_\alpha),\ (\alpha = i, e). \tag{9}$$

This corresponds to the generally accepted "isothermic approximation", in which one assumes that the temperatures are not perturbed ($\delta T_\alpha = 0$). It is convenient to modify (9) by omitting the $\boldsymbol{E}_0$-term and by substituting at the same time ϕ_α by some other function ϕ_α^*, chosen in such a manner that the ensuing equation:

$$\frac{\partial\, \delta f_\alpha}{\partial t} + \boldsymbol{v} \cdot \frac{\partial\, \delta f_\alpha}{\partial \boldsymbol{r}} + \frac{e_\alpha}{m_\alpha}(\delta \boldsymbol{E} + \boldsymbol{v} \times \delta \boldsymbol{B}) \cdot \frac{\partial f_\alpha^0}{\partial \boldsymbol{v}} + \\ + \frac{e_\alpha}{m_\alpha}(\boldsymbol{v} \times \boldsymbol{B}_0) \cdot \frac{\partial\, \delta f_\alpha}{\partial \boldsymbol{v}} = -\nu_\alpha (\delta f_\alpha - \delta n_\alpha \Phi_\alpha^*),\ (\alpha = i, e), \tag{10}$$

remains equivalent to eq. (9) in the sense that they both yield the same equations of mass and momentum transfer [14]. With non–parallel fields $\boldsymbol{B}_0$ and $\boldsymbol{E}_0$, one thus obtains:

$$\Phi_\alpha^* = \frac{1}{(2\pi)^{3/2} v_{T\alpha}^3} \exp\left[-\frac{(\boldsymbol{v} - \boldsymbol{U}_\alpha)^2}{2 v_{T\alpha}}\right], \quad \boldsymbol{U}_\alpha = \frac{e_\alpha \boldsymbol{E}_0}{m_\alpha \nu_\alpha}. \tag{11}$$

Using cylindrical coordinates in the velocity space, and carrying out Fourier transformations with respect to both space coordinates and time, one may rewrite (10) as:

$$\frac{\partial \delta \tilde{f}_\alpha}{\partial \varphi} + i \frac{\omega + i\nu_\alpha - k_z v_z - k_\perp v_\perp \cos\varphi}{\omega_{B\alpha}} \delta \tilde{f}_\alpha = \frac{e_\alpha}{m_\alpha \omega_{B\alpha}} (\delta \tilde{\boldsymbol{E}} + \boldsymbol{v} \times \delta \tilde{\boldsymbol{B}}) \cdot \frac{\partial f_\alpha^0}{\partial \boldsymbol{v}} - \frac{\nu_\alpha}{\omega_{B\alpha}} \Phi_\alpha^* \delta \tilde{n}_\alpha, \tag{12}$$

the tildes denoting the double Fourier transforms mentioned. The integration of (12) is performed with the aid of the expansion with respect to Bessel functions [15], and the components of the plasma dielectric tensor are evaluated following the well–known standard procedure. It will not be necessary to give all these components here.

3. QPESIC WAVES

The dispersion equation for the QPESIC waves reads [16]:

$$\delta\varepsilon_i + \delta\varepsilon_e = 0. \tag{13}$$

For QP propagation, and for the modal frequencies around that of the Rth ($R =$

1, 2, 3, ...) IC harmonic, one obtains in the "one-harmonic approximation" [13]:

$$\delta\varepsilon_i = \frac{\omega_{pi}^2}{k^2 v_{Ti}^2}\left\{\left[1 - \frac{\omega A_R(\mu)}{\omega - R\omega_{Bi}} + \sin\chi\frac{u_\perp}{v_{Ti}}\Gamma_R\right] + \right.$$
$$\left. +i\left[\frac{\omega\nu_i A_R(\mu)}{(\omega - R\omega_{Bi})^2} + \sqrt{\frac{\pi}{2}}\frac{\omega A_R(\mu)}{\omega - R\omega_{Bi}}\xi_R \exp\left(-\frac{1}{2}\xi_R^2\right) + \cos\chi\frac{u_\perp}{v_{Ti}}\Delta_R\right]\right\}, \tag{14}$$

where $\mu = k_\perp^2 v_{Ti}^2/\omega_{Bi}{}^2$, $A_R(\mu) = I_R(\mu)e^{-\mu}$, $\xi_R = \dfrac{\omega - R\omega_{Bi}}{|k_z|v_{Ti}}$, and ω_{pi} denotes the ion plasma frequency, while:

$$\Gamma_R = \frac{\omega_{Bi}}{\omega - R\omega_{Bi}}\sqrt{\mu}\left[(2\mu+1)A_R'(\mu) - \left(\frac{R^2}{\mu} - 1\right)A_R(\mu)\right] - \frac{RA_R(\mu)}{\sqrt{\mu}}, \tag{15}$$

$$\Delta_R = \frac{\omega A_R'(\mu)}{\omega - R\omega_{Bi}}\sqrt{\mu} - \frac{R\omega_{Bi}}{\omega - R\omega_{Bi}}\frac{A_R(\mu)}{\sqrt{\mu}}. \tag{16}$$

As far as the electronic contribution to (13) is concerned, there are two domains of weak damping of the IC waves. One of these is at small wavelengths (the "shortwave region", $\nu_e \ll |k_z|v_{Te}$), and the other at large ones ($\nu_e \gg |k_z|v_{Te}$, the "longwave region"; here, one must have $\omega\nu_e \ll k_z^2 v_{Te}^2$). Using superscripts (s) and (l) to designate the two regions, one has [13,14]:

$$\delta\varepsilon_e = \frac{\omega_{pe}^2}{k^2 v_{Te}^2}\left(1 + \frac{i}{a}\frac{\omega - k_z u_z}{|k_z|v_{Te}}\right), \tag{17}$$

$$a^{(s)} = \sqrt{\frac{2}{\pi}}, \qquad a^{(l)} = \frac{|k_z|v_{Te}}{\nu_e}. \tag{18}$$

Inserting (14) and (17) into (13), and equating the real part of the resulting dispersion equation to zero, one obtains the spectrum of the IC mode around the Rth harmonic. The effects of non-parallel fields are very small [13], so that one may use the simplified result:

$$\omega = R\omega_{Bi}\frac{T}{T - A_R(\mu)}, \qquad T = 1 + \frac{1}{z_i\tau}, \tag{19}$$

obtained by omitting the Γ_R-term in the real part of (14). With the aid of (19), one may simplify (16) and obtain:

$$\Delta_R(\mu, T) = T\sqrt{\mu}\frac{A_R'(\mu)}{A_R(\mu)} - \frac{T - A_R(\mu)}{\sqrt{\mu}}, \tag{20}$$

In the derivations that follow, the functions

$$Z_R(\mu, T) = \Delta_R(\mu, T)A_R(\mu). \tag{21}$$

will have an important role. They all have a minimum, the abscissa of which coincides with that of the maximum of the corresponding function $A_R(\mu)$, and with the negative ordinate depending on T. The functions (21) were tabulated for the purpose of the analysis that follows.

Equating with zero the imaginary part of the dispersion equation obtained by inserting (14) and (17) into (13), one arrives at the condition of marginal instability. It can be solved explicitly for u/v_{Ti} (u being the electron drift magnitude), defining thereby the critical drift required for the spontaneous excitation of the QPESIC instability pertaining to the Rth IC harmonic in the given direction of wave propagation (which is specified by the angle χ and by the variables μ and ξ_R connected with $k_\perp$ and $k_\parallel = |k_z|$). The threshold drift is the minimum of those critical drifts, and it corresponds to a direction of wave propagation not known beforehand. In the shortwave region, the final result for the threshold drift reads:

$$\left(\frac{u^*}{v_{Ti}}\right)_R^{(s)} = T\frac{G\left[(\xi_R^*)^{(s)}\right] + \sqrt{\frac{2}{\pi}}z_i\left(\frac{m_i}{m_e}\right)^{1/2}\tau^{3/2}\frac{\nu_i}{\omega_{Bi}}\frac{T-(A_R)_{max}}{R}}{(A_R)_{max} + \sqrt{\frac{2}{\pi}}z_i^2\tau^2\frac{\sigma_{en}}{\sigma_{in}}\frac{\nu_i}{\omega_{Bi}}|\tan\gamma||(Z_R)_{min}|}, \tag{22}$$

where $G(x) = x^3/(x^2-1)$, and $(\xi_R^*)^{(s)}$ is the larger of the two solutions of the equation $(x^2-1)e^{-x^2/2} = a_R$, with $a_R = z_i(m_i/m_e)^{1/2}\tau^{3/2}(A_R)_{max}$. In the longwave region one similarly has:

$$\left(\frac{u^*}{v_{Ti}}\right)_R^{(l)} = \frac{2T}{(A_R)_{max}}(\xi_R^*)^{(l)}, \qquad (\xi_R^*)^{(l)} = \left(b_R + c_R^2\right)^{1/2} - c_R, \tag{23}$$

where

$$b_R = a_R\frac{\sigma_{in}}{\sigma_{en}}, \qquad c_R = z_i^2\tau^2|\tan\gamma|\frac{R|(Z_R)_{min}|}{T-(A_R)_{max}}. \tag{23a}$$

In both regions, the spontaneous excitation takes place for $\chi = 0^0$ first.

A more detailed analysis of the results (22) and (23) discloses certain regularities in the threshold drifts for the QPESIC instability. Firstly, the threshold drifts increase with the harmonic number R, in both shortwave and longwave regions. This feature is specific for the "one–harmonic approximation". Secondly, the threshold drifts increase with the non–isothermality, as τ is raised above 1. This is an effect amenable to ion collisions in the shortwave region [with $\nu_i = 0$, eq. (22) would yield a function which decreases as τ increases], and to electron collisions in the longwave region (with $\nu_e = 0$, this region vanishes altogether). Thirdly, the threshold drifts in both wavelength regions decrease as γ, the angle between $\boldsymbol{B_0}$ and $\boldsymbol{E_0}$, is augmented. The augmentation of γ enhances the perpendicular drift, hence, the presence of this drift component largely facilitates the excitation of the instabilities considered [7,8], although the accompanying perpendicular current is absent.

On the other hand, the longwave threshold drifts are almost invariably considerably smaller than the shortwave ones. Consequently, the QPESIC instability in the vicinity of the given (R, th) harmonic will appear in the longwave region first, if the drift(viz. the magnitude of the applied external d.c. electric field) is gradually increased. More precisely, the longwave threshold drifts do not depend on magnetization (this is actually due to the fact that the ionic Landau damping could be neglected in this region). The shortwave ones, on the other hand, decrease as ω_{Bi}/ν_i increases, as shown by (22). In non isothermal plasmas with $\tau > 1$ it may happen, therefore, that the shortwave drift becomes smaller than the longwave one, provided that the magnetization is exceptionally high ($\omega_{Bi}/\nu_i \geq 20$).

It should be underlined that the QPESIC threshold drifts do not depend on gas pressure (or, alternatively, on neutral number density n_n) if R, τ, γ and ω_{Bi}/ν_i are regarded as the fundamental parameters. As far as the QPESIC instabilities are concerned, the numerical value of n_n only fixes the minimum magnetic field magnitude required to magnetize the plasma (it is obtained from the condition $\omega_{Bi}/\nu_i = 1$).

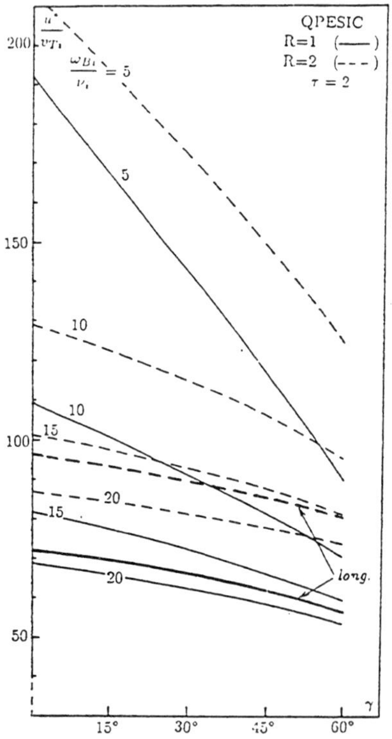

Figure 3.

Figure 3 shows the QPESIC threshold drifts evaluated from (22) and (23) for $R = 1$ (unbroken lines) and $R = 2$ (dashed lines) in the moderately non-isothermal ($\tau = 2$) weakly ionized hydrogen ($z_i = 1$, $\sqrt{m_i/m_e} = 42.8$) plasma with $\sigma_{in} = \sigma_{en}$ (the so-called "strict" billiard–bal model); using typical gas-discharge parameters ($T_i \approx 10^3 K$, $\sigma_{in} \approx 10^{-20} m^2$, $n_n \approx 3 \times 10^{23} m^{-3}$), one obtains that the minimum magnetic field ensuring the magnetization is about $900\, Ga$. Bold lines (both unbroken and dashed) show the longwave threshold drifts. The numbers attached to the shortwave curves indicate the pertaining values of ω_{Bi}/ν_i. It is easily noticed that the shortwave drifts pertaining to $\omega_{Bi}/\nu_i = 20$ fall below the longwave drift curve. It is also easily noticed that the rate of decrease of the threshold drifts with the angle γ is very remarkable with $\omega_{Bi}/\nu_i = 5$ (comparatively low magnetization), and then decreases as the magnetization is augmented.

4. QPEMIC WAVES

Foe low–pressure plasmas with moderate non–isothermalities (we shall assume $\tau \leq$

5), the dispersion equation for QPEMIC waves reads [16]:

$$\delta\varepsilon_i + \delta\varepsilon_e - \frac{\omega^2}{c^2k^2}\varepsilon_{xx}\varepsilon_{zz} = 0, \tag{24}$$

Here, $\delta\varepsilon_i$ and $\delta\varepsilon_e$ have the same meaning as in eq. (13) of the preceding paragraph, and

$$\varepsilon_{xx} = 1 + \delta\varepsilon_{xx}^{(i)} + \delta\varepsilon_{xx}^{(e)}, \quad \varepsilon_{zz} = 1 + \delta\varepsilon_{zz}^{(i)} + \delta\varepsilon_{zz}^{(e)}; \tag{25}$$

are the two relevant components of the plasma dielectric tensor. They are evaluated by appropriate integrations of eq. (12). For the electronic contributions the results are:

$$\delta\varepsilon_{xx}^{(e)} = 0, \quad \delta\varepsilon_{zz}^{(e)} = \frac{\omega_{pe}^2}{|k_z|^2 v_{Te}^2}\left(1 + \frac{i}{a}\frac{\omega - k_z u_z}{|k_z| v_{Te}}\right); \tag{26}$$

the coefficient a being again given by (18). For the ionic contributions one obtains:

$$\delta\varepsilon_{xx}^{(i)} = -\frac{\omega_{pi}^2}{\omega^2}\frac{R^2 T}{\mu}\left\{1 - i\left[\frac{\nu_i}{\omega}\frac{T}{A_R(\mu)} + \sqrt{\frac{\pi}{2}}\,\xi_R \exp\left(-\frac{1}{2}\xi_R^2\right)\right]\right\} + \ldots, \tag{27a}$$

$$\delta\varepsilon_{zz}^{(i)} = -\frac{\omega_{pi}^2}{\omega^2}T\left\{1 - i\left[\frac{\nu_i}{\omega}\frac{T}{A_R(\mu)} + \sqrt{\frac{\pi}{2}}\,\xi_R^3 \exp\left(-\frac{1}{2}\xi_R^2\right)\right]\right\} + \ldots \tag{27b}$$

the dots denoting the terms which are proportional to the small parameter u_i/v_{Ti} and which can be neglected in (24), in view of multiplication with another small parameter ω^2/c^2k^2.

The spectrum, obtained by equating the real part of (24) to zero is very little affected by the new terms proportional to ω^2/c^2k^2, and the result (19) can be adopted here again. The condition of marginal instability is also little altered by the new terms. It can be inferred, therefore, that the direction of propagation of QPEMIC waves of growing amplitude, at the onset of the instability, is practically the same as in the QPESIC case. This yields the following threshold drifts for the Rth harmonic:

$$\left(\frac{u^*}{v_{Ti}}\right)_R^{(s)} = T\frac{(\xi_R^*)^{(s)} + \dfrac{C^{(s)}}{D^{(s)}}z_i\left(\dfrac{m_i}{m_e}\right)^{1/2}\tau^{3/2}Q_R}{(A_R)_{max} + \dfrac{1}{D^{(s)}}\sqrt{\dfrac{2}{\pi}}z_i^2\tau^2\dfrac{\sigma_{en}}{\sigma_{in}}\dfrac{\nu_i}{\omega_{Bi}}|\tan\gamma||(Z_R)_{min}|}, \tag{28}$$

where

$$Q_R = (A_R)_{max}(\xi_R^*)^{(s)}\exp\left\{-\frac{1}{2}\left[(\xi_R^*)^{(s)}\right]^2\right\} + \sqrt{\frac{2}{\pi}}\frac{\nu_i}{\omega_{Bi}}\frac{T - (A_R)_{max}}{R}, \tag{28a}$$

and

$$\left(\frac{u^*}{v_{Ti}}\right)_R^{(l)} = T\frac{\left[(\xi_R^*)^{(l)}\right]^2 + \dfrac{C^{(l)}}{D^{(l)}}z_i\left(\dfrac{m_i}{m_e}\right)^{1/2}\tau^{3/2}(A_R)_{max}\dfrac{\sigma_{in}}{\sigma_{en}}}{(\xi_R^*)^{(l)}(A_R)_{max} + \dfrac{1}{D^{(l)}}z_i^2\tau^2\dfrac{R(A_R)_{max}}{T - (A_R)_{max}}|\tan\gamma||(Z_R)_{min}|}, \tag{29}$$

respectively in the shortwave and longwave regions. Here $(\xi_R^*)^{(s)}$ and $(\xi_R^*)^{(l)}$ have the same meaning as in (22) and (23), and

$$C = 1 - \frac{v_{Ti}^2}{v_A^2}\frac{T^2}{z_i\tau}\frac{(\xi_R^*)^2}{\tilde{\mu}_R (A_R)_{max}^2}, \quad D = 1 + \frac{v_{Ti}^2}{v_A^2}T^3\frac{(\xi_R^*)^2}{\tilde{\mu}_R (A_R)_{max}^2}, \tag{30a,b}$$

with $\tilde{\mu}_R$ denoting the abscissa of the maximum of $A_R(\mu)$. The above two correction coefficients have different values in the shortwave and the longwave regions [this is indicated by superscripts (s) and (l) in (28) and (29)], since (ξ_R^*) is not the same in the two regions. As C is somewhat smaller and D somewhat larger than 1, the threshold drifts for the QPEMIC instabilities are generally somewhat smaller than the corresponding QPESIC ones.

A closer analysis of (28) and (29) reveals that the QPEMIC threshold drifts exhibit a behaviour largely analogous to that of the QPESIC ones. Thus, the QPEMIC threshold drifts increase if R or τ are augmented, and decrease if γ or ω_{Bi}/ν_i increase, always remaining somewhat below the QPESIC threshold drift values for the same set of the above parameters. Also, the threshold drifts in the longwave region are almost invariably smaller than the shortwave ones, for the same R, τ, γ and ω_{Bi}/ν_i, provided that the magnetization is not too high.

Certain novel features in the QPEMIC threshold drift behaviour are amenable to the presence of the ratio $(v_{Ti}/v_A)^2$ in the correction coefficients (30a,b). If this ratio is neglected (i.e. if $C = D = 1$), the QPEMIC results (28) and (29) coincide with (22) and (23) for the QPESIC instabilities. On the other hand, the ratio considered may be written as:

$$\frac{v_{Ti}^2}{v_A^2} = \frac{\nu_i^2}{n_n^2\sigma_{in}^2}\frac{\mu_0 m_i n_i}{B_0^2} = 1.91 \times 10^{-17}\frac{z_i X}{A n_n \sigma_{in}^2}\left(\frac{\nu_i}{\omega_{Bi}}\right)^2 \tag{31}$$

where X is the degree of ionization ($n_i = X n_n$), and A is the atomic mass number ($m_i = A m_H$, m_H being the mass of the hydrogen atom). Hence, the correction coefficients (30a,b) depend on magnetization (i.e. on ω_{Bi}/ν_i) and on the gas pressure in the discharge tube (by way of n_n), and this is reflected upon the QPEMIC threshold drift values. For low magnetizations (say, ω_{Bi}/ν_i not much above 5) and low gas pressures (say, in the range of $1\,mmHg$) they differ considerably from the QPESIC values; if the magnetization or the gas pressure are augmented, C and D gradually tend to unity, and the differences between the QPEMIC and QPESIC threshold drifts become very small. These differences also gradually diminish as γ, the angle between B_0 and E_0, is increased.

To illustrate the above characteristics of the drifts, one may take again the gas-discharge hydrogen plasma ($z_i = 1$, $A = 1$) with $T_i \approx 10^3 K$ and $\sigma_{in} \approx 10^{-22} m^2$. In order to ensure noticeable effects, one takes $n_n \approx 3 \times 10^{22} m^{-3}$ and $n_n \approx 10^{22} m^{-3}$ (the gas pressures approximately $3\,mmHg$ and $1\,mmHg$, respectively); the QPESIC results remain unaffected by this change in the gas density, so that the results shown in Fig. 3 ar still valid. The results for $R = 1$, $\tau = 2$, and $\omega_{Bi}/\nu_i = 5$ and 15 are presented in Fig.

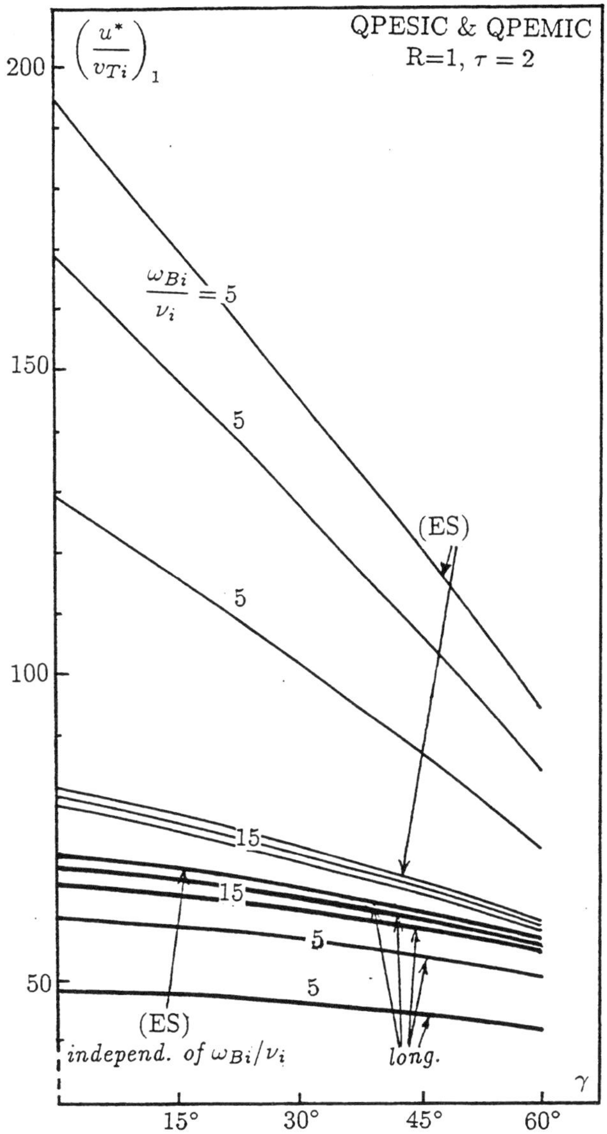

Figure 4.

4, under the assumption that the "strict" billiard–bal model holds for the collisions. The basic features of the threshold drift behaviour can be clearly seen. For sake of comparison, the QPESIC curves pertaining to $R = 1$, $\tau = 2$ and $\omega_{Bi}/\nu_i = 5$ and 15 from Fig. 3 are also given, marked (ES). In both short- and longwave domains, the lower of the two QPEMIC curves pertaining to the same value ω_{Bi}/ν_i corresponds to lower pressure.

REFERENCES

[1] A.M. LEVINE, A.F. FUCKES, *Phys. Fluids* **9**, 2263 (1966).

[2] J. JUUL–RASMUSSEN, R. SCHRITTWIESER, Å. SKØELV, *Physica Scripta* **34**, 821 (1986).

[3] D.M. SUSZCYNSKY, N.D'ANGELO, R.L. MERLINO, *J. Geophys. Res.* **94**, 8966 (1989).

[4] R.E. WALTZ, R.R. DOMINGUEZ, *Phys. Fluids* **24**, 1575 (1981).

[5] M.J. ALPORT, P.J. BARRETT, M.A. BEHRENS, *Plasma Physics* **25**, 1059 (1983).

[6] S. CUPERMAN, L. GOMBEROFF, *J. Plasma Phys.* **18**, 391 (1977).

[7] A.V. VOLOSEVICH, V.A. SHAFTAN, *Geomagnetism and Aeronomy* (USSR) **28**, 959 (1988).

[8] P. SATYANARAYANA, P.K. CHATURVEDI, M.J. KESKINEN, J.D. DUBA, S.L. OSAKOW, *J. Geophys. Res.* **90**, 12209 (1985).

[9] T.H. STIX, *"The Theory of Plasma Waves"*, McGraw–Hill, New York, 1962.

[10] A.B. MIKHAILOVSKI, *"The Theory of Plasma Instabilities"* (in Russian), Atomizdat, Moscow, 1970.

[11] A. FRUCHTMAN, W.F. CUMMINS, *Phys. Fluids* **30**, 1569 (1987).

[12] M.A. GREENSPAN, R. PAL, D. HAMMER, S. HUMPHRIES, *Appl. Phys. Lett.* **37**, 248 (1980).

[13] B.S. MILIĆ, D.Ž. GAJIĆ, *Physica Scripta* **45**, 264 (1992).

[14] B.S. MILIĆ, N.R. BRAJUŠKOVIĆ, *J. Plasma Phys.* **29**, 21 (1983).

[15] I.B. BERNSTEIN, *Phys. Rev.* **109**, 10 (1958).

[16] A.I. AKHIEZER, I.A. AKHIEZER, R.V. POLOVIN, A.G. SITENKO, K.N. STEPANOV, *"Plasma Electrodynamics"* (in Russian), Nauka, Moscow, 1974.

Low Temperature and
General Plasmas
M. Milosavljevic and Z. Petrovic

ISBN 1-56072-399-8

PLASMA VORTICES

D. Jovanović
Institute of Physics
P.O. Box 57, Yu-11001 Belgrade, Yugoslavia

Abstract:

Recent developments in the theory of plasma vortices are reviewed. Vortices are "generic" nonlinear, coherent, stationary and spatially localized solutions of magnetohydrodynamic (MHD) and fluid equations in the regime $\frac{\partial}{\partial t} \ll \Omega_i$. They have the form of pairs of vortex tubes, rotating in opposite directions, slightly tilted relative to the magnetic field. Main vortex modes are associated with nonlinear drift waves, shear-Alfvén waves, and short wavelength drift-Alfvén waves, with the typical perpendicular scalelengths of the order of the ion inertial length, collisionless skin depth, and the ion Larmor radius, respectively.

In this review main vortex types are studied to some detail. Stability of drift- and shear-Alfvén-type vortices under small perturbations is discussed, and linear stability of drift-wave vortices propagating in the direction of the ion diamagnetic drift is proved in a numerical experiment. Although stability of shear-Alfvén vortices could not be shown analytically, a strong evidence is found indicating that an eventual instability would be followed by a self-organization process which may yield to a new coherent structure. Finally, the interaction between vortices and particles is studied, and the creation of a quasi-three dimensional electron hole/vortex structure by particle trapping is demonstrated.

1 Introduction

The dominance of the convective nonlinearity, with the leading order velocity being given by $\vec{v}_E = \vec{E} \times \vec{B}/B^2$, may lead to the formation of low-frequency convective structures in plasmas. Such convective nonlinearities give rise to vector-product, or Poisson bracket, terms in the corresponding nonlinear partial differential equations, which are responsible for the self-organization in the direction perpendicular to the magnetic field, and the creation of two-dimensional vortex structures. A large number of model equations, describing plasma dynamics in various geometries and regimes belong to this family [1] - [4], and possess double vortex (also called modon) solutions.

Plasma vortices are not solitons in the mathematical sence, since the corresponding equations generally have a finite number of conserved quantities [5]. However, they are very robust objects, and it is expected that a turbulent plasma state can be represented as a superposition of coherent vortex structures and weakly correlated wave-like fluctuations [6] - [8]. Such coherent vortices, generated in some kind of self-organization process are expected to play a major role in the particle transport in radio-frequency (RF) heated magnetically confined plasmas, by the mechanism of particle detrappind-trapping during inelastic collisions between vortices in the fully developed turbulence [10]. Emergence of self-organization was observed experimentally for the case of flute mode turbulence in the presence of externally injected plasma vortices [9], while computer simulations and analytical studies indicate vortex creation because of driving forces of density and temperature gradients (in the drift-wave instability) [11], [12], by the perpendicular electric field (Kelvin-Helmholtz instability) [13], parallel electric fields [14] and the ponderomotive force [15]. All these processes occur on the spatial scale which is much longer than the ion Larmor radius, and they are connected with various ideal and resistive MHD instabilities. However, the energy and particle transport is often affected by turbulent fluctuations with the scalelength below the ion Larmor radius, which are driven by various microinstabilities. Typical example of such processes is the stochatization of the magnetic field. Vortices on the microturbulence scale, associated with nonlinead drift-Alfvén waves are also possible, excited by e. g. inhomogeneous parallel shear flows [16].

The paper is organized as follows. In Sections 2 and 3 plasma vortices associated with drift waves, and torsional Alfvén waves are constructed, and their linear stability is studied using Lyapunov method. Gyrokinetic theory of three-dimensional electron holes, arising from the electron trapping in the direction of the magnetic field, and coupled with usual hydrodynamic vortices, are constructed as Bernstein-Green-Kruskal (BGK) solutions of the drift-kinetic equation in Section 4.

2 Stability of drift-wave modons

Particularly important vortex-type equation is the Charney-Hasegawa-Mima equation, which describes both nonlinear drift waves, and Rossby waves in rotating shallow fluids. In the description of plasmas, it is derived from the ion continuity equation using the standard expression for the ion hydrodynamic velocity $\vec{v}_i = \vec{v}_E + \vec{v}_P$ where the $\vec{E} \times \vec{B}$ and ion polarization drift velocities, respectively are given by

$$\vec{v}_E = \frac{\vec{E} \times \vec{B}}{B^2} \quad , \quad \vec{v}_P = \frac{1}{\Omega_i}\left(\frac{\partial}{\partial t} + \vec{v}_E \nabla\right)\frac{\vec{E}_\perp}{B}. \tag{1}$$

In the purely electrostatic case with a slowly time-varying electrostatic potential $\phi(x,y,t)$, assuming weakly inhomogeneous plasma density $n_0(x)$, quasineutrality $n_i = n_e$, and Boltzmann distributed electrons

$$n_e = n_0(x) \exp\left(\frac{e\phi}{T_e}\right) \tag{2}$$

the ion continuity readily gives

$$\left[\frac{\partial}{\partial t} + \frac{1}{B_0}(\vec{e}_z \times \nabla\phi) \cdot \nabla\right]\left[\left(1 - \rho_s^2 \nabla_\perp^2\right)\phi - B_0 v_{*e} x\right] = 0 \tag{3}$$

Here v_{*e} and ρ_s denote the electron diamagnetic drift velocity and the ion inertial length, respectively

$$v_{*e} = -\frac{T_e}{eB_0 n_0}\frac{dn_0}{dx} \quad , \quad \rho_s^2 = \frac{T_e}{m_i \Omega_e^2} \tag{4}$$

and B_0, T_e, $n_0(x)$ are the unperturbed magnetic field, electron temperature, and plasma density, respectively.

For a solution ϕ_S which is stationary in the reference frame moving with the velocity $\vec{e}_y u_y$ Eq. (3) is readily integrated one time to give

$$\phi_S - B_0 u_y x = G\left[\left(1 - \rho_s^2 \nabla_\perp^2\right)\phi_S - B_0 v_{*e} x\right] \tag{5}$$

where G is an arbitrary function. In the usual vortex scenario it is taken to be a linear function, $G(\xi) = \xi G$, having different slopes G^{in}, G^{out} inside and outside of the vortex core, which is adopted as a circle in the x, y plane with the radius equal to R. A finite solution for $x \to \infty$ is possible only if $G^{out} = u_y / v_{*e}$.

Matching both the potential ϕ_S and its Laplacian at the core edge $r = R$, we may write the vortex solution in the form

$$\phi_S = B_0 u_y R \cos\theta \begin{cases} \left(1 + \dfrac{\mu^2}{\kappa^2}\right)\dfrac{r}{R} - \dfrac{\mu^2}{\kappa^2}\dfrac{J_1(r\kappa)}{J_1(R\kappa)}, & r < R \\ \dfrac{K_1(r\mu)}{K_1(R\mu)}, & r > R \end{cases} \tag{6}$$

where $r = (x^2 + y^2)^{1/2}$, $\theta = \arctan y/x$, J_1, K_1 are the first order ordinary and modified Bessel functions, and the inside wavenumber κ (and also the slope G^{in}) is determined from the continuity of the gradient $\frac{\partial}{\partial R}\phi_S(R)$, yielding the following nonlinear dispersion relation

$$\kappa \frac{J_1(R\kappa)}{J_2(R\kappa)} = -\mu \frac{K_1(R\mu)}{K_2(R\mu)}, \tag{7}$$

where the outside and inside wavenumbers μ, κ are defined by

$$\mu^2 = \frac{1}{\rho_s^2}\left(1 - \frac{1}{G^{out}}\right) \quad , \quad \kappa^2 = \frac{1}{\rho_s^2}\left(\frac{1}{G^{in}} - 1\right) \tag{8}$$

Well localized vortex for $r > R$ is, obviously, obtained only if $\mu^2 > 0$, which indicates the existance of two vortex modes, propagating in the direction of the electron and the ion diamagnetic drift, respectively

$$G^{out} = \frac{u_y}{v_{*e}} > 1 \qquad \text{and} \qquad G^{out} < 0 \tag{9}$$

An approximative solution of Eq. (7) is $\kappa = j_{1,m}/R$, where $j_{n,m}$ denotes the m-th zero of the Bessel function J_n. In the following we will study only the ground state modon, which corresponds to the smallest solution for κ obtained from Eq. (7).

A stationary solution ϕ_S is considered to be linearly stable if any initially infinitezimally small perturbation remains finite for all times. In a broader sence, stability may be defined allowing even an infinite growth of perturbations in a finite number of isolated singular points, but requiring that the perturbations are square integrable across these singularities.

Stability of the vortex ϕ_S, Eq. (6) in the presence of small perturbations $\delta\phi \ll \phi_S$ is studied in detail in [17]. The vortex evolution equation is obtained after linearization of the Hasegawa-Mima equation (3) around the stationary solution ϕ_S, yielding

$$\frac{\partial}{\partial t}\hat{p}\delta\phi = \hat{A}\hat{H}\hat{p}\delta\phi \tag{10}$$

where the following notation is used

$$\hat{A} = -\frac{1}{B_0}\left[\vec{e}_z \times \nabla\left(\hat{p}\delta\phi - B_0 v_{*e} x\right)\right]\cdot\nabla \quad , \quad \hat{p} = 1 - \rho_s^2\nabla_\perp^2 \quad , \quad \hat{H} = G - \hat{p}^{-1} \tag{11}$$

Eq. (10) is linear, but it is practically impossible to find analytically its general solution. Instead, we will investigate its stability applying the Lyapunov analysis, making use of an obvious quadratic conserved quantity of Eq. (10)

$$L = \int_{-\infty}^{\infty} dx dy \hat{p}\delta\phi \hat{H} \hat{p}\delta\phi \equiv \langle \hat{p}\delta\phi | \hat{H} | \hat{p}\delta\phi \rangle = const. \tag{12}$$

Perturbation $\hat{p}\delta\phi$ will be expanded as

$$\hat{p}\delta\phi\left(x,y,t\right) = \sum_n e^{in\theta}\left[\sum_k \mathcal{P}_{n,k}\left(t\right) g_{n,k}\left(r\right) + \int dk \mathcal{P}_n\left(k,t\right) g_n\left(k,r\right)\right] \tag{13}$$

where the complete basis of ortogonal and normalized functions g_n is determined from the following eigenvalue problem [18]

$$\hat{H} e^{in\theta} g_n = k e^{in\theta} g_n \qquad \Leftrightarrow \qquad \left[\nabla_\perp^2 + \frac{1}{\rho_s^2}\left(\frac{1}{G-k} - 1\right)\right] e^{in\theta} g_n = 0 \tag{14}$$

The continuous spectrum $g_n\left(k,r\right)$ is obtained from the above equation if

$$\xi_n^2\left(k\right) \equiv \frac{1}{\rho_s^2}\left(\frac{1}{G^{in} - k} - 1\right) > 0 \quad , \quad \eta_n^2\left(k\right) \equiv \frac{1}{\rho_s^2}\left(\frac{1}{G^{out} - k} - 1\right) > 0 \tag{15}$$

yielding

$$g_n\left(k,r\right) = a_n\left(k\right)\begin{cases} \dfrac{J_n\left(r\xi_n\left(k\right)\right)}{J_n\left(R\xi_n\left(k\right)\right)} \ , & r < R \\[2ex] \dfrac{J_n\left(r\eta_n\left(k\right)\right) + d_n\left(k\right) Y_n\left(r\eta_n\left(k\right)\right)}{J_n\left(R\eta_n\left(k\right)\right) + d_n\left(k\right) Y_n\left(R\eta_n\left(k\right)\right)} \ , & r > R \end{cases} \tag{16}$$

and the coefficient $d_n\left(k\right)$ is determined from the smoothness at $r = R$

$$d_n\left(k\right) = -\frac{\xi_n\left(k\right) J_{n-1}\left(R\xi_n\left(k\right)\right) J_n\left(R\xi_n\left(k\right)\right) - \eta_n\left(k\right) J_{n-1}\left(R\eta_n\left(k\right)\right) J_n\left(R\eta_n\left(k\right)\right)}{\xi_n\left(k\right) J_{n-1}\left(R\xi_n\left(k\right)\right) J_n\left(R\xi_n\left(k\right)\right) - \eta_n\left(k\right) Y_{n-1}\left(R\eta_n\left(k\right)\right) Y_n\left(R\eta_n\left(k\right)\right)}. \tag{17}$$

The continuous spectrum, according to Eq. (15) exists only for the eigenvalues k satisfying

$$\min\left(G^{in}, G^{out}\right) \geq k \geq \max\left(G^{in} - 1, G^{out} - 1\right) \tag{18}$$

For the two vortex modes defined in Eq. (9), continuous spectra exist if

$$G^{in} \geq k \geq G^{out} - 1 \geq 0 \qquad \text{and} \qquad G^{in} - 1 \leq k \leq G^{out} \leq 0 \tag{19}$$

respectively. Similarly, the discrete spectrum is obtained from Eq. (14) for

$$\xi_{n,k}^2 \equiv \frac{1}{\rho_s^2}\left(\frac{1}{G^{in} - k} - 1\right) > 0 \quad , \quad q_{n,k}^2 \equiv \frac{1}{\rho_s^2}\left(1 - \frac{1}{G^{out} - k}\right) > 0 \tag{20}$$

which gives

$$g_{n,k}\left(r\right) = a_{n,k}\begin{cases} \dfrac{J_n\left(r\xi_{n,k}\right)}{J_n\left(R\xi_{n,k}\right)} \ , & r < R \\[2ex] \dfrac{K_n\left(rq_{n,k}\right)}{K_n\left(Rq_{n,k}\right)} \ , & r > R \end{cases} \tag{21}$$

Continuity of $\frac{\partial}{\partial R} g_{n,k}(R)$ yields the following dispersion relation

$$\xi_{n,k}\frac{J_{n-1}(R\xi_{n,k})}{J_n(R\xi_{n,k})} = -q_{n,k}\frac{K_{n-1}(Rq_{n,k})}{K_n(Rq_{n,k})}. \tag{22}$$

which has an infinite number of solutions. The eigenvalue k corresponds to the k-th solution of Eq. (22).

We will adopt the parameters $a_{n,k}$, $a_n(k)$ from the normalization

$$\int_0^\infty drrg_{n,k}(r)\,g_{n,k'}(r) = \frac{\delta_{k,k'}}{2\pi} \quad , \quad \int_0^\infty drrg_n(k,r)\,g_n(k',r) = \frac{\delta(k-k')}{2\pi} \tag{23}$$

which permits us to rewrite the conserved quantity, Eq. (12) as

$$L = L^{(+)}(t) - L^{(-)}(t) = const. \tag{24}$$

where

$$L^{(+)}(t) = \sum_n \left[\int_{k>0} dkk\mathcal{P}_n^2(k,t) + \sum_{k>0} k\mathcal{P}_{n,k}^2(t)\right]$$

$$L^{(-)}(t) = -\sum_n \left[\int_{k<0} dkk\mathcal{P}_n^2(k,t) + \sum_{k<0} k\mathcal{P}_{n,k}^2(t)\right] \tag{25}$$

Obviously, both functionals $L^{(+)}$ and $L^{(-)}$ are positive definite. Noting that they are defined as integrals over the whole x, y plane of some positive definite functions, their boundedness implies that the potential $\delta\phi$ remains finite for all times and on the whole x, y plane, except possibly some finite size regions with finite perturbations, and of well behaved (i. e. square integrable) singularities.

A careful inspection of the dispersion relation Eq. (22) reveals that there are only two discrete modes, $g_{0,1}$ and $g_{1,1}$, whose eigenvalues are negative. These modes correspond to the lowest order monopole and dipole perturbations. Likewise, the continuous spectrum, according to Eq. (19), corresponds to positive eigenmodes k in the case of the zero-order vortex propagating in the electron diamagnetic drift direction, $u_y/v_{*e} > 1$, while for the vortices propagating fast enough in the ion diamagnetic drift direction

$$\frac{u_y}{v_{*e}} \leq G^{in} - 1 < 0 \tag{26}$$

the continuous spectrum is absent. Hence, the linear instability of a vortex is possible only if the modes $g_{0,1}$, and $g_{1,1}$, corresponding to an asymmetry and tilt of the vortex, respectively, are unstable.

Vortex behavior under the above described critical perturbations can easily be tested in a numerical experiment. We demonstrate that vortices propagating in the electron diamagnetic drift direction, faster than linear drift waves $u_y > v_{*e}$ are unstable. Applying a dipole-type perturbation, we find that a small initial tilt is growing exponentially, leading to the splitting of the cyclone and anticyclone parts, which propagate independently and eventually disperse. Conversely, vortices propagating in the ion diamagnetic drift direction, with $u_y < 0$, are stable for tilt perturbations. They oscillate around the equilibrium position and preserve their shape for a long time (we were able to follow two complete cycles around the computational box). Similar

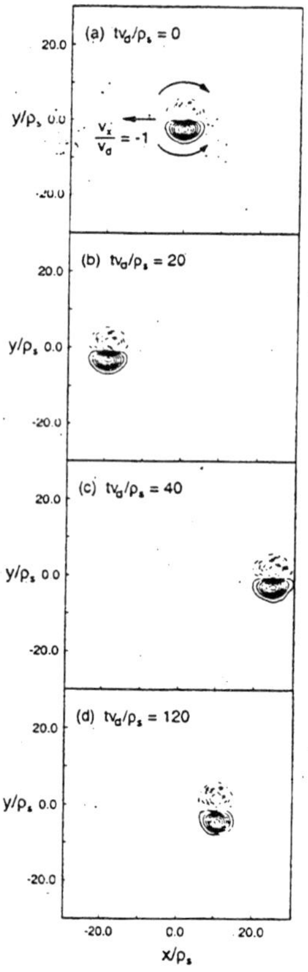

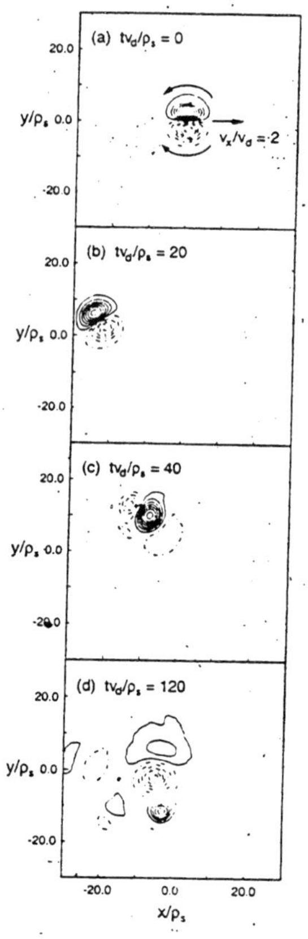

Figure 1: Unstable motion of vortex propagating in the electron diamagnetic drift direction, due to a 2% monopolar perturbation. Angular rotation frequency is $\Omega_E \approx 3$ at frame (a). In frame (b), at $tv_{*e}/\rho_s = 20$, the position of the tilt is at 45% and the center is at $\Delta x \approx 40$ and $\Delta y \approx R \sim 6$. In frame (c), at $tv_{*e}/\rho_s = 40$, anticyclone $(\phi > 0)$ dominates and $\Delta y \approx 9\rho_s$. In frame (d), $tv_{*e}/\rho_s = 120$, about 50 rotations of the core, dispersion is setting in.

Figure 2: Stable motion of the vortex propagating in the ion diamagnetic drift direction. The same 2% monopolar perturbation is applied. In frame (a) the rotation rate is $\Omega_E \approx 3$ and the initial speed is $u_x/v_{*e} = -1$. In frame (b), $tv_{*e}/\rho_s = 20$ and $\Delta x \approx -20$, $\Delta y = 0$ after ten rotations. Frames (c) and (d) show the stable structure at times $tv_{*e}/\rho_s = 40$ and 120, respectively. The vortex is slowly losing speed probably because of numerical effects.

behavior was observed in the case of perturbation by a small monopole. Trajectory of unstable vortices, with $u_y > v_{*e}$ was curving away from the straight line, while for the stable ones with $u_y < 0$ the trajectory was always curving back toward the straight one and exibited only small oscillations. Thus, from their resilience in the presence of the only two critical perturbations, we conclude that the vortices propagating in the ion diamagnetic drift direction are linearly stable.

3 Shear-Alfvén vortices and their stability

In this section we study electromagnetic perturbations in a homogeneous, very low-β plasma

$$\beta = \frac{2n_o T_e}{c^2 \epsilon_0 B_0^2} \ll \frac{m_e}{m_i}. \tag{27}$$

which are slowly varying in time, compared to the ion gyrofrequency Ω_i, and weakly z-dependent

$$\frac{\partial}{\partial t} \ll \Omega_i \quad , \quad \frac{\partial}{\partial z} \ll \nabla_\perp, \tag{28}$$

In that case, the ion and electron hydrodynamic velocities are given by

$$\vec{v}_i = \vec{v}_E + \vec{v}_P \quad , \quad \vec{v}_e = \vec{v}_E + \frac{\vec{B}}{B} v_{\parallel e}. \tag{29}$$

Here subscripts $\parallel$, $\perp$ denote vector components parallel, resp. perpendicular to the zero-order magnetic field, and $\vec{v}_E$, $\vec{v}_P$ are the $\vec{E}\times\vec{B}$ and ion polarization drift velocities, respectively, defined in Eq. (1). We assume small perturbations of the electron density and of the magnetic field, and neglect the compressional component of the magnetic field perturbation, $\delta B_z = 0$, which is justified in the low β regime. Then, subtracting the electron and ion continuity equations, and making use of the quasineutrality, $n_e \approx n_i$, we readily obtain

$$\nabla_\perp \vec{v}_P - \frac{\vec{B}}{B} \cdot \nabla v_{\parallel e} = 0, \tag{30}$$

while the parallel electron momentum equation takes the form

$$\left(\frac{\partial}{\partial t} + \vec{v}_E \nabla\right) v_{\parallel e} = -\frac{e}{m_e} \frac{\vec{B} \cdot \vec{E}}{B}. \tag{31}$$

Here e and m_e are are the electron charge and mass, respectively.

Finally, expressing the electric and magnetic fields in terms of the electrostatic potential ϕ and the z-component of the vector potential $\vec{e}_z A_z$ and making use of the parallel component of the Ampere's law $j_\parallel = -en_0 v_{\parallel e} = -c^2 \epsilon_0 \nabla_\perp^2 A_z$ we may rewrite Eqs. (30), (31) in the following convinient form

$$\left[\frac{\partial}{\partial t} + \frac{1}{B_0}(\vec{e}_z \times \nabla\phi) \cdot \nabla\right] \nabla_\perp^2 \phi = -c_A^2 \left[\frac{\partial}{\partial z} - \frac{1}{B_0}(\vec{e}_z \times \nabla A_z) \cdot \nabla\right] \nabla_\perp^2 A_z \tag{32}$$

$$\left[\frac{\partial}{\partial t} + \frac{1}{B_0}(\vec{e}_z \times \nabla\phi) \cdot \nabla\right] \left(1 - \lambda_s^2 \nabla_\perp^2\right) A_z = -\frac{\partial \phi}{\partial z}. \tag{33}$$

Here $c_A = c\Omega_i/\omega_{pi}$ is the Alfvén speed, $\lambda_s = c/\omega_{pe}$ is the collisionless skin depth, and ω_{pi}, ω_{pe} are the ion and electron plasma frequencies, respectively. Equations (32), (33) constitute our basic system describing nonlinear shear-Alfvén perturbations. The dipole vortex, [20], [21] is found as a time stationary, z-independent solution of Eqs. (32), (33), in the reference frame travelling with the velocity $\vec{u} = u_y\vec{e}_y + u_z\vec{e}_z$, $\mid u_z \mid \gg \mid u_y \mid$, i.e. we assume that we have a two-dimensional elongated structure moving with the velocity u_y along the y axes, and making a pitch angle u_y/u_z to the z-axis. Using the properties of the Poisson brackets we can integrate Eqs. (32), (33) one time, yielding

$$\left(1 - \lambda_s^2\nabla_\perp^2\right)(u_zA_z - B_0u_yx) = H(\phi - B_0u_yx) \tag{34}$$

$$\frac{c_A^2}{u_z^2}H'(\phi - B_0u_yx)\cdot\nabla_\perp^2u_zA_z - \nabla_\perp^2\phi = G(\phi - B_0u_yx), \tag{35}$$

where $H(\zeta)$, $G(\zeta)$ are arbitrary functions of their arguments, and primes denote derivatives of these functions. Same as before, in the vortex scenario, $H(\zeta)$, and $G(\zeta)$ are taken to be linear functions, $H(\zeta) = \zeta H$, $G(\zeta) = \zeta G$, allowing for different slopes G^{out}, H^{out} and G^{in}, H^{in} outside and inside the vortex core, respectively. The core is taken to be a circle in the x, y plane, with the radius R. Obviously, a localized solution is possible only if $H^{out} = 1$, $G^{out} = 0$. Equations (34), (35) may be decoupled to give

$$\left(\nabla_\perp^2 + \kappa_1^2\right)\left(\nabla_\perp^2 + \kappa_2^2\right)(\phi - B_0u_yx) = 0, \tag{36}$$

where the wavenumbers κ_1, κ_2 are related with the slopes H, G through

$$\kappa_1^2\kappa_2^2 = -\frac{G}{\lambda_s^2} \quad , \quad \kappa_1^2 + \kappa_2^2 = \frac{1}{\lambda_s^2}\left(\frac{c_A^2}{u_z^2}H^2 - 1\right) + G. \tag{37}$$

In the outer region $r > R$ we have

$$\kappa_1^{out} = 0 \quad , \quad \kappa_2^{out\,2} = -\frac{1}{\lambda_s^2}\left(1 - \frac{c_A{}^2}{u_z^2}\right).$$

The fourth order wave equation (36) separates variables in the cylindrical frame. Every localized solution must contain the first cylindrical harmonic due to the presence of the term B_0u_yx. Thus, we can readily write the simplest localized solution in the standard form of a double vortex

$$\phi = B_0u_yR\cos\theta\begin{cases}\dfrac{r}{R} + \alpha_1J_1\left(r\kappa_1^{in}\right) + \alpha_2J_1\left(r\kappa_2^{in}\right), & r < R\\[2ex] \beta_1\dfrac{R}{r} + \beta_2K_1(r\mu), & r > R\end{cases} \tag{38}$$

where $r = \left(x^2 = y^2\right)^{1/2}$, $\theta = \arctan y/x$, J_1, K_1 are the ordinary and modified Bessel functions of the first order. For simplicity, we introduced

$$\mu^2 \equiv -\kappa_2^{out\,2} = \frac{1}{\lambda_s^2}\left(1 - \frac{c_A^2}{u_z^2}\right).$$

Obviously, the continuity of the function $G(\phi - B_0u_yx)$ requires that at the core edge we have $\phi - B_0u_yx\mid_{r=R} = 0$. Coefficients α_1, α_2, β_1, β_2 are determined from the continuity of the potentials ϕ, A_z and their radial derivatives at the core edge, which after a somewhat lengthy, but straightforward algebra (for details see Ref. [21], reveals the existence of two distinct vortex modes:

1. Drift-vortex mode

This mode (called also the Larichev-Reznik mode [21]) can be identified as the generalization of the vortex associated with nonlinear drift-wave perturbations to the case of large parallel phase velocities u_z, i. e. to the regime when the electrons are adiabatic, rather than Boltzmann distributed as studied in the preceeding section. For this mode, parameter $H = 1$ is constant on the whole x, y plane, while the constant of integration G^{in} is determined from the following dispersion relation

$$\frac{\lambda_s^2 \kappa_1^{in\,2}}{1+\lambda_s^2 \kappa_1^{in\,2}} \frac{\frac{\partial}{\partial R} J_1\left(R\kappa_1^{in}\right)}{J_1\left(R\kappa_1^{in}\right)} = \frac{\lambda_s^2 \kappa_2^{in\,2}}{1+\lambda_s^2 \kappa_2^{in\,2}} \frac{\frac{\partial}{\partial R} J_1\left(R\kappa_2^{in}\right)}{J_1\left(R\kappa_2^{in}\right)} \tag{39}$$

and the coefficients α_1, α_2, β_1, β_2 are given by

$$\alpha_m = \frac{2}{R\frac{\partial}{\partial R} J_1\left(R\kappa_m^{in}\right)} \frac{c_A^2}{u_z^2} \frac{\kappa_n^{in\,2}}{\kappa_m^{in\,2} - \kappa_n^{in\,2}} \frac{1}{1+\lambda_s^2 \kappa_n^{in\,2}}, \qquad m = 1,2$$

$$\beta_1 = 1 - \frac{2J_1\left(R\kappa_1^{in}\right)}{R\frac{\partial}{\partial R} J_1\left(R\kappa_1^{in}\right)} \frac{c_A^2}{c_A^2+u_z^2} \frac{\lambda_s^2 \kappa_2^{in\,2}}{1+\lambda_s^2 \kappa_2^{in\,2}}$$

$$\beta_2 = \frac{2J_1\left(R\kappa_2^{in}\right)}{R\frac{\partial}{\partial R} J_1\left(R\kappa_2^{in}\right)} \frac{c_A^2}{c_A^2+u_z^2} \frac{\lambda_s^2 \kappa_1^{in\,2}}{1+\lambda_s^2 \kappa_1^{in\,2}} \frac{1}{K_1(R\mu)} \tag{40}$$

Note that vortices of this type are well localized ($\mu^2 > 0$) only if they propagate faster than the Alfvén speed, $u_z^2 > c_A{}^2$. They are produced by nonlinearities acting within the vortex core, while the outer solution is an evanescent linear Alfvén wave, whose dispersion relation is given by

$$k_\perp^2 \left(\lambda_s^2 k_\perp^2 + 1 - \frac{c_A^2 k_z^2}{\omega^2}\right) = 0 \tag{41}$$

In the MHD limit $\lambda_s \to 0$, the drift-vortex becomes infinitezimally thin, $1/\kappa_1^{in} \sim 1/\kappa_2^{in} \sim R \sim \lambda_s$, i.e. it reduces to a line (or filament). One can easily verify that finite size MHD drift vortices are forbidden. Namely, Eqs. (37) with $\lambda_s^2 \kappa_1^{in\,2} \ll 1$ yield $\lambda_s^2 \kappa_2^{in\,2} = c_A^2/u_z^2 - 1 < 0$, and consequently it is impossible to have a finite solution at $r = 0$.

2. Convective-cell mode

This vortex type arises from the coupling between electrostatic and magnetostatic convective-cell modes. It is characterized by the discontinuity at $r = R$ of both constants of integration G and H. In the outer region it decays as $\cos\theta/r$, and physically it is just a $k_\perp = 0$ linear quasimode. The "inside" wavenumbers are related through the following dispersion relation

$$\alpha_1 J_1\left(R\kappa_1^{in}\right) = \alpha_2 J_1\left(R\kappa_2^{in}\right) = 0. \tag{42}$$

Coefficients α_1, α_2 in Eq. (38), when different than zero, are given by

$$\alpha_m = -\frac{2\left(1+\lambda_s^2\kappa_m^{in\,2}\right)^{\frac{1}{2}}}{R\frac{\partial}{\partial R} J_1\left(R\kappa_m^{in}\right)} \times$$
$$\left\{\left[\left(1+\lambda_s^2\kappa_1^{in\,2}\right)^{\frac{1}{2}} + \left(1+\lambda_s^2\kappa_2^{in\,2}\right)^{\frac{1}{2}}\right]^{-1} + \frac{c_A - u_z}{u_z} \frac{\left(1+\lambda_s^2\kappa_n^{in\,2}\right)^{\frac{1}{2}}}{\lambda_s^2\left(\kappa_n^{in\,2} - \kappa_m^{in\,2}\right)}\right\},$$
$$m = 1,2 \tag{43}$$

while the "outside" coefficients β_1, β_2 are given by

$$\beta_1 = 1 \quad , \qquad \beta_2 = 0. \tag{44}$$

Particularly interesting is the convective-cell vortex given by only one Bessel function inside

$$\alpha_1 = \frac{-2}{R\frac{\partial}{\partial R}J_1\left(R\kappa_1^{in}\right)} \quad , \qquad \alpha_2 = 0 \tag{45}$$

which corresponds to

$$\kappa_1^{in} = \frac{j_{1,m}}{R} \quad , \qquad \kappa_2^{in} = \frac{1}{\lambda_s}\left[\frac{c_A^2}{u_z^2}\left(1+\lambda_s^2\kappa_1^{in\,2}\right)-1\right]^{\frac{1}{2}} \tag{46}$$

where $j_{1,m}$ is the m-th zero of the Bessel function J_1. In this special case we have $\phi = u_z A_z$, its z-component of the electric field is equal to zero, and consequently it can not be Landau damped by electrons propagating along the magnetic field lines.

In the MHD limit $\lambda_s \to 0$, the convective-cell type vortex becomes singular, i.e. reduces to a line (or filament), similarly to the drift-vortex. It is obvious from Eq. (43) that a vortex with $u_z \neq c_A$ has an infinite amplitude if $\lambda_s \to 0$. In the special case $u_z = c_A$, solution of Eqs. (34), (35) with a finite perpendicular scalelength $\lambda_s^2\nabla_\perp^2 \sim \lambda_s^2/R^2 \ll 1$ is possible only if $G^{out} = G^{in} = 0$, $H^{out} = H^{in} = 1$. However, in that case we have essentially linear Alfvén modes, and localized vortices are not possible.

We study the stability of the vortex solution, Eq. (38) in the presence of small perturbations of the potentials ϕ, A_z. Linearizing our starting system of equations (32), (33) around the stationary solution $\phi^{(0)}$, $A_z^{(0)}$, given by Eq. (38), and using Eqs. (34), (35) we obtain the following set of equations for the perturbations

$$u_z\frac{\partial}{\partial z}\delta\phi + \frac{\partial}{\partial t}\left(1-\lambda_s^2\nabla_\perp^2\right)\delta\psi = \vec{V}\nabla\left[H\delta\phi - \left(1-\lambda_s^2\nabla_\perp^2\right)\delta\psi\right] \tag{47}$$

$$\frac{\partial}{\partial t}\nabla_\perp^2\delta\phi + \frac{c_A^2}{u_z^2}\frac{\partial}{\partial z}\nabla_\perp^2\delta\psi = -\vec{V}\nabla\left[\left(\nabla_\perp^2+G\right)\delta\phi - H\frac{c_A^2}{u_z^2}\nabla_\perp^2\delta\psi\right] - \vec{W}\nabla\left[H\delta\phi - \left(1-\lambda_s^2\nabla_\perp^2\right)\delta\psi\right]. \tag{48}$$

Here we used the notations $\delta\psi = u_z\delta A_z$ and

$$\vec{V} = \frac{1}{B_0}\vec{e}_z \times \nabla\left(\phi^{(0)} - B_0 u_y x\right) = \vec{v}_E^{(0)} - \vec{e}_y u_y$$

$$\vec{W} = \frac{1}{B_0}\vec{e}_z \times \nabla\frac{u_z^2}{c_A^2}\frac{\lambda_s^2}{H}\left(\nabla_\perp^2 + G\right)\left(\phi^{(0)} - B_0 u_y x\right) = -u_z\lambda_s^2\nabla_\perp^2\delta\vec{B}^{(0)} \tag{49}$$

where $\vec{v}_E^{(0)}$ and $\delta\vec{B}^{(0)}$ are the leading order fluid velocity and the magnetic field perturbation, associated with the zero-order vortex.

Equations (47), (48) for the perturbations separate Fourier components only in the ω and k_z domains, vhile the $\vec{k}_\perp$ components are coupled via the inhomogeneous flows $\vec{V}$ and $\vec{W}$. Thus, any initial perturbation with a localized $\vec{k}_\perp$ spectrum will eventually, given enough time, spread over the entire range of perpendicular wavenumbers k_x, k_y.

One can easily verify that the drift- (or Larichev-Reznik-) type vortex, Eqs. (38) - (40) is unstable. For perturbations whose characteristic perpendicular wavelength is much shorter than the vortex radius $R^2\nabla_\perp^2 \gg 1$, in the vicinity of an arbitrary point in the x, y plane we may solve Eqs. (47), (48) in the local approximation, using $\vec{V}$, $\vec{W} = const$. Thus, in the outer region $x^2 + y^2 > R^2$, using $H^{out} = 1$, $G^{out} = 0$, we obtain from Eqs. (47), (48) the following linear dispersion relation

$$\frac{(1+\lambda_s^2 k_\perp^2)\left(\omega - \vec{k}_\perp \vec{V}\right)}{\lambda_s^2 k_\perp^2 \left(k_z u_z - \vec{k}_\perp \vec{V}\right) + (1+\lambda_s^2 k_\perp^2)\vec{k}_\perp \vec{W}} = \frac{k_z u_z - \vec{k}_\perp \vec{V}}{\lambda_s^2 k_\perp^2 \left(\omega - \vec{k}_\perp \vec{V}\right) + \frac{c_A^2}{u_z^2}\vec{k}_\perp \vec{W}}, \tag{50}$$

Particularly simple is the case of short wavelength perturbations $\lambda_s^2 k_\perp^2 \gg 1$ (which is in a good agreement with the assumption $R^2\nabla_\perp^2 \gg 1$ used in the derivation of the above dispersion relation) when from Eq. (50) we obtain the following condition for instability

$$\frac{\vec{k}_\perp \vec{W}}{\vec{k}_\perp \vec{V} - k_z u_z} > 1 \qquad \Leftrightarrow \qquad -u_z \frac{\vec{k}_\perp \cdot \lambda_s^2 \nabla_\perp^2 \delta \vec{B}^{(0)}}{\vec{k} \cdot \left(\vec{v}_E^{(0)} - \vec{u}\right)} > 1. \tag{51}$$

Obviously, perturbations with an arbitrary wavevector $\vec{k}$ are unstable in the spatial region where the fluid velocity associated with the vortex, calculated in the reference frame moving with the vortex, is perpendicular to the wave vector

$$\vec{k} \cdot (\vec{v}_E^{(0)} - \vec{u}) \approx 0.$$

For a given k_z this condition may not be fulfilled for all values of the perpendicular wavevector $\vec{k}_\perp$. However, as we have discussed above, the perpendicular wavenumber spectrum of perturbations is broad, $(k_x, k_y) \in (-\infty, \infty)$ due to the presence of inhomogeneous flows $\vec{V}$, and $\vec{W}$. As a consequence, for an arbitrary value of the parralel wavenumber k_z, the spectrum will always contain some unstable modes, whose wavevector satisfies Eq. (51). Thus, with an arbitrary initial perturbation, the maximum characteristic time of the vortex destruction is determined by the rate of spectrum cascade towards these unstable modes.

For convective-cell vortices, Eqs. (38), (43), (44), we have $\vec{W}^{out} = 0$, and the dispersion relation, Eq. (50) does not indicate any short-wavelength instability in the exterior region. Thus, their stability analysis requires a more detailed inspection of the interior ($r < R$) solution, and of the perturbation with longer scale lengths, $k_\perp R \leq 1$. We will restrict ouselves to the "one Bessel function" convective cell vortex, Eqs. (38), (45), (46), in order to exclude the Landau damping and the consequent vortex dispersion. Then, our evolution equations (47), (48) for small perturbations $\delta\phi$, $\delta\psi$ simplify to

$$u_z \frac{\partial}{\partial z}\delta\phi + \frac{\partial}{\partial t}\left(1 - \lambda_s^2 \nabla_\perp^2\right)\delta\psi = \vec{V}\nabla\left[\left(1 + \lambda_s^2 \kappa_1^2\right)\delta\phi - \left(1 - \lambda_s^2 \nabla_\perp^2\right)\delta\psi\right] \tag{52}$$

$$\frac{\partial}{\partial t}\nabla_\perp^2 \delta\phi + \frac{c_A^2}{u_z^2}\frac{\partial}{\partial z}\nabla_\perp^2 \delta\psi = -\vec{V}\nabla\left(\nabla_\perp^2 + \kappa_1^2\right)\left(\delta\phi - \frac{c_A^2}{u_z^2}\delta\psi\right). \tag{53}$$

Here κ_1 is the wavenumber of the zero-order vortex, Eq. (45), whose values are $\kappa_1^{out} = 0$, $\kappa_1^{in} = j_{1,m}/R$.

We proceed applying the Lyapunov stability analysis to the above system. In the construction of the Lyapunov functional we need to have at least one quadratic quantity conserved by Eqs. (52), (53). For that purpose we use the following integrals.

First, we multiply Eq. (52) by $\left[\left(1+\lambda_s^2\kappa_1^2\right)\delta\phi-\left(1-\lambda_s^2\nabla_\perp^2\right)\delta\psi\right]$, and integrate within the vortex core $r=\left(x^2+y^2\right)^{1/2}<R$.

Next, we multiply Eqs. (52), (53) respectively by $\left(\nabla_\perp^2+\kappa_1^2\right)\left(\delta\phi-\frac{c_A^2}{u_z^2}\delta\psi\right)$, $\left[\left(1+\lambda_s^2\kappa_1^2\right)\delta\phi-\left(1-\lambda_s^2\nabla_\perp^2\right)\delta\psi\right]$, add, and integrate for the whole x,y plane.

Finally, for the special case of zero-order vortices which satisfy

$$\lambda_s^2\kappa_1^{in\,2}\equiv\frac{\lambda_s^2}{R^2}j_{1,m}^2=\frac{u_z^2}{c_A^2}-1 \tag{54}$$

we may write down the sought for quadratic conserved quantity, performing a proper linear combination of the above defined integrals, and the integration in z,t. Thus, after a somewhat lengthy but straightforward algebra we obtain the following conserved quantity L, which is a quadratic function of the perturbations $\delta\phi,\delta\psi$

$$\begin{aligned}L&=const.\\&=\int_{-\infty}^{\infty}dxdydz\left\{\left[\nabla_\perp\left(\delta\phi-\frac{c_A}{u_z}\delta\psi\right)\right]^2+\lambda_s^2\left[\nabla_\perp^2\left(\delta\phi-\frac{c_A}{u_z}\delta\psi\right)\right]^2+\right.\\&\left.\frac{c_A^2}{u_z^2}\left(1+\lambda_s^2\kappa_1^2\right)\lambda_s^2\left(\nabla_\perp^2\delta\psi\right)^2-\delta\phi\lambda_s^2\nabla_\perp^2\left(\nabla_\perp^2+K\right)\delta\phi-\frac{c_A^2}{u_z^2}\delta\psi\lambda_s^2\nabla_\perp^2\left(\nabla_\perp^2+K\right)\delta\psi\right\}\end{aligned} \tag{55}$$

where the parameter K is defined by

$$\lambda_s^2K=\frac{c_A}{u_z}\left(1+\lambda_s^2\kappa_1^2\right)-1 \tag{56}$$

or, with the use of Eq. (54)

$$\lambda_s^2K^{in}=\frac{u_z}{c_A}-1>0\quad,\quad\lambda_s^2K^{out}=\frac{c_A}{u_z}-1<0. \tag{57}$$

Obviously, the first three terms in the integrand of Eq. (55) are positive definite, and in the case of an instability they can grow only at the expence of the last two. Thus, the sufficient and necessary stability condition is that the (possibly negative) sum of the last two terms remains bounded from below for all times, and for all possible realizations of $\delta\phi,\delta\psi$, which satisfy the evolution equations (52), (53). If this was fulfilled, the above defined integral of motion L would be the Lyapunov functional for our system.

For a detailed analysis of the functional L, Eq. (55), we will expand the potentials $\delta\phi,\delta\psi$ on the complete basis of orthogonal and normalized functions g, similarly to Eq. (13)

$$\delta\phi\left(t,z,\vec{r}_\perp\right)=\sum_n e^{in\theta}\left[\sum_{k>0}\widehat{\delta\phi}_{n,k}\left(t,z\right)g_{n,k}\left(r\right)+\int_{-\infty}^{0}dk\widehat{\delta\phi}_n\left(k,t,z\right)g_n\left(k,r\right)\right] \tag{58}$$

with an analogous expression for $\delta\psi$.

Eigenfunctions $g_{n,k}\left(r\right)$, $g_n\left(k,r\right)$ are obtained from the Schrödinger equation describing a particle in a two-dimensional cylindrical potential well, whose radius is equal to the vortex core radius R, and whose depth is s^2.

$$\left[\nabla_\perp^2+S^2\left(r\right)\right]e^{in\theta}g_n\left(r\right)=ke^{in\theta}g_n\left(r\right)\quad,\quad S^2\left(r\right)=\begin{cases}s^2\ , & r<R\\0\ , & r>R\end{cases} \tag{59}$$

Discrete and continuous eigenfunctions $g_{n,k}(r)$, $g_n(k,r)$ and their spectra are given by Eqs. (21), (22) and (16), (17), respectively, with the notation $q_{n,k}^2 = k > 0$, $\xi_{n,k}^2 = s^2 - k$ and $\eta_n(k)^2 = -k > 0$, $\xi_n(k)^2 = s^2 - k$.

The depth of the potential well s^2 will be conviniently adopted as

$$s^2 = \frac{1}{2}\left(K^{in} - K^{out}\right) = \frac{1}{2\lambda_s^2}\frac{c_A}{u_z}\left(\frac{u_z^2}{c_A^2} - 1\right), \tag{60}$$

which permits us to write down the following identity

$$\nabla_\perp^2\left(\nabla_\perp^2 + K\right)e^{in\theta}g_n(r) = \left[k\left(k + K^{out}\right) - \frac{1}{4}\left(K^{in2} - K^{out2}\right)h(R-r)\right]e^{in\theta}g_n(r) \tag{61}$$

where h denotes the Heaviside unit step function.

Finally, using the expansions, Eq. (58), orthogonality, Eq. (23), and the above identity, we may rewrite the functional L, Eq. (55) in the following form

$$L = L^{(+)}(t) - L^{(-)}(t) = const., \tag{62}$$

where

$$L^{(+)}(t) = \int_{-\infty}^{\infty} dxdydz\left\{\left[\nabla_\perp\left(\delta\phi - \frac{c_A}{u_z}\delta\psi\right)\right]^2 + \lambda_s^2\left[\nabla_\perp^2\left(\delta\phi - \frac{c_A}{u_z}\delta\psi\right)\right]^2 + \right.$$
$$\left.\frac{c_A^2}{u_z^2}\left(1 + \lambda_s^2\kappa_1^2\right)\lambda_s^2\left(\nabla_\perp^2\delta\psi\right)^2 + \frac{\lambda_s^2}{4}\left(K^{in2} - K^{out2}\right)\left(\delta\phi^2 + \frac{c_A^2}{u_z^2}\delta\psi^2\right)h(R-r)\right\} -$$
$$\int_{-\infty}^{\infty} dz\sum_{k>0,n}\left(|\,\widehat{\delta\phi}_{n,k}\,|^2 + \frac{c_A^2}{u_z^2}\,|\,\widehat{\delta\psi}_{n,k}\,|^2\right)k\left(k - K^{out}\right), \tag{63}$$

and

$$L^{(-)}(t) = \int_{-\infty}^{\infty} dz\sum_n\int_{-\infty}^{0} dk\left(|\,\widehat{\delta\phi}_n(k)\,|^2 + \frac{c_A^2}{u_z^2}\,|\,\widehat{\delta\psi}_n(k)\,|^2\right)k\left(k - K^{out}\right). \tag{64}$$

Noting that, according to Eq. (54) our zero-order vortex is propagating faster than the Alfvén speed, $u_z^2 > c_A^2$, we have $K^{out} < 0$, and consequently the functional $L^{(-)}(t)$, Eq. (64) is positive definite. Similarly, from

$$K^{in2} - K^{out2} = \frac{1}{\lambda_s^4}\left(1 - \frac{c_A^2}{u_z^2}\right)\left(\frac{u_z}{c_A} - 1\right)^2 > 0.$$

follows that the functional $L^{(+)}(t)$ will be positive definite if $k + K^{out} < 0$ or equivalently, making use of Eqs. (54), (57)

$$\frac{u_z}{c_A}\left(\frac{u_z}{c_A} + 1\right) > \left(\frac{u_z}{c_A} - 1\right)D_{n,k} \quad , \quad D_{n,k} = \frac{\kappa_1^{in2}}{2\xi_{n,k}^2}. \tag{65}$$

We will study the lowest order stationary vortex solution, Eq. (46), i.e. we adopt $\kappa_1^{in} = j_{1,1}/R$ (where $j_{n,m}$ denotes the m-th zero of the Bessel function J_n). The "inside" eigenvalues $\xi_{n,k}$ may be estimated from the dispersion relation, Eq. (22), whose smallest solution is somewhat below $j_{0,1}/R$, but very close to it, which sets the numerical value of the parameter $D_{n,k}$ in Eq. (65) below 1.3:

$$\xi_{n,k} \geq \xi_{0,1} \sim \frac{j_{0,1}}{R}. \quad \Rightarrow \quad D_{n,k} \leq \frac{\kappa_1^{in2}}{2\xi_{0,1}^2} \sim 1.2688 \tag{66}$$

Inequality Eq. (65) is always satisfied with the above value of $D_{n,k}$, which implies that the functional $L^{(+)}(t)$ is positive definite.

In order to complete our stability proof, it would be necessary to show that either $L^{(+)}(t)$, or $L^{(-)}(t)$ is bounded from above (then, from Eq. (62) the other functional automatically remains finite). Noting that $L^{(+)}(t)$ and $L^{(-)}(t)$ are defined as integrals for the entire space of some positive definite functions, their boundedness implies that their integrands, and consequently also the perturbations $\delta\phi$, $\delta\psi$, remain infinitezimally small for all times, and in the whole space. Possible exceptions of finite size spatial regions with finite perturbations, and of "well behaved" (i. e. square integrable) isolated singularities of $\delta\phi$, $\delta\psi$ are not regarded as instabilities in the usual sense, as defined above.

It is important to note at this point that only the continuum spectrum of perturbations is contributing to the functional $L^{(-)}(t)$. Furthermore, it can be shown that modes belonging to the high $n, |k|$ end of the continuum spectrum are stable, and that consequently their contribution to the functional $L^{(-)}(t)$ is bounded. Namely, if the characteristic scalelength is much shorter than the vortex size

$$n \sim kr \sim kR \gg 1 \tag{67}$$

the eigenfunctions $e^{in\theta} g_n(kr)$ may be replaced by simple Fourier components. In this regime we also consider $\vec{V} \approx const.$, and apply the local approximation to Eqs. (52), (53), which permits us to calculate the frequency from the corresponding linear dispersion relation. As it has already been shown, in the outer region, setting $\kappa_1^{out} = 0$ in Eqs. (52), (53) linearly stable Fourier components are obtained

$$\left(\omega - \vec{k}_\perp \vec{V}\right)^2 = \frac{\left(k_z u_z - \vec{k}_\perp \vec{V}\right)^2}{1 + \lambda_s^2 k_\perp^2}. \tag{68}$$

Within the vortex core, and in the short-wavelength limit $\lambda_s^2 k_\perp^2 \sim R^2 k_\perp^2 \gg 1$ the linear dispersion relation derived from Eqs. (52), (53) has the form

$$\frac{u_z^2}{c_A^2} \lambda_s^2 k_\perp^2 \left(\frac{\omega - \vec{k}_\perp \vec{V}}{k_z u_z - \vec{k}_\perp \vec{V}} \right)^2 = 1 - \lambda_s^2 \kappa_1^{in\,2} \frac{\vec{k}_\perp \vec{V}}{k_z u_z - \vec{k}_\perp \vec{V}}. \tag{69}$$

and there exists a linear stability, Im $\omega = 0$, provided

$$\frac{k_z u_z}{\vec{k}_\perp \vec{V}} < 1 \qquad \text{or} \qquad \frac{k_z u_z}{\vec{k}_\perp \vec{V}} > 1 + \lambda_s^2 \kappa_1^{in\,2} = \frac{u_z^2}{c_A^2}. \tag{70}$$

For a broad range of parallel wavenumbers k_z these conditions are satisfied by short wavelength perturbations. A simple example are the perturbations with a small enough value of the parallel wavenumber $\lambda_s k_z \leq \frac{u_y}{u_z}$ which are unstable, according to Eq. (70), only if $\lambda_s k_\perp \leq 1$. However, this is outside of the validity region of the dispersion relation Eq. (69), and the existence of the instability remains uncertain. For a conclusive proof of either stability, or instability, extensive numerical investigation is required, solving three-dimensional evolution equations in the presence of initial perturbations belonging to the continuum spectrum $g_n(k,r)$, whose the scalelength is comparable to the vortex size $n \sim kR \sim 1$.

From Eqs. (62) - (64) it should be noted that an eventual instability of the continuum modes, having a wave-like structure far from the vortex, is followed by a simultaneous growth of the discrete modes $\widehat{\delta\phi}_{n,k}(t,z)\, g_{n,k}(r)$, which have a better spatial localization than the zero-order vortex. Such a behavior bears a close resemblence with the dual cascade in the weak shear-Alfvén turbulence [19], [20], towards both smaller and larger wavenumbers. This may possibly indicate a transition to a new nonlinear structure, different from the vortex Eqs. (38), (43), (44). However, to prove this a detailed numerical solution is required of the fully nonlinear, three-dimensional evolution equations describing convective-cell vortices.

4 Shear-Alfvén electron holes

Holes, or phase-space vortices, consist of particles trapped in a self-consistent potential well. They can be found as a spatially localized Bernstein-Green-Kruskal (BGK) solution of the Vlasov-Poisson system. Since a local potential minimum can trap only particles whose potential energy is not exceeding the depth of the well, holes are localized in the velocity space as well. The presence of the magnetic field, however, gives rise to the usual $\vec{E} \times \vec{B}$ drift in the perpendicular direction, and strong coupling between holes and plasma vortices takes place. A unified theory of electron holes and vortices is developed in Ref. [22], applying the gyrokinetic description of electrons and the usual fluid description of two-dimensional ions.

In the low-β, slowly t, z-varying regime described in the preceeding section, electron drift-kinetic equation can be written as

$$\frac{\partial f}{\partial t} + \left(\frac{\vec{B}}{B} v_\parallel + \frac{\vec{E} \times \vec{B}}{B^2} \right) \cdot \nabla f - \frac{e}{m_e} \frac{\vec{E} \cdot \vec{B}}{B} \frac{\partial f}{\partial v_\parallel} = 0 \tag{71}$$

where f is the electron distribution, and $v_\parallel$ is the velocity component parallel to the magnetic field. Similarly to the case of shear-Alfvén vortices, we seek a solution which is stationary in the reference frame moving with the velocity $\vec{u} = \vec{e}_y u_y + \vec{e}_z u_z$, when the drift-kinetic equation takes the form

$$(\vec{e}_z \times \nabla \psi) \cdot \nabla f + \frac{\partial f}{\partial v_\parallel} (\vec{e}_z \times \nabla \psi) \cdot \nabla \frac{e}{m_e} \frac{\varphi}{v_\parallel - u_z} = 0 \tag{72}$$

where ϕ, ψ denote the electrostatic potential in the moving frame, and the potential "seen" by the particle moving with the parallel velocity $v_\parallel$

$$\varphi = \phi - u_z A_z \quad , \quad \psi = \phi - v_\parallel A_z + (v_\parallel - u_z) \frac{u_y}{u_z} B_0 x \tag{73}$$

Characteristics of the stationary drift-kinetic equation Eq. (72)

$$\frac{dx}{-\frac{\partial \psi}{\partial y}} = \frac{dy}{\frac{\partial \psi}{\partial x}} = \frac{m_e (v_\parallel - u_z) \, dv_\parallel}{e (\vec{e}_z \times \nabla \psi) \cdot \nabla \varphi} \tag{74}$$

yield two integrals of motion, which can be identified as the electron energy and its parallel momentum in the moving frame

$$W = \frac{m_e}{2} (v_\parallel - u_z)^2 - e\varphi \quad , \quad P = m_e (v_\parallel - u_z) + e \left(\frac{u_y}{u_z} B_0 x - A_z \right) \tag{75}$$

Distribution of particles with positive values of the energy is adopted in the form

$$f = f_0 (X, V_\parallel) \tag{76}$$

where $f_0(x, v_\parallel)$ is the unperturbed distribution function, and the characteristics X, $V_\parallel$ are the initial particle position and velocity, defined in terms of the characteristics W, P as

$$X = \frac{u_z}{u_y \Omega_e}\left[\operatorname{sign}(v_\parallel - u_z)\left(\frac{2}{m_e}W\right)^{1/2} - \frac{P}{m_e}\right] \tag{77}$$

$$V_\parallel = u_z + \operatorname{sign}(v_\parallel - u_z)\left(\frac{2}{m_e}W\right)^{1/2} - \frac{e}{m_e}(1+H)\left(\frac{u_y}{u_z}B_0 x\right) \tag{78}$$

In the above, H is the function introduced in Eq. (34).

One may easily verify that if only the contribution of nonresonant perticles $\frac{m_e}{2}(v_\parallel - u_z)^2 \gg e\varphi$ is retained in the distribution function Eq. (76) the hydrodynamic results of the preceeding section are recovered.

In the cese of positive potentials φ, there exists a region of resonant particle velocities $v_\parallel$

$$|v_\parallel - u_z| \le \left(\frac{2e}{m_e}\varphi\right)^{1/2} \tag{79}$$

for which the the distribution Eq. (76) is not applicable. Distribution function of resonant particles is adopted in the form of a hole in the velocity space

$$f = (1-a)\, f_0(X_{Res}, V_{Res}) \tag{80}$$

$$X_{Res} = -\frac{u_z}{u_y}\frac{P}{m_e \Omega_e} \quad , \quad V_{Res} = u_z - \frac{e}{m_e}(1+H)\left(\frac{u_y}{u_z}B_0 X_{Res}\right) \tag{81}$$

The depth of the hole, given by the parameter a, is determined by the processes which contribute to its "filling up", such as collisions, turbulent diffusion, etc, which are not studied in this context; in an ideal case of perfectly absorbing boundaries at $z = \pm\infty$ and the absence of dissipation, the hole would be empty, $a = 1$.

Now, calculating the parallel current with the use of Eqs. (76), (80), and applying the same procedure as in the preceeding section, the following equation for the potential φ is obtained

$$\left(\nabla_\perp^2 + \kappa_1^2\right)\left(\nabla_\perp^2 + \kappa_2^2\right)\varphi + \gamma h(\varphi)\left(c_1 - c_2\nabla_\perp^2\right)\varphi^{1/2} = 0 \tag{82}$$

where h is the Heaviside unit step function, and

$$c_1 = G + \frac{c_A^2}{u_z^2}(1+H)^2 \quad , \quad c_2 = \frac{c_A^2}{u_z^2} - 1 \quad , \quad \gamma = 2a f_0(u_z)\frac{u_z^2}{n_0\lambda_s^2}\left(\frac{2m_e}{e}\right)^{1/2} \tag{83}$$

and the wavenumbers κ_1, κ_2 are defined in Eq. (37).

In the presence of electron trapping, $a \neq 0$ the wave equation (82) is difficult to solve due to the presence of the nonlinear term $\varphi^{1/2}$. It can be seen, however that a simple cylindrically symmetric and spatially localized solution does not exist, since the nonlinearity is not strong enough to produce the localization (nonlinearities with the power $\ge 3/2$ could produce it).

However, there is one particular case which can be colved analytically. The "one Bessel function" dipole, Eq. (45) has the electrostatic potential in the moving frame equal to zero,

$\varphi^{(1)} = \phi^{(1)} - u_z A_z^{(1)} = 0$, and consequently it does not interact with the particles. This permits us to calculate the cylindrically symmetric (monopole) component independently, from

$$\left(\nabla_\perp^2 + K\right) \varphi^{(0)}(r) + \gamma \left(\frac{c_A^2}{u_z^2} - 1\right) \sqrt{\varphi^{(0)}(r)} = 0 \quad , \quad K = \begin{cases} \frac{c_A^2}{u_z^2} - 1 \quad , & r > R \\ {\kappa_2^{in}}^2 \quad , & r < R \end{cases} \tag{84}$$

while κ_2^{in} is defined in Eq. (46). A localized solution of the ordinary diferential equation (84) can easily be found numerically; it is also instructive to present the folowing approximative solution, obtained after the linearization of Eq. (84) around $\varphi^{(0)} = 0$, and $\varphi^{(0)} = \gamma^2 \mu^4 / (W {\kappa_2^{in}}^4)$ outside and inside the vortex core, respectively

$$\varphi^{(0)}(r) = \frac{\gamma^2}{W} \frac{\mu^4}{{\kappa_2^{in}}^4} \begin{cases} W + \mu K_1(\mu R) J_0(\kappa^{(\bullet)} r) , & r < R \\ \kappa^{(\bullet)} J_1(\kappa^{(\bullet)} R) K_0(\mu r) , & r > R \end{cases} \tag{85}$$

where $\mu = (1 - c_a^2/u_z^2)/\lambda_s^2$, $\kappa^{(\bullet)} = \kappa_2^{in}/\sqrt{2}$ and W is the Wronskian

$$W = \kappa^{(\bullet)} J_1(\kappa^{(\bullet)} R) K_0(\mu R) - \mu K_1(\mu R) J_0(\kappa^{(\bullet)} R) \tag{86}$$

With the above solution, the ratio of monopole and dipole components can be estimated as

$$M = \frac{\gamma^2}{B_0 u_y R} \approx 16\pi a^2 \sqrt{\frac{m_e}{m_i}} \frac{\lambda_s}{R} \frac{u_z^2}{u_y c_A} \frac{u_z^2}{v_{Te}^2} \exp\left(\frac{-u_z^2}{2 v_{Te}^2}\right) \tag{87}$$

and one may see that the monopoles dominate if the vortex size is of the order of collisionless skin depth, $R \sim \lambda_s$, provided that the plasma is not too cold, i. e. that the resonant velocity is not too deep in the tail of the distribution function.

Monopolar hole, Eq. (85) appears as a "rider" superimposed on the hydrodynamic vortex, which determines the size and the speed of the entire structure, while the monopole provides the self-organization from the parallel phase space trapping.

Since the vortex contains a separatrix dividing the electron orbits between the trapped and passing, wave fields and other perturbations will lead to a stochastic layer. Besides that, the presence of trapped electrons in the kinetic structure raises the possibility of significant detrapping, or separatrix crossing, during collisions between such structures, which reduces their resilience as compared to the corresponding hydrodynamic sructures. We speculate that these new three-dimensional coherent localized structures are robust and stable to vortex-vortex interaction, like the two-dymensional hydrodynamic vortices. Clearly, new hybrid ion fluid and drift-kinetic codes will be required to fully investigate the new structures presented here. Vortex/hole structures are expected to be stable on the electron bounce timescale, $\tau_{be} \sim (1/k_\parallel)(m_e/e\phi)^{1/2}$ and to play an important role in the anomaluos transport in open-endend confinement devices.

References

[1] A. B. Mikhailovskii, V. P. Lakhin, L. A Mikhailovskaya, and O. G. Onischenko, Zh. Eksp. Teor. Fiz., **86**, 2061, (1984) [Sov. Phys. JETP **59**, 1198, (1984)]

[2] V. I. Petviashvili, and O. A. Pokhotelov, Fiz. Plazmy, **12**, 1127, (1986) [Sov. J. Plasma Phys., **12**, 651, (1987)]

[3] M. V. Nezlin, Usp. Fiz. Nauk, **150**, 3, (1986) [Sov. Phys. Usp., **29**, 807, (1987)]

[4] V. I. Petviashvili and V. V. Yankov, in *Reviews in Plasma Physics,* edited by B. B. Kadomtsev (Consultants Bureau, New York, 1989), vol. 14, pp. 39–55

[5] L. I. Piteborg and E. I. Shulman, Phys. Lett., **A 140** 29, (1989)

[6] W. Horton, X. N. Su and P. J. Morrison; Fiz. Plazmy **16**, 969, (1990) (in Russian)

[7] A. Hasegawa and M. Wakatani, Phys. Rev. Lett., **59**, 1581, (1987)

[8] T. D. Kaladze and V. I. Petviashvili, Fiz. Plazmy **12**, 1335, (1986) [Sov. J. Plasma Phys., **12**, 772, (1986)]

[9] S. Iizuka, T. Huld, H. L. Pecseli and J. J. Rasmusen, Phys. Rev. Lett., **60**, 1026, (1988)

[10] W. Horton, Phys. Fluids, **B 1**, 524, (1989)

[11] W. Horton, Phys. Fluids, **29**, 1491, (1986)

[12] M. Kono and E. Miyashita, Phys. Fluids, **31**, 326, (1988)

[13] W. Horton, T. Tajima and T. Kamimura, Phys. Fluids, **30**, 3485, (1987)

[14] D. Jovanović, H. L. Pecséli and J. J. Rasmussen, Phys. Fluids, **B 2**, 2035, (1990)

[15] D. Jovanović, P. K. Shukla and U. de Angelis, Phys. Fluids, **B 1**, 715, (1989)

[16] D. Jovanović, P. K. Shukla, U. de Angelis and W. Horton, Phys. Fluids, **B 3**, 45, (1991)

[17] D. Jovanović and W. Horton, Phys. Fluids, **B 5**, 9, (1993)

[18] E. W. Laedke and K. H. Spatschek, Phys. Fluids, **31**, 1492, (1988)

[19] A. Bondesson, Report NSF-ITP-85-16

[20] D. Jovanović, H. L. Pécseli, J. Juul Rasmussen and K. Thomsen, J. Plasma Phys., **37**, 81, (1987)

[21] D. Jovanović, H. L. Pécseli, J. Juul Rasmussen, Phys. Fluids, **B2**, 2035, (1990)

[22] D. Jovanović, W. Horton, Phys. Fluids, **B 5**, 433, (1993)

Low Temperature and
General Plasmas
M. Milosavljevic and Z. Petrovic

ISBN 1-56072-399-8

RAY TRACING NEAR THE ELECTRON CYCLOTRON RESONANCE IN THERMAL PLASMAS

S. Pešić and A. Stojić

Institute of Nuclear, Sciences "VINČA", Physics Laboratory,
P.O.B. 522, Belgrade, Yugoslavia

Abstract. The ray tracing in weakly dispersive and dissipative inhomogeneous media and in particular, near and through the electron cyclotron resonance (ECR) is investigated. First, the wave behaviour near the ECR is analysed qualitatively and the source of differences between the thermal plasma and cold plasma ray descriptions is identified. This is followed by a detailed numerical study of ray paths in tokamak plasmas in which a fully three dimensional code is applied to trace ray paths initialized to the low or high field launched ordinary or extraordinary modes. A toroidal (i) cold plasma, (ii) hot Maxwellian plasma and (iii) weakly relativistic Maxwellian plasma is considered. It is found that in the ECR region thermal effects may induce a substantial ray refraction, ray reflection and change of the wave polarization. This has important implications on the propagation and absorption characteristics of both, the low and the high field side launched electromagnetic waves. The preliminary numerical results obtained within the new, complex treatment of the dispersion equation confirm the afore mentioned conclusions.

I. INTRODUCTION

The ray tracing technique is widely used to study the wave propagation, absorption and emission in weakly dispersive and dissipative inhomogeneous media. In general, the paths of the energy in the wave packet are determined by using the ray equations in which the wave properties are described by the the local cold plasma dispersion relation. Namely, it is generally accepted that the wave dispersion relation in the cold plasma limit provides a satisfactory approximation for the real part of the wave refractive index, the ray path and the wave polarization. The cold plasma dispersion relation is real and the ray path represents the position and the wave vector of a ray in a real, physical space. When finite-temperature plasma is considered however, the plasma dielectric tensor is no longer Hermitian and consequently, the dispersion relation becomes complex. In order to restrict the ray path to real values then the wave absorption is assumed to be weak[1], i.e. $|Im\mathrm{N}| \ll |Re\mathrm{N}|$ where N is the wave refractive index. Furthermore, it has been shown[2] that if the wave damping at the electron cyclotron resonance (ECR) is weak, the ray paths can be defined and can be traced through the resonance by using the cold plasma dispersion relation. We have reconsidered the problem of the influence of thermal effects on the ray path near and through the resonance and found that they are essential, in particular in plasma regions in which efficient wave - particle interaction takes place (particle resonances, wave resonances, etc.). The results obtained in weakly dispersive and dissipative inhomogeneous hot plasma indicate indeed that thermal effects may induce a substantial ray refraction, ray reflection and/or a change of the wave polarization near the ECR. Obviously, the wave power deposition along the ray is also altered.

The paper is organized as follows. In Section II we discuss the ray description in dissipative plasmas and present a new treatment which offers the possibility to obtain meaningful physical informations in these plasmas. In Section III ray behaviour near the ECR is analysed qualitatively and the source of differences between the thermal plasma and cold plasma descriptions is identified. The fully three dimensional code which is applied to trace the rays in toroidal (i) cold plasma, (ii) hot Maxwellian plasma and (iii) weakly relativistic Maxwellian plasma is presented in Section IV. The paths of rays initialized to the low and high field side launched ordinary and extraordinary modes are analysed in Section V. Finally, our conclusions are summarized in Section VI.

II. RAY DESCRIPTION IN DISSIPATIVE PLASMAS

Within the geometrical optics approximation the path of the energy in a wave packet is given by the ray equations,

$$\frac{d\mathbf{r}}{ds} = - \frac{\partial D/\partial \mathbf{N}}{|\partial D/\partial \mathbf{N}|}, \tag{1}$$

$$\frac{d\mathbf{N}}{ds} = \frac{\partial D/\partial \mathbf{r}}{|\partial D/\partial \mathbf{N}|}. \tag{2}$$

Here $\mathbf{r}$ is the spatial coordinate, s is the arc length along the ray, $\mathrm{N} = (c/\omega)\mathbf{k}$ is the wave refractive index and $D \equiv D(\mathrm{N},\omega,\mathbf{r},t)$ is the local plasma dispersion relation. We note that the presented form of ray equations is more convenient in applications than the familiar one in which the wave vector $\mathbf{k}$ is used as a variable. The pair of ray equations (1) - (2) are identical to the Hamilton's equations of motion for a particle with momentum **k** and Hamiltonian $D(\mathbf{N},\omega,\mathbf{r},t)$.

The electromagnetic properties of the considered plasma are completely characterized by the local dielectric tensor whose calculation varies depending on the nature of the plasma. Depending on the wave - particle interaction (and on the degree of the approximation used to describe this interaction) three various physical situations can occur:

(i) The dielectric tensor is everywhere Hermitian. If the wave - particle interaction is weak, one can disregard the thermal scatter of particles and formally treat the dielectric tensor as Hermitian. In this case the ray equations yield paths in the real space. We note that in the cold plasma approximation, the plasma is degenerate at the cyclotron resonance surface, that is, the corresponding dispersion relation has multiple roots on this surface. Since the dispersion relation changes its sign as one passes through the resonance, it represents in fact a degenerate surface separating non-degenerate plasma regions.

(ii) There are plasma regions in which the dielectric tensor is non-Hermitian. This is the most frequently encountered case. The ray equations (1) - (2) initialized with real initial values yield ray paths which stay in the real space in the outer, low-temperature plasma region. So, the rays describe the path of the energy of the two familiar (cold plasma) electromagnetic modes: the extraordinary and the ordinary mode. In the hot plasma core however, the dielectric tensor is no longer Hermitian and consequently, the dispersion relation is complex. This implies the pair of real equations,

$$Re\{D(\mathbf{N},\omega,\mathbf{r},t)\} \equiv D_R(\mathbf{N},\omega,\mathbf{r},t) = 0 \tag{3}$$

$$Im\{D(\mathbf{N},\omega,\mathbf{r},t)\} \equiv D_I(\mathbf{N},\omega,\mathbf{r},t) = 0 \tag{4}$$

which in general, will not be satisfied simultaneously by the same root $\omega = \omega(\mathbf{N},\mathbf{r},t)$. So, equations (3) - (4) describe complex ray paths which nevertheless, in low-dissipative plasmas may remain close to the real world. Obviously, the distance of the ray path from the real world is directly related to the wave decay or growing in time (or space). Let us introduce the complex frequency $\Omega = \omega + i\nu$ where ν is assumed to be small compared to ω. If ν is small the dispersion equation can be approximated by

$$D(\mathbf{N},\omega+i\nu,\mathbf{r},t) \simeq D_R - \nu\partial D_I/\partial\omega + i(D_I + \nu\partial D_R/\partial\omega) = 0. \quad (5)$$

Instead of the dispersion equation $D(\mathbf{N},\omega,\mathbf{r},t) = 0$ then the equation

$$\Delta(\mathbf{N},\omega,\mathbf{r},t) = D_R\partial D_R/\partial\omega + D_I\partial D_I/\partial\omega = 0, \quad (6)$$

can be regarded as a dispersion relation yielding ω [3] while ν can be determined from

$$\nu = D_R/(\partial D_I/\partial\omega) = - D_I/(\partial D_R/\partial\omega) \quad (7)$$

So, as long as the rays do not propagate far from the real world, this treatment offers the possibility to obtain meaningful physical information. Note that similar results can be obtained by using a fully complex formulation of the geometric optics and by expanding the complex ray path back to a neighboring point in the real world[4].

(iii) The dielectric tensor is everywhere non-Hermitian. After the initialization at the plasma edge, in this case the ray quickly propagate far from the real world and thus equations (1) - (2) give no further information about the solution at points in the real space.

III. QUALITATIVE ANALYSIS OF THE RAY BEHAVIOUR NEAR THE ECR

Before presenting the ray paths obtained by numerical analysis of the ray equations, let us briefly review the main features of the dispersion relation in the considered frequency range. Local symmetry around the direction of the confining magnetic field implies that the dispersion relation is a function of the wave refractive index only through $N_\perp^2$ and $N_\parallel$. So, using the local reference frame where the magnetic field is along one of the orthogonal axis, the dispersion relation can be written as,

$$D = \epsilon_{11}N_\perp^4 + 2\epsilon_{13}N_\parallel N_\perp^3 + \left[\epsilon_{13}^2 + \epsilon_{11}(N_\parallel^2 - \epsilon_{22}) - \epsilon_{12}^2 + \epsilon_{33}(N_\parallel^2 - \epsilon_{11})\right]N_\perp^2 +$$

$$2\left[\epsilon_{13}(N_\parallel^2 - \epsilon_{22}) + \epsilon_{12}\epsilon_{23}\right]N_\parallel N_\perp + \epsilon_{11}\epsilon_{23}^2 - \epsilon_{22}\epsilon_{13}^2 + 2\epsilon_{12}\epsilon_{13}\epsilon_{23} + N_\parallel^2(\epsilon_{13}^2 - \epsilon_{23}^2) +$$

$$\epsilon_{33}(N_\parallel^2 - \epsilon_{11} - \epsilon_{22})N_\parallel^2 + \epsilon_{33}(\epsilon_{12}^2 + \epsilon_{11}\epsilon_{22}) = 0, \quad (8)$$

where ϵ_{ij} are the components of the corresponding dielectric tensor. For frequencies close to the fundamental electron cyclotron (EC) frequency, the dielectric tensor components of a nonrelativistic Maxwellian plasma can be approximated by,

$$\epsilon_{11} = \epsilon_{22} = 1 - i\epsilon_{12} = 1 + i\epsilon_{21} \simeq 1 - \frac{X}{2(1 + Y)} + \eta Z(z_1) \qquad (9)$$

$$\epsilon_{13} = \epsilon_{31} = - i\epsilon_{23} = i\epsilon_{32} \simeq \frac{XN_\perp}{2Y\,|N_{\|}|}\left[1 + z_1 Z(z_1)\right] \qquad (10)$$

$$\epsilon_{33} \simeq 1 - X + i\,\frac{XN_\perp{}^2\gamma}{2Y^2\,|N_{\|}|}\, z_1{}^2\, \mathrm{Im}\{Z(z_1)\} \qquad (11)$$

where $Z(z)$ is the plasma dispersion function, $\eta = z_0 X/2$, $z_0 = (\gamma\,|N_{\|}|)^{-1} \simeq 16/|N_{\|}|\,T^{1/2}$ (keV), $\gamma = v_t/c$, $z_1 = z_0(1 - Y)$, $X = \omega_p{}^2/\omega^2$, $Y = \omega_c/\omega$ with ω_p and ω_c being the plasma and EC frequency, v_t being the electron thermal velocity and c is the speed of light in vacuum. For $\gamma\,|N_{\|}| \ll 1$ one has $|z_0 Z(z_1)| \gg 1$, and after some algebraic manipulations the dispersion relation can be approximated by,

$$D \simeq \epsilon_{11} Q(N_\perp{}^2, N_{\|}, X, Y) + S(N_\perp{}^2, N_{\|}, X, Y) \qquad (12)$$

where Q and S are real algebraic functions of $N_\perp{}^2$, $N_{\|}$, X and Y (see also Eq. (32) in [5]). It is convenient to split the susceptibility into two terms, $\chi_{11} = \chi_{1c} + \chi_{1t}$, where $\chi_{1c} = -X/(1 - Y^2)$ represents the susceptibility in cold plasma while $\chi_{1t} = X/2(1 - Y) + \eta Z(z_1)$ describes the contribution of the resonant term in thermal plasma. The parameter η can be considered as a quantity which weights the influence of plasma effects in the wave propagation. In most practical cases, $\gamma \ll 1$ and $X \lesssim 1$ so that one finds oneself deeply inside the finite density regime, $\eta > 1$. On the other hand, for equilibrium plasma parameters such that $\eta < 1$, the plasma (and thermal) contribution to the dielectric tensor is small compared to 1, and one is in the tenuous plasma regime[6]. For frequencies which are not very close to the EC frequency or more precisely, for $|z_1| \gg 1$ (and $|z_0 Z(z_1)| > 1$) one has $\mathrm{Re}\{\eta Z(z_1)\} \simeq - X/2(1 - Y)$ and $\mathrm{Im}\{\eta Z(z_1)\} \simeq 0$, and the dispersion relation (12) reduces to that obtained in the cold plasma limit, $D_R \simeq (1 + \chi_{1c})Q \simeq D_c$ and $D_I \simeq 0$. We recover the fact that in this frequency range the anti-Hermitian part of the dispersion relation is vanishingly small while the Hermitian part is well described by the cold plasma model. For near-resonant frequencies however, the plasma dispersion function starts to deviate significantly from its asymptotic form and thermal effects affect the wave behaviour. In the middle of this frequency range the imaginary part of $\eta Z(z_1)$ comes into but $|\mathrm{Im}\{\eta Z(z_1)\}| \ll |\mathrm{Re}\{\eta Z(z_1)\}|$ for $|z_1| > 1.5$. Therefore, in a large part of the frequency range around the ECR (let's say, for $1.5 \lesssim$

$|z_1| \lesssim 6$) the dispersion relation is basically a real function. Only in a narrow frequency band (for $|z_1| \lesssim 0.15$) the anti-Hermitian part of the dispersion relation may become larger than the Hermitian part but in many cases of practical interest the inequality $|Im\mathrm{N}| \ll |Re\mathrm{N}|$ still applies. Here, we recall that the previous qualitative analysis is based on the nonrelativistic dispersion relation. In the weakly relativistic description the wave damping is asymmetrical and the discussed variation of the Hermitian and anti-Hermitian part of the dispersion relation is modified. Nevertheless, within this description the dispersion relation remains a real function affected by thermal effects in a relatively large frequency range, also.

Turning now to the spatial derivative of the dispersion relation whose general form can be written as,

$$\frac{\partial D}{d\mathbf{r}} = \frac{\partial D}{\partial X}\frac{\partial X}{\partial \mathbf{r}} + \frac{\partial D}{\partial Y}\frac{\partial Y}{\partial \mathbf{r}} + \frac{\partial D}{\partial \gamma}\frac{\partial \gamma}{\partial \mathbf{r}} + \frac{\partial D}{\partial N_\perp^2}\frac{\partial N_\perp^2}{\partial \mathbf{r}} + \frac{\partial D}{\partial N_\parallel}\frac{\partial N_\parallel}{\partial \mathbf{r}}, \tag{13}$$

we note that in thermal plasma it contains an additional term, $(\partial D/\partial\gamma)(\partial\gamma/\partial\mathbf{r})$. The contribution of this term becomes appreciable in thermal plasma but it is small at not too high electron temperatures ($T \leq 20$ keV). Away from the ECR region, the derivatives of the weakly relativistic dispersion relation $\partial D_r/\partial X$, $\partial D_r/\partial Y$, $\partial D_r/\partial N_\perp^2$ and $\partial D_r/\partial N_\parallel$ nearly coincide with those obtained in the cold plasma limit so that $\partial D_r/d\mathbf{r} \simeq \partial D_c/d\mathbf{r}$. In the cold plasma approximation the aforementioned derivatives are proportional to $(1-Y)^{-1}$. Thus, as the wave approaches the resonance, the absolute value of $\partial D_c/\partial X$, $\partial D_c/\partial Y$, $\partial D_c/\partial N_\perp^2$ and $\partial D_c/\partial N_\parallel$ increases. All these derivatives are of the same order of magnitude except $\partial D_c/\partial Y$, the absolute value of which for the O mode and X mode is smaller and larger than the other derivatives, respectively.

The most temperature dependent partial derivative of the weakly relativistic D_r or hot plasma dispersion relation D_h is that with respect to the static magnetic field. In the ECR layer, these spatial derivatives are proportional to $Xz_0^2 = 4\eta^2/X$. For equilibrium plasma parameters such that $\eta < 1$, i.e. in the tenuous plasma limit, $|\partial D_r/\partial Y|$ (or $|\partial D_h/\partial Y|$), are relatively small, and thermal plasma ray paths tend to the cold plasma ones. As the parameter η is increased, the thermal contribution to the dielectric tensor enlarges and the ray paths start to deviate from those obtained in the cold plasma limit. Actually, the relatively large value of the derivative $|\partial D_r/\partial Y|$ in the resonance region causes changes of sign of the equations governing the poloidal wave number N_θ which in turn induces further modifications of the ray path as compared to the cold one: $d\phi/ds$, N_θ and $d\theta/ds$ change sign (see Eqs. (15),(16) and (18)).

In previous ray tracing studies of the wave propagation and absorption near the ECR, thermal effects have been neglected and the ray paths are determined by using the cold plasma dispersion relation, $D_c = 0$. The wave power deposition along the ray is usually determined within the weak damping approximation. The foregoing discussion however, indicate that thermal effects are substantial for the ray behaviour in particular, in plasma regions in which an effective wave-particle interaction takes place (for instance, near particle or wave resonances). The results presented in this work demonstrate indeed that thermal effects induce additional ray refraction, or reflection in the ECR layer.

IV. THE RAY TRACING CODE

To verify the previous qualitative analysis we have developed a fully three-dimensional ray tracing code. The code is written in the usual toroidal coordinates $\{r,\theta,\phi\}$ where r and $R = R_0 + r\cos\theta$ are the minor and major radial positions, θ and ϕ are the poloidal and toroidal angles, respectively. We note that right-handed Cartesian coordinate systems are used throughout this paper. By applying a canonical transformation and by implicit differentiation of the dispersion relation we obtained the ray equations in the following form,

$$\frac{dr}{ds} = - F \frac{\partial D}{\partial N_r} = - F\, 2N_r \frac{\partial D}{\partial N_\perp^2} \tag{14}$$

$$\frac{rd\theta}{ds} = - F \frac{\partial D}{\partial N_\theta} = - F \left[2b_\phi N_u \frac{\partial D}{\partial N_\perp^2} + b_\theta \frac{\partial D}{\partial N_\parallel} \right] \tag{15}$$

$$\frac{Rd\phi}{ds} = - F \frac{\partial D}{\partial N_\phi} = - F \left[-2b_\theta N_u \frac{\partial D}{\partial N_\perp^2} + b_\phi \frac{\partial D}{\partial N_\parallel} \right] \tag{16}$$

$$\frac{dN_r}{ds} = F \frac{\partial D}{\partial r} = F \left\{ \frac{\partial D}{\partial X}\frac{\partial X}{\partial r} + \frac{\partial D}{\partial \gamma}\frac{\partial \gamma}{\partial r} - \frac{Y\cos\theta}{R}\frac{\partial D}{\partial Y} + \right.$$

$$\left. \frac{df}{dr} \left[Y_\phi b_\theta \frac{\partial D}{\partial Y} + N_u b_\phi^2 \left(-2N_\parallel \frac{\partial D}{\partial N_\perp^2} + \frac{\partial D}{\partial N_\parallel} \right) \right] \right\} + N_\phi \cos\theta \frac{d\phi}{ds} + N_\theta \frac{d\theta}{ds} \tag{17}$$

$$\frac{d(rN_\theta)}{ds} = F \frac{\partial D}{\partial \theta} = r \sin\theta \left[\frac{FY}{R}\frac{\partial D}{\partial Y} + N_\phi \frac{d\phi}{ds} \right] \tag{18}$$

$$\frac{d(RN_\phi)}{ds} = F\frac{\partial D}{\partial \phi} = 0. \tag{19}$$

Here $F \equiv 1/|\partial D/\partial \mathbf{N}|$, $\mathbf{b} = \mathbf{Y}/Y = \{0,b_\theta,b_\phi\}$, and $f = B_\theta/B_\phi = r/(R_0q(r))$. The magnetic surfaces are assumed to have circular and concentric meridian cross-sections. The equilibrium tokamak magnetic induction has a toroidal component $B_\phi(r,\theta) = B_{\phi 0}R_0/R$ and a poloidal component $B_\theta(r,\theta) = fB_\phi(r,\theta)$, where the safety factor is $q(r) = q(0) + [q(a) - q(0)]r^2/a^2$, and a is the plasma radius. Furthermore, the density and temperature are assumed to be constant on the flux surfaces: $n(r) = n(0)\ (1 - r^2/a^2)$ and $T(r) = T(0)\ (1 - r^2/a^2)^2$. The components of the wave refractive index in the orthogonal frame of reference tied to $\mathbf{B}$ are determined by the following relations: $N_{||} = b_\theta N_\theta + b_\phi N_\phi$, $N_u = -b_\theta N_\phi + b_\phi N_\theta$ and $N_\perp = (N_r^2 + N_u^2)^{1/2}$.

We have assumed that the waves are launched from the plasma boundary at the position $\{a,\theta_a,\phi_a\}$ in a direction characterized by the angles χ and ψ. χ is the angle between the wave refractive index N and the unit vector $\mathbf{e}_\phi$, while ψ is the angle between $\mathbf{e}_r$ and the projection of N in the poloidal plane. Since the present study is not intended to match any existing device, we chose parameters characterizing a medium-size tokamak device (for instance, RTP): $a = 0.24$m, $R_0 = 0.72$m, $B_{\phi 0} = 2.14$T, $q(0) = 1$ and $q(a) = 3$ and varied the central electron density and temperature in wide ranges: $n(0) = 10^{16} - 10^{20}$ m^{-3} and $T(0) = 1$eV - 20keV, respectively. For simplicity, we chose the frequency of the launched waves such that the ECR takes place at the plasma centre. This leads to a pump frequency f = 60GHz.

By using the local dispersion equation (3) the code is applied to trace rays in toroidal **(i)** cold plasma, **(ii)** hot Maxwellian plasma, and **(iii)** weakly relativistic Maxwellian plasma. The wave power deposition along the ray path in this case is determined from the complex roots of the complete **(ii)** hot plasma or **(iii)** weakly relativistic plasma dispersion relation. Furthermore, the modified local dispersion equation (6) and the relation (7) are used and the code is apllied to trace **(iv)** rays in toroidal weakly relativistic Maxwellian plasma. The dielectric tensor components ϵ_{ij} of cold plasma and hot Maxwellian plasma are defined elsewhere[7,8]. In the relativistic treatment, we used the dielectric tensor of weakly relativistic thermal plasma[9] since in most cases which are of interest in current fusion research, electron temperatures are not too high ($T \lesssim 20$ keV).

V. RAY PATHS IN TOKAMAKS

First, we consider rays injected from the low magnetic field side of the torus. The paths of the rays initialized to the ordinary mode have two different topologies, depending on the value of the parallel wave refractive index. At large values of $N_{\parallel}$, $Y/(1+Y) \lesssim N_{\parallel}^2 \leq 1$, that is, for $\chi \lesssim 60°$ the rays suffer radial reflection with $N_r = \left[N_{\perp}^2 - N_u^2\right]^{1/2} = 0$. The rays reflect before the high-density (or left-hand) cutoff $X_c = (1 - N_{\parallel}^2)(1 + Y)$ is reached, that is, at an electron density X_r which is smaller than X_c. If $N_{\parallel} \rightarrow 1$ the ray reflection occurs near the high-density cutoff. As $N_{\parallel}$ is decreased however, the perpendicular wave refractive index of the ordinary mode increases and the rays reflect well before $N_{\perp}^2$ itself vanishes. In the considered large $N_{\parallel}$-value range the rays do not traverse the ECR layer, and consequently, the ray paths obtained in the cold plasma limit coincide with the weakly relativistic ones. This is graphically illustrated in Fig. 1 where we show the poloidal and toroidal projections of rays launched from the points $\{a,\theta = -80°(20°)80°,\phi=0°\}$ in direction of the plasma centre

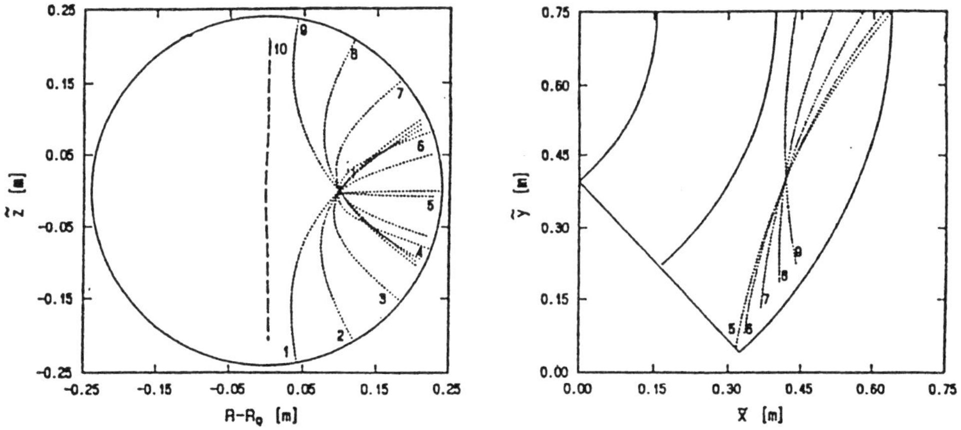

FIG. 1. Poloidal (a) and toroidal (b) views of the O mode ray paths for θ = -80°(20°)80°, ψ = 180°, χ = 40°, n(0) = 2x10^{19} m^{-3} and T(0) = 1keV.

(ψ = 180°) at χ = 40°. The represented ray paths are obtained in weakly relativistic plasma with central electron density $n(0) = 2\text{x}10^{19}\text{m}^{-3}$ and central electron temperature $T(0)$ = 1keV. The cold plasma ray paths practically coincide with those shown in Fig. 1. Here we note that in all figures which illustrate our discussion the ray paths obtained in weakly-relativistic and cold plasma are represented by lines consisting of dots and short dashes, respectively. The line consisting of long dashes shows the ECR position. It should be pointed out that the condition of weak absorption ($|Im\text{N}| \ll |Re\text{N}|$) is satisfied along all the ray paths shown in the present paper. Now, coming back to the ray paths in Fig. 1, we note that the radial component of the wave refractive index vanishes precisely at the point where the ray becomes tangent to a magnetic surface. Notice the strong focusing which brings the reflected rays to a real focal point in the midplane. For small values of the angle θ the initially cylindrical wavefront is strongly distorted due to the radial ray reflection. The polarization of the ordinary mode projected into the plane perpendicular to **B** deviates only little from circular. This quasi-circular wave rotates opposite to the Larmor motion of electrons.

In the range of intermediate values of the angle χ (50° $\leq \chi \leq$ 60°), the rays refract near the plasma centre and reflect in the high magnetic field side. Simultaneously, the length of the ray paths in toroidal direction increases. The wave launching in this parameter range has the advantage that the rays cross twice the absorption region. The ray paths in finite density plasma obtained within the weakly relativistic and cold plasma description start here to differ. Finally, for small $N_{\|}$ - values ($N_{\|}^2 \leq Y/(1+Y)$) that is, for $\chi \geq 60°$ the radial ray reflection disappears and the rays reach the plasma edge on the high magnetic field side. In Fig. 2(a,b) we show the poloidal projections of ray paths in (a) cold plasma and (b) weakly relativistic thermal plasma with $T(0)$ = 1keV. Here χ = 70° while the values of the other

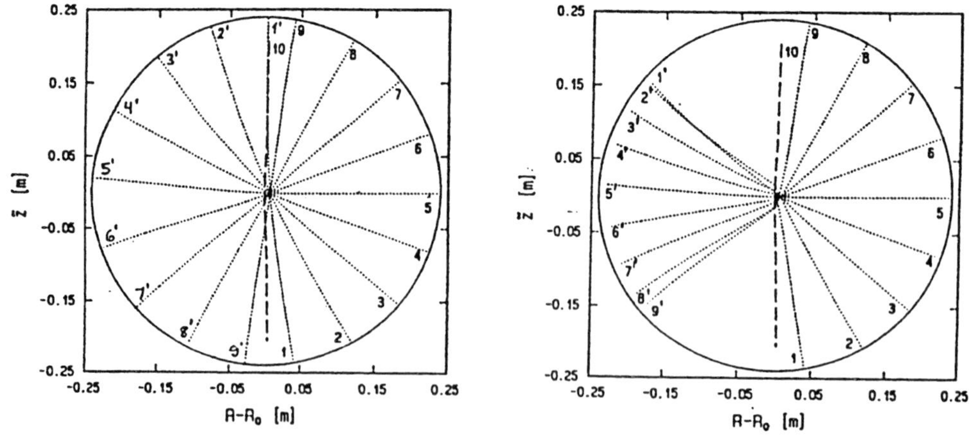

FIG. 2. Poloidal views of the O mode ray paths for χ = 70° in (a) cold plasma and (b) weakly relativistic plasma with T(0) = 1keV. The values of the other parameters are same as in Fig. 1.

parameters are same as in Fig. 1. In both represented cases the incident rays converge to a real focus. In the weakly relativistic plasma description, however, the rays emerging from the focus are substantially refracted near the ECR. This strong defocusing and the toroidal effects slightly affect the wavefront. For small $N_{||}$-values the ordinary mode tends to be linearly polarized in the direction of **B**. In the plane perpendicular to **B** the ordinary mode is elliptically polarized with the large axe in the $\mathbf{e}_y$-direction.

In Fig. 3(a,b) the ray paths obtained for various (a) central electron densities and (b) central electron temperatures are compared with those obtained in the cold plasma limit. As one can see, up to the resonance layer the ray paths coincide.

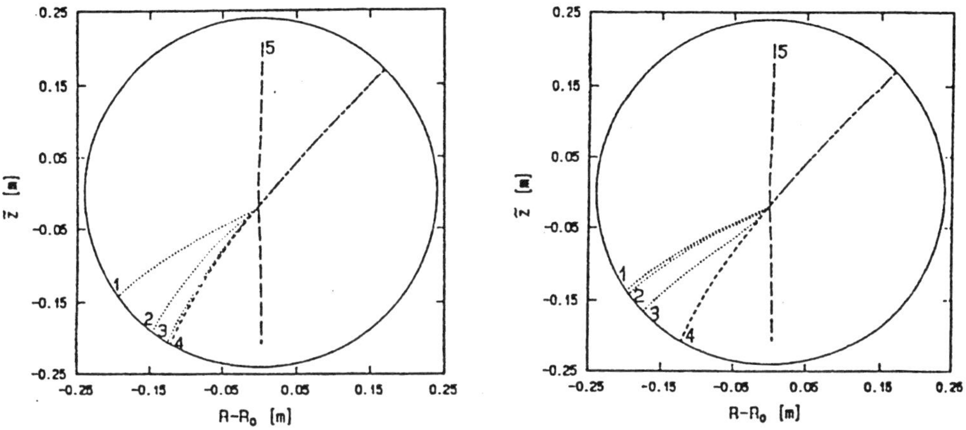

FIG. 3. Poloidal views of the O mode ray paths for $\theta = 45^\circ$, $\chi = 60^\circ$, $\psi = 180^\circ$ and (a) $T(0) = 1$kev, (1) $n(0) = 10^{19}m^{-3}$, (2) $n(0) = 10^{18}m^{-3}$, (3) $n(0) = 10^{17}m^{-3}$, (4) $T(0) = 0$eV, $n(0) = 10^{17}m^{-3} - 10^{19}m^{-3}$, and (b) $n(0) = 10^{19}m^{-3}$, (1) $T(0) = 100$eV, (2) $T(0) = 1$keV, (3) $T(0) = 10$keV and (4) $T(0) = 0$eV.

There, the ray refraction enlarges for increasing the parameter $\eta = X/2N_{||}\gamma$.

As is well known, the ECR is not directly accessible to the fast extraordinary mode propagating in direction of increasing magnetic field. This mode can reach the ECR only by tunelling through the evanescent region between the low-density (or right-hand) cutoff and the upper hybrid resonance or by mode conversion of O-waves into X-waves at the plasma cutoff layer. We have investigated the latter scheme and found a fairly good agreement with the results reported in[10]. The launching of extraordinary waves with frequency significantly smaller than the EC frequency is also examined. It is found that at large frequency down-shifts and low electron temperatures ($T_e \leq 5$ keV) the obtained ray paths almost coincide with those determined by using the cold plasma approximation[11].

Let us now examine the ray paths of high field side launched waves. The equation governing the poloidal wave number in this case changes sign at ECR layer. This induces further modifications of the ray path, as compared to the cold one: $d\phi/ds$, N_θ and $d\theta/ds$ change sign. Furthermore, the wave polarization in the plane perpendicular to **B** changes in the ECR region. For waves initialized to the O mode it changes from left to right. In general, when the wave leaves this layer its polarization comes back to the left one. In dense plasma the wave with large parallel wave refractive index suffers radial reflection with $N_r = 0$. As in the case of ordinary wave launching from the low field side, the ray reflection takes place at an electron density which is significantly smaller than the high-density plasma cutoff. After reflection, in cold plasma the rays cross the resonance. In tenuous thermal plasma the ray reflection disappears and the ray paths nearly coincide with those obtained in cold plasma. In dense thermal plasma however, the rays do not cross the resonance: they are bent back at the ECR layer. The wave is polarized quasi-circularly rotating in direction of the electron gyration. Similar ray behaviour is obtained at smaller $N_{\|}$-values that is, at larger values of the angle χ.

In Fig. 4(a,b) we show the poloidal views of ray paths initialized to the ordinary mode in weakly relativistic (a) tenuous and (b) finite-density plasmas. The

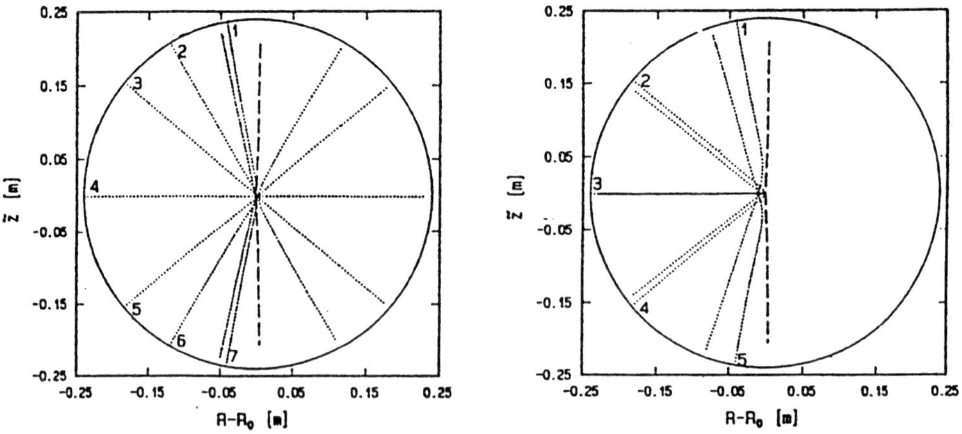

FIG. 4. Poloidal views of the O mode ray paths for $\chi = 80°$, $\psi = 180°$, T(0) = 1keV, (a) n(0) = $10^{17}m^{-3}$, θ = 100°(20°)140°(40°)220°(20°)260°, and (b) n(0) = $2x10^{19}m^{-3}$, θ = 100°(40°)260°.

waves are launched from the high magnetic field side (θ = 100°(20°)140°(40°)220°(20°)260°) in direction of the plasma centre (ψ = 180°) at an angle χ = 80°. For $n(0) = 10^{17}m^{-3}$ (Fig. 4(a)), that is in tenuous plasma, the rays cross the ECR and their paths nearly coincide with those obtained using the cold plasma dispersion equation. This does not apply for the rays launched from θ = 100° and θ = 260° which suffer reflection at the ECR layer. In this layer the wave is elliptically (right) polarized with a very elongated axe in the e_y direction. In contrast to the other ray paths represented in Fig. 4(a), the rays launched from θ = 100° and θ = 260°, change their polarization when they leave the ECR layer. Away from this layer their polarization becomes a quasi-linear one in the direction of **B**. Now, if the central electron density is increased to $n(0) \geq 10^{18}m^{-3}$, one goes over to the finite density regime and the ray path changes completely. As one can see, the rays reflect at the ECR layer. This is due to the aforementioned changes of N_θ and $d\theta/ds$ in this layer. The O-mode tends to be linearly polarized in the direction of **B**. In the plane perpendicular to **B**, for 100° ≤ χ ≤ 260° it rotates in the direction opposite to the electron gyration. Here we recall that it is the wave polarization that determines the strength of the ECR interaction.

We turn now to the rays initialized to the extraordinary mode. In the cold plasma description, the X mode which propagates in decreasing magnetic field, first encounters the ECR layer where, in general, it is partially absorbed, and then possibly the upper hybrid resonance (UHR) layer. Its polarization changes from right to left at the layer in which $N^2 = \epsilon_{11}$. This density layer moves towards the (high-field) plasma edge when the peak electron density is increased and/or the parallel wave refractive

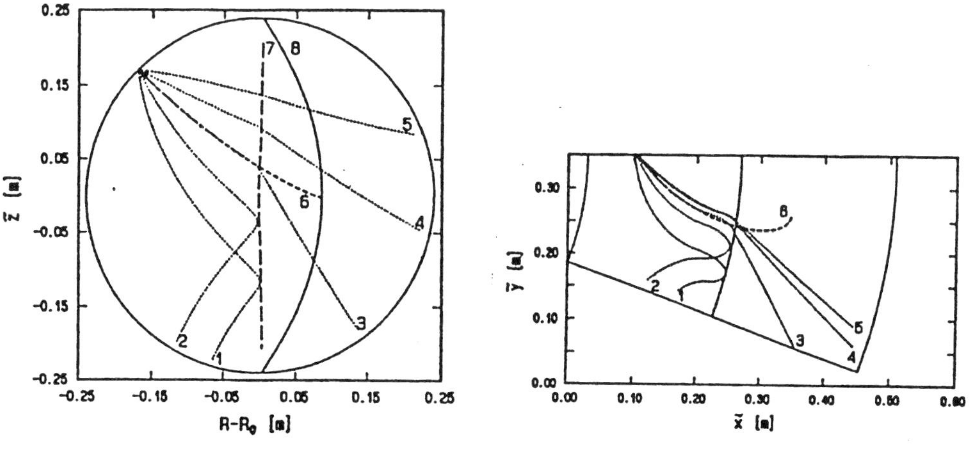

FIG. 5. Poloidal (a) and toroidal (b) views of the X mode ray paths for $n(0) = 10^{19}m^{-3}$, θ = 135°, χ = 45°, ψ = 140° (20°) 220°, T(0) = 1keV, and for (6) ψ = 180°, T(0) = 0eV.

index is decreased. Thus, for quasi-parallel propagation this mode is almost circularly (left) polarized except at the ECR layer where it is longitudinal. As $N_{||}$ decreases its polarization gradually becomes quasi-elliptical. A typical ray path obtained within the cold plasma approximation is shown in Fig. 5 (the curve labelled "6"). Now, if thermal efffects are taken into account the wave behaviour in the ECR layer changes completely. As the peak electron density increases and/or $N_{||}$ decreases the X-mode tends to be longitudinal in the direction of **B**. In the plane perpendicular to **B** it is elliptically polarized rotating in the direction opposite to the gyration of electrons. Depending on the values of the parameter η and the launching angles θ and ψ, the X-mode refracts or reflects at the resonance layer. This ray behaviour is illustrated in Fig. 5(a,b) where we show the poloidal and toroidal views of ray paths obtained in weakly relativistic plasma (with the exception of the ray path labelled "6"). The X mode is launched from the high field side at the point {a,135°,0°} in the direction χ = 45°. From the cold plasma wave theory we should expect that the X mode has a resonance at the UHR region. As a consequence of the change of polarization in the ECR layer, however, in tenuous plasma the X-mode does not "feel" the presence of the UHR, and thus it passes straight through. In finite density plasma the rays which cross the ECR behave similarly. In contrast to the tenuous plasma case however, they are also significantly refracted at the ECR layer. It turns out that in thermal plasma the UHR is not directly accessible to the X wave obliquely propagating in decreasing magnetic field. This finding supports experimental results in which the coupling of the X mode to the quasi-electrostatic electron cyclotron mode (Bernstein mode) has not been observed[12].

VI. CONCLUSIONS

We have investigated the ray tracing in the EC frequency range with the main objective of clarifying the influence of thermal effects on the ray paths. First, we have analysed qualitatively the wave behaviour near the ECR and we have identified the principal source of differences between the thermal plasma and cold plasma ray descriptions. This has been followed by a detailed numerical study of the ray paths in tokamak like plasmas in which a fully three dimensional code is applied to trace paths of rays initialized to the low or high field side launched ordinary or extraordinary modes in the ECR frequency range. This study has demonstrated the importance of thermal effects on the ray paths. Namely, it is found that thermal effects may induce a substantial ray refraction, ray reflection and/or change of the wave polarization near the ECR. Thus, the ray paths determined within the cold plasma approximation are considerably modified. Obviously, the wave power deposition along the ray is also altered. In a wide range of plasma parameters, the waves are efficiently damped in one pass through the ECR. The condition of weak absorption $|Im\mathrm{N}| \ll |Re\mathrm{N}|$ has been satisfied along most of the considered cases. The preliminary numerical results obtained within the new, complex treatment of the dispersion equation (see (6) - (7)) confirm the preceding conclusions.

As it was noted, some experimental results support the paths of the energy obtained in thermal plasmas. Nevertheless, the importance of the considered problem certainly suggests the undertaking of further experimental steps to check if the presented theoretical predictions have an impact on reality.

REFERENCES

[1] I.B.Bernstein, Phys. Fluids **18** (1975) 320.

[2] H.Weitzner and D.B.Batchelor, Phys. Fluids **23** (1980) 1359.

[3] I.B.Bernstein and L.Friedland, in *Handbook of PLasma Physics*, edited by A.A.Galeev and R.N.Sudan (North-Holand, Amsterdam, 1984), Vol. I, p. 367.

[4] A.Bravo-Ortega, Phys. Fluids **3** (1991) 529.

[5] I.Fidone, G.Granata and R.L.Meyer, Phys. Fluids **25** (1982) 2249.

[6] M.Bornatici, R.Cano, O.DeBarbieri and F.Engelmann, Nucl. Fusion **23** (1983) 1153.

[7] S.Pešić, Phys. Fluids **31** (1988) 115.

[8] S.Pešić and A.Stojić, J. Phys. Soc. Jpn **62** (1993) 2359.

[9] I.P.Shkarofsky, J.Plasma Phys. **35** (1986) 319.

[10] F.R.Hansen, J.P.Lynov and P.Michelsen, Plasma Phys. **27** (1985) 1077.

[11] E.Mazzucato, I.Fidone, G.Giruzzi and V.Krivenski, Nucl. Fusion **26** (1986) 3.

[12] R.Wilhelm,V.Erckman,G.Jansen,W.Kasparek,G.Muller,E.Rauchle,P.Schuller,K.Schworer M.Thumm and W.VII-A Team, Plasma Phys. and Contr. Fusion **26** (1984) 1433.

Low Temperature and
General Plasmas
M. Milosavljevic and Z. Petrovic

ISBN 1-56072-399-8

ON TEMPORAL VARIATIONS OF SOLAR SPECTRAL LINES

Ištvan Vince
Astronomical Observatory, Volgina 7, 11000 Belgrade

1. INTRODUCTION

The main objective of this paper is to outline the results on the observations and theoretical interpretations of the variations of solar spectral lines. The discussion will be limited to the Fraunhofer lines, i.e., to the absorption spectral lines in the visible region of solar spectra.

Almost all these spectral lines originate in the photosphere. Consequently, they carry information about the physical state and its variation in this tiny layer of the Sun. Since the photosphere is intimately connected with the convective zone, the variations of physical conditions in this layer must have an influence on Fraunhofer lines, too. Magnetohydrodynamical and thermal modulation of the internal structure of the Sun can be manifest in photosphere, among others, as changes in spectral lines. The energy generated in the solar core is transported through the photosphere. Therefore, the photosphere must also have an indirect connection with energy production in the core and its transport through the interior of the Sun.

The spectral lines observations show a variability on all time scales: from minutes to years and decades. The variations in the range from minutes to days (short-term variations) relate to the convective motions (granulation and supergranulation) and to the solar oscillations. The variations lasting for years and decades (long-term variations) are connected with dynamical and thermal modulation of the internal layers and are also linked to the evolution of magnetic fields over a 11-year activity cycle of the Sun.

This paper will present results of spectral line variations on these two very different time scales.

2. THEORETICAL BACKGROUND

A question that arises is what can be the essential cause of the temporal variation of solar spectral lines? Perhaps the best way to answer this question, briefly, is the study of the formal solution of the line profile formation process in photosphere.

The temporal variations of a spectral line profile are, in fact, the variations of intensity at a given frequency in a spectral line with time. These variations can yield the shift of the whole spectral line profile or the changes of its particular parts.

The emergent intensity at frequency ν can be given by following formal solution of the equation of the radiative transfer:

$$I_\nu(0,\mu) = \int_0^\infty \frac{1}{\mu} S(\tau_\nu) e^{-\tau_\nu/\mu} d\tau_\nu, \tag{1}$$

where $I_\nu(0,\mu)$ is the emergent intensity, $S(\tau_\nu)$ is the source function, τ_ν is the total optical depth and μ is the cosine of the heliocentric angle.

The source function (S_ν) is the ratio between emissivity and absorptivity and consists of contribution from the spectral line (S_ν^l) and continuum (S_ν^c)

$$S_\nu = \frac{k_l \cdot S_\nu^l + k_c \cdot S_\nu^c}{k_l + k_c}. \tag{2}$$

k_l and k_c are the absorption coefficients for the spectral line and the continuum radiation, respectively.

Usually spectral line profiles cover a relative small frequency interval, within which only the line source function changes considerably, while the continuum source function, practically, remains constant.

The source function for radiation in the line may be written as

$$S_\nu^l = \frac{2 \cdot h \cdot \nu^3}{c^2} \cdot \frac{1}{\frac{\mathrm{N_l}}{\mathrm{N_u}} \cdot \frac{g_u}{g_l} \cdot \frac{\varphi_\nu}{\Psi_\nu} - 1} \cdot \frac{\sigma_\nu}{\Psi_\nu}, \tag{3}$$

where N_l and N_u are the number of atoms per unit volume in lower (l) and upper (u) levels, g_l and g_u are the statistical weights of these levels. h is Planck's constant, c is the velocity of light, σ_ν is the spontaneous emission profile, Ψ_ν is the induced emission profile and φ_ν is the absorption profile. In local thermodynamic equilibrium

$$\frac{\mathrm{N_l}}{\mathrm{N_u}} \cdot \frac{g_u}{g_l} = e^{\frac{-h \cdot \nu}{k \cdot T}}, \tag{4}$$

where k is Boltzmann's constant, $\mathbf{T}$ is the temperature.

The total optical depth (τ_ν) has two components: the optical depth in the continuum (τ_ν^c) and in spectral line (τ_ν^l):

$$\tau_\nu = \tau_\nu^l + \tau_\nu^c. \tag{5}$$

The optical depth in the spectral line can be given by equation:

$$\tau_\nu^l(z) = \int_z^\infty N_a \cdot \alpha_o \cdot V(a, b) \cdot dz', \tag{6}$$

where z and z' are the geometrical depths, $\mathbf{N}_a$ is the number density of absorbers, and α_0 is the total absorption by one atom. The total absorption coefficient per unit volume ($\mathbf{N}_a\alpha_0$) is broadened, simultaneously, by Doppler effect (thermal, turbulent and other random motions) and different kinds of damping (radiation and different collision damping) over a finite frequency interval. At temperatures and pressures of solar photosphere one anticipate that the impact theory of collision broadening is applicable. Therefore, this combine broadening can be given by Voigt-function, $V(a, b)$, which is defined by two dimensionless parameters:

$$a = \frac{\gamma}{\Delta\nu_D}, \tag{7}$$

$$b = \frac{\nu - \nu_0 + d}{\Delta\nu_D}, \tag{8}$$

where

$$\Delta\nu_D = \frac{\nu}{c}\sqrt{\frac{2 \cdot k}{m} \cdot \mathbf{T} + \sum_i^n v_i^2}, \tag{9}$$

γ and d are the half-half width and the shift, respectively, caused by damping effects, $\Delta\lambda_D$ is the Doppler width and v_i are different random influences like microturbulent velocity, small scale magnetic field etc.

The broadening parameters are given as

$$\gamma = \sum_i^n \gamma_i \tag{10}$$

$$d = \sum_i^n d_i \tag{11}$$

where γ_i are different damping constants (e.g., radiation damping constant (γ_n), damping constant caused by collisions with neutral perturbers (γ_{pn}), damping constant caused by collisions with charged perturbers (γ_{pe}) etc.), d_i are different shifts (e.g., Doppler shift, shift caused by collisions with neutral or charged perturbers, by Zeeman effect of large-scale magnetic fields etc.)

The broadening and shift of the absorption coefficient due to absorber's collisions with neutral perturbers (γ_{pn} and d_{pn}) are usually related to different interaction potentials (e.g., van der Waals, Lennard-Jones, Smirnov-Roueff etc. potentials). Which of them one use depends on their validity for concrete transition and on availability of relevant atomic data for their calculation. The broadening and shift caused by charged particles can be calculated from the Stark broadening theories.

Generally these parameters can be given by a complex number:

$$\gamma_\diamond + i \cdot d_\diamond = 2 \cdot \pi \mathrm{N}_\diamond \int \mathbf{v} \cdot f(\mathbf{v}) \cdot d\mathbf{v} \cdot \int \prod(\rho) \cdot \rho \cdot d\rho \tag{12}$$

$$\diamond = pn, pe, ...$$

$$\mathrm{N}_\diamond = \mathbf{N_H}, \mathbf{N_e}, \mathbf{N_{H^+}}, \mathbf{N_{He}}...$$

Here $\mathbf{v}$ is the relative velocity of the absorbers with respect to the perturbers, f($\mathbf{v}$) is the velocity distribution function, $\prod(\rho)$ is a function that describes the interactions between absorbers and perturbers and depends on the collision parameter (ρ).

The velocity depends on the temperature:

$$\mathbf{v} = f(\mathbf{T}). \tag{13}$$

In thermodynamic equilibrium the degrees of ionization are given by Saha equation:

$$\mathbf{N_e} \cdot \frac{\mathbf{N_{r+1}}}{\mathbf{N_r}} = \frac{U_{r+1}(\mathbf{T})}{U_r(\mathbf{T})} \cdot g_e \cdot \frac{(2 \cdot \pi \cdot m_e \cdot k \cdot \mathbf{T})^{\frac{3}{2}}}{h^2} \cdot e^{-\frac{\chi_r}{k \cdot \mathbf{T}}}, \tag{14}$$

where $\mathbf{N}_r$, $\mathbf{N}_{r+1}$ and $\mathbf{N}_e$ are the number densities of ions in rth state of ionization, in the next higher state and electrons, respectively, m_e is the mass of the electron, χ_r is the ionization energy of the ion in rth state of ionization. U_r and U_{r+1} are the partition functions for $\mathbf{N}_r$ and $\mathbf{N}_{r+1}$, and g_e is the statistical weight for electrons.

In Equations 3 to 14. the solar photospheric quantities that influence the spectral line profiles are designed by bold face letters. The temporal variations of these quantities yield the temporal changes of spectral line profiles. Evidently the spectral line profiles are influenced by temperature, by abundance of chemical elements, by velocity, by magnetic and electric fields.

3. SHORT-TERM VARIATIONS OF LINE PROFILES

The smallest spatially resolved photospheric feature is the solar granulation. It is a cellular pattern which cover the entire solar surface and composes of bright elements (the granules) isolated by dark intergranular lanes. The bright granules are hot upwards-moving gas bubbles, while dark intergranular lanes are cooler, downwards-moving material. The granula, in fact, is the upper visible part of a convective cell a few hundred km in size, with a central upflow velocity of about 400 m/s having a lifetime of about 8 min and a temperature difference between granules and intergranular space about hundred K. The combination of high spectral dispersion with high spatial resolution spectrograms show a wiggly spectral line structure hundred different granules at different states of their evolution and at random positions, i.e. places with different brightness and velocities. Two parallel photometric scans along the opposite of a spectral line gives the Doppler velocity distribution along the slit position on the solar surface across the spectral line give the line profile variation along the slit. Comparison of the line profiles of granules and intergranular lanes shows that the granular lines are blushifted with respect to intergranular lines, in average, about 0.8 km/s, and that the intergranular profiles are always broader and line cores are brighter than the granular ones (Holweger and Kneer, 1989). This wiggly structure of spectral lines , due to time evolution of granules, varies with time.

The probability that at a given place on the solar surface the velocity-shift belongs to a given interval is taken as Gaussian. In other words, it means that the spectral line profile is varying randomly with time and in space too (along the spectrograph slit). Consequently, the time averaged or the spatially unresolved spectral line profile has a Gaussian component. In synthetic line profile calculations this component is considered through microturbulent velocity (eq. 9).

It has to be pointed out that in low spatial resolution spectral observations, due mainly to the influence of intensity and velocity distribution in garnulation, the spectral lines are blueshifted with respect to their laboratory wavelengths. Namely, the emerging (blueshifted) hot granules are brighter and larger than the descending (redshifted) intergranular lanes, i.e., the granules contribute more to the formation of the line profile than the intergranular lanes. Of course this is true only in vicinity of the center of solar disk. Observations show that this blueshift decreases from the center towards the limb of solar disk (this phenomenon is called the limb effect). The limb effect is mainly due to the decrease of the granular contrast and the line of sight component of vertical velocity in granules towards the limb. The transition from coherent to noncoherent flow that occurs about 150 km height in the photosphere (Hanslmeier et al., 1993) has an influence on the limb effect too.

Beside blueshift the spatially unresolved spectral line profiles show an asymmetry with the well known C-shaped bisectors. This asymmetry in great part originates also in granulation. There are other processes that can affect the line profile as, e.g., atomic collisions (Vince and Dimitrijević, 1989), but they are not so prominent. There are two in principle different theoretical models for explanation of asymmetry of spectral line profiles. According to the first one the asymmetry of spatially unresolved line profile is the result of the combination of symmetric line profiles originating in granule and in intergranular lane (Dravins et al., 1981). According to the second model the asymmetry of a spatially unresolved profile consists of more, highly asymmetric profiles (Steffen, 1989). Observations (e.g., Mattig et al., 1989) show that the asymmetry of the spectral line profile increases with enhanced spatial resolution in both the granular and in intergranular lane. This behaviour of observed spectral line profiles supports the second model. But, on the other hand, even the high-resolution observations are spatial averages over many different unresolved elements. So, the observed high-resolution line asymmetries may be a function of the spatial resolution and not the peculiarity of individual element (e.g., depth dependence of the velocity field in an element).

The global variation of granulation parameters (e.g., size, temperature, velocity etc.) with time will be manifested as a temporal variation of spectral line profiles.

Another type of short-term variation of spectral lines is caused by the 5-minute photospheric oscillations. The 5-minute oscillations are acoustic waves in a subphotospheric convection zone cavity with a large number (more than 10^7) of discrete modes. The amplitudes of a single modes are about 30 cm/s or less. These modes have periods of oscillation between about 3 and 12 minutes. This periodic velocity field manifests itself as a local Doppler shift variation of spectral lines with amplitude of about 0.5-1 km/s resulting from the interference of those 10^7 modes. Areas on the solar surface as large as 30000 km in size oscillate almost in phase. On the basis of measurements of spectral line shift variation a new method of study the Sun has been developed during the last few decades: the helioseismology.

The two different types of velocity fluctuations, the granulation and the 5-minute oscillations, contribute simultaneously to the observed line shifts. In principle, there are two possibilities to separate granular motions from 5-minute oscillations. The best method is to calculate the velocity power spectrum in the wavenumber-frequency ($k-\omega$) space. Namely, in $k-\omega$ plane the velocity power of 5-minute oscillation spectrum is distributed so that equal power contours form certain ridges, i.e., the power spectrum is not randomly distributed in this plane. Analysis of this distribution yields to components of oscillatory motions, while the residual noise of the velocity signal contains the granular velocity component. For this kind of analysis it is necessary to have observation time series. When time series are not available then a filtering process has to be introduce in spatial wavenumber with a cut-off at a given wavenumber ($k_0 \approx 2.5 Mm^{-1}$). 5-minute oscillations have smaller, while granules have larger wavenumbers than the cut-off value.

The 5-minute oscillations have also been observed even in the flux spectrum, i.e., in spatially unresolved sunlight. These oscillations often are called global oscillation. Global oscillation is low-degree mode of such oscillations, because the high-degree oscillations tend to cancel each other due to existence of larger number of areas oscillating in the opposite phase.

Recently the global low-degree oscillations are used to get information on the solar interior. For this purpose it is necessary to have high frequency resolution. This resolution is determined by the duration of observation.

High-spectral resolution observations show that the 5-minute oscillations cause not only the shift of the whole spectral line profile but they influence its particular parts too. So, the well-known C-shape of spectral line bisectors changes during 5-minute period. The variation of the asymmetry of the spectral line profile anticorrelates with the oscillations of the whole line profile. The 5-minute oscillation does not affect equally the two line flanks: at each intensity level lower than 70% the rms velocity of the red wing is smaller than that one of the blue wing. Amplitude of the profile oscillation is increasing with decreasing residual intensities.

The observed differences between the oscillations of the line flanks can be explained by phase lag greater than 90 degree between the temperature and velocity wave. This phase-lag between the temperature and velocity can be caused by radiative damping in photosphere (Gomez et al. 1987). The radiative relaxation time in the upper photosphere is the same order as the wave period and the oscillation is nearly adiabatic (the phase difference between temperature and velocity is about 90 degree), while in the low photosphere the radiative relaxation time is short (few seconds) and the temperature perturbations are strongly damped (the phase difference between temperature and velocity is nearly 180 degree).

4. LONG-TERM VARIATIONS OF LINE PROFILES

The long-term variations of line profiles are usually connected with the 11-year solar activity cycle. But, spectroscopic measurements during a time interval longer than the solar activity cycle indicate a possible secular variations in the spectrum line parameters and recent high quality spectral measurements show rotation modulation of spectral line parameters, too.

In the past several decades many authors have been studied the activity-dependent variations of spectral line profiles. Kharadze (1935) was, probably, the first investigator who had measured some activity-dependent changes of the widths of H_δ and H_γ spectral line profiles.

Later Derviz et al. (1961), observing 31 Fraunhofer lines of the integrated solar radiation in the period 1951-1959, found an activity-dependent increase of the central depth of spectral lines of various origin. No changes of equivalent widths have been found.

Based on spectral observation of the center of the solar disk, Mitchell (1969) found a very weak connection of the central intensity in a spectral line with Wolf sunspot number.

Zhukova and Mitrofanova (1973) established such a dependence for period 1964-1972. However, covering a part of this time interval (1969-1972) Krat et al. (1975) did not corroborate the results of Zhukova and Mitrofanova.

Studying the Sun as a star at two epoch with different levels of floccular activity and general magnetic field, Stepanyan and Shcherbakova (1979) extracted 63 spectral lines, out of 1000 observed, showing considerable changes of their central depths, half-widths and equivalent widths. About 75% of lines from that sample showed an increase, and 25% a decrease of, at least, one of the observed three parameters. The authors' attempt to find a connection among changes of equivalent widths, excitation potential of the lower level of the electronic transitions and the mean optical depths of formation of spectral lines gave no results.

Livingston and Holweger (1982) analysed the sensitivity of seven spectral lines to photospheric parameters based on the full disk spectrophotometric data from the period 1976-1980. They found a secular decrease of equivalent widths with time up to 2.3 % and conclude that the lines' weakening are due to global variations of surface properties (these results have recently not been confirmed).

By observations from 1969 till 1979 Kokhan (1987) found that the central depth, half-width and equivalent width of spectral lines change in dependence on the solar activity. The most sensitive parameter was the equivalent width. A certain dependence of all three observed quantities on the excitation potential of lower transition levels has been noticed: the lower the excitation potential, the larger activity-dependent changes of spectral line parameters are.

Livingston and Wallace (1987) applied improved corrections for water vapour and an instrument slit change on full disk spectrophotometric records for selected spectral lines observed during the period 1976-1985. They found that the equivalent width of the MnI 539.47 nm line show the greatest variability (about 2%) and anticorrelates with the CaK_3 393.3 nm line intensity changing. The CI 538.03 nm line after these corrections becomes almost constant. Other lines among the observed ones showed little or no change with time what is an opposite result to that given by Livingston and Holweger (1982). Recent results by Livingston (1992) show that strong metallic lines (with equivalent widths greater than 10 pm) have little or no cycle variations, while the intermediate strong metallic lines (with equivalent widths from 4 to 8 pm) vary about 1%. The CI 538.03 nm line show a monotonic increases in equivalent width from 1979 to 1992 (0.03%/year). No solar cycle modulation was detected. Similar, but not so prominent secular change was observed in the case of the FeI 5397.9 nm line (0.01%/year). This behaviour of the temperature sensitive CI line (it is formed almost at continuum level) and the FeI 537.9 nm line suggest that a component of solar irradiance variation exists which is independent of the solar activity. Namely, such a secular change of equivalent width may be caused by the change of the temperature gradient in solar photosphere, implying a change in the convective energy transport causing a solar irradiance change. There are other observational data suggesting that there may be secular irradiance variations that are not correlate with solar activity (e.g., changes in limb darkening, changes in solar limb brightness etc.). If the solar irradiance is varying in a really secular manner, it could have some effect on the climate of the Earth.

Babij (1991) found that the scattering of observed central depth of moderately strong spectral lines decrease with an increase of excitation potential, while for weak lines such a dependence is absent. This behaviour of spectral lines could be explained by existence of one unstationary photospheric layer above a stationary one. The variation of this unstationary layer causes time variation of the central depth of some spectral lines.

Theoretical works and model calculations of the heat transfer process from the solar interior show that the presence of magnetic fields in the convective layer depresses convection. That can be seen as a changeable asymmetry of spectral line profiles. For example, the FeI 525.06 nm line in active regions is more symmetric than out of them (Livingston 1982). He also found the same effect observing the Sun as a star at two different activity phases. However, then the asymmetry of the observed spectral line kept decreasing even after the activity maximum. The phenomenon is explained by the author as a diffusion of local magnetic fields (highly concentrated during the maximum activity phase) into considerably large areas after the maximum. This diffuse, or general solar magnetic field disturbs the line asymmetry of the integrated sunlight more than the local fields in the active regions. Changes in line parameters as a function of the magnetic filling factor have been investigated by Brandt and Solanki (1990). The filling factor was determined by them from a statistical analysis of Zeeman-sensitive line widths, assuming that the magnetic field strength in fluxtubes was independent of the filling factor. The presence of small magnetic fluxtubes yields to inhibition of granular convection (the granule becomes smaller and convectively less effective). The line depth decrease, while the line width increases with the increasing filling factor. The lower part of the bisector is straightening out. The equivalent width of some lines (mostly strong or temperature insensitive lines) remains unchanged, while decreases for the other ones (weak lines). Similar results have been found by Immerschitt and Schröter (1989) who analysed the behaviour of line profile parameters of the photospheric FeI 557.61 nm line (g=0) in regions of different magnetic activity.

A similar conclusion has been derived by Cavallini et al. (1986) from annual mean values of the observed asymmetries of FeI 629.78 nm, 630.15 nm and 630.25 nm spectral lines at the center of the solar disk during 1983 and 1984. It should be noticed that these two years belong to the descending phase of solar activity cycle when the number of tiny magnetic flux tubes (facular points), varying in antiphase with the sunspot number, increases. Besides, the authors pointed out the increasing of the width of both the 630.15 nm and the 630.25 nm spectral lines, especially near the continuum level. This result is in accordance with activity cycle dependence of the line equivalent widths found by Livingston and Holweger (1982).

Calculating the influence of convective motion on the spectral line profiles, Dravins et al. (1981) concluded that the main cause of their asymmetry is a specific distribution of plasma velocities and temperature within the convective cells. A decrease of convective velocity leads to smaller line profile asymmetry. It should be noticed that a part of the asymmetry in some spectral lines may be caused by interatomic collisions of absorbers and perturbing particles (Vince and Dimitrijević, 1989). In such cases a change of plasma parameters will introduce a change in the asymmetry of spectral line.

Since the question of correlation of spectral line parameters with the solar activity has not been finally solved, a long-term program of full solar disk observations of photospheric spectral lines was initiated in 1986 at Belgrade Astronomical Observatory for the study of variations of equivalent widths, half-widths and central residual intensities of spectral lines during the 22nd solar activity cycle (Arsenijević et al., 1987). Routine observations are in progress since 1987.

The observations are performed with the equatorial solar Litrow-type spectrograph of 9 m effective focal length and dispersion about 70 mm/nm in 5th order, with a Baush & Lomb replica grating (154x206 mm^2, 600 lines/mm)(Kubičela, 1975). Using three flat mirrors an unimaged beam of sunlight is fed through the 100-150 μm wide entrance slit to give the solar integrated flux spectrum. For the purpose the spectrograph was converted into a scanning monochromator in 1986 (Arsenijević et al., 1988). The scanning is realized by optically shifting the image of the spectrum with respect to the exit slit.

The choice of the spectral lines to be observed was, in general, governed by several criteria according to which a list of 31 spectral lines has been compiled (Vince et al., 1988).

The first results for 10 spectral lines observed in period from 1987-1990 were published by Skuljan et al. (1992) and it was found that some equivalent widths are constant and some change progressively. In another paper (Skuljan et al., 1993) the observational results of equivalent width variations of 18 selected Fraunhofer lines of solar flux during the years 1987–1992 (so far, our observations cover the rise from short time after the minimum through the maximum and a part of the decreasing wing of the 22nd solar activity cycle) were presented. After the elimination of systematic day-to-day errors using a set of spectral lines that do not change with time, a second degree polynomial fit to the data had been drawn for each line from where the moments of maximum equivalent width (T) were found. Then, taking 3.8 years as the mean solar activity rise time (Coffey, 1993), a cosine–wave fit, starting at T for each respective spectral line, has been applied to the same data and the activity minimum-to-maximum changes of the equivalent widths have been evaluated. Using the triple r.m.s. error of the single cosine-wave amplitude as the criterion, significant changes have been found in 8 spectral lines, out of 18. All these equivalent widths reach their maximum values near the time of solar activity maximum. The relative changes range between 5% and 14%. The other Fraunhofer lines do not change its equivalent widths or the changes are not reliably measured now.

So far we do not have a proper theoretical explanation of the behaviour of these lines. However, some further steps toward its understanding have been undertaken. Erkapić and Vince (1993a) theoretically studied changes of the equivalent widths, full half-widths and central fluxes of a set of Fraunhofer lines (comprising the observed ones) as functions of variations of photospheric vertical temperature gradient. They found two types of equivalent width changes. (i) All single-ionized atom lines and the CI 538.03 nm line increased the equivalent width with increasing temperature gradient. (ii) Spectral lines of all neutral atoms, except the mentioned CI line, exhibited the opposite changes. On the other hand, Erkapić and Vince (1993b) examining the same sample of Fraunhofer lines, found a negligible dependence of the spectral line profile parameters on variations of the photospheric pressure gradient.

One might expect that further studies of temperature dependence of the spectral line profile parameters on excitation and ionization energy can lead toward understanding of global time-dependent spectral variations.

From this briefly reviewed results of various authors it has been seen that the central intensity, half-width, equivalent width and asymmetry of Fraunhofer lines change in dependence on the solar activity. The results of various authors are very different and some of them are even contradictory (e.g., Zhukova and Mitrofanova (1973), and Krat et al. (1975)). Since the question of correlation of spectral line parameters with solar activity has not been finally solved, further studies of variability of Fraunhofer lines can be taken as scientifically interesting.

REFERENCES

Arsenijević J., Karabin M., Kubičela A., Vince I.: 1987, in Solar and Stellar Physics, Fifth European Meeting on Solar Physics (April 27–30, 1987, Titisee), P5.

Arsenijević J., Kubičela A., Vince I., and Jankov S.: 1988, *Bull. Obs. Astron. Belgrade* **138**, 1.

Babij B.T.: 1991, *Kinematika i fizika nebesnyh tel* **712**, 16.

Brandt P. N. and Solanki S. K.: 1990, Astron. and Astrophys. **231**, 221.

Cavallini F., Cepatelli G. and Righini A: 1986, Astron. and Astrophys. **158**, 275.

Coffey, E.: 1993, Solar-Geophysical Data **584**, 31.

Derviz T.E., Kuprevich N.F., and Mitrofanova I.A.: 1961, *Astron. Astrophys.* **38**, 448.

Dravins D., Lindegren L., and Nordlund A.: 1981, Astron. and Astrophys. **96**, 345.

Erkapić, S. and Vince, I.: 1993a, *Publ. Obs. Astron. Belgrade* **44**, 29.

Erkapić, S. and Vince, I.: 1993b, *XVI Summer School and International Symposium on the Physics of Ionized Gases (SPIG)*, (ed. M. Milosavljević), Contributed papers, p. 369, Belgrade.

Gomez M.T., Marmolino C., Roberti., and Severino G.: 1987, Astron. Astrophys. **270**, 169-177.

Hanslmeier A., Nesis A., and Mattig W.: 1993, Astron. Astrophys. **270**, 516-521.

Holweger H. and Kneer F.: 1989, in Solar and stellar granulation (eds. R.J. Rutten and G. Severino), Kluwer Academic Publishers, Dordrecht, Boston, London, p. 173-186.

Immerschitt and Schröter .: 1989, *Astron. and Astrophys.* **208**, 307.

Kharadze E.K.: 1935, *Zeitshr. Astrophys.* **10**, 339.

Kokhan E.K.: 1987, *Izv. Glav. Astron. Obs. Pulkove* **204**, 81.

Krat V.A., Kokhan E.K., and Pechinskaya N.T.: 1975, *Izv. Glav. Astron. Obs. Pulkove* **193**, 3.

Kubičela A.: 1975, *Publ. Astron. Obs. Belgrade* **20**, 47.

Livingston W.: 1982, *Nature* **297**, 208.

Livingston W.: 1992, in Proceedings of the Workshop on the Solar Electromagnetic Radiation Study for Solar Cycle 22 (ed. R.F. Donnelly) p. 11.

Livingston W. and Holweger H.: 1982, *Astrophys. J.* **252**, 375.

Livingston W. and Wallace L.: 1987, *Astrophys. J.* **314**, 808.

Mattig W., Hanslmeier A., and Nesis A.: 1989, in Solar and Stellar Granulation (eds. R.J. Rutten and G. Severino), NATO ASI Series: Mathematical and Physical Sciences, Vol. 263, Kluwer Academic Publishers, Dodrecht, Boston , London, p. 187.

Mitchell W.E.: 1969, *Astrophys. J.* **155**, 709.

Skuljan J., Karabin M., Vince I., and Kubičela A.: 1992, *Bull. Astron. Belgrade* **145**, 1.

Skuljan, J., Kubičela, A., Vince, I., Arsenijević, J. and Popović, L. Č., 1993, *Publ. Obs. Astron. Belgrade* **44**, 37.

Steffen M.: 1989, in Solar and Stellar Granulation (eds. R.J. Rutten and G. Severino), NATO ASI Series: Mathematical and Physical Sciences, Vol. 263, Kluwer Academic Publishers, Dodrecht, Boston , London, p. 425.

Stepanyan N.N. and Shcherbakova Z.A.: 1979, *Izv. Krym. Astrophys. Obs.* **59**, 91.

Vince I. and Dimitrijević M.: 1989, in Solar and Stellar Granulation (eds. Rutten R.J. and Severino G.), Kluwer Academic Publishers, p. 93-97.

Vince I., Kubičela A., and Arsenijević J.: 1988, *Bull. Obs. Astron. Belgrade* **139**, 25.

Zhukova L.N. and Mitrofanova L.A.: 1973, *Sol. Dann.* **6**, 65.

Low Temperature and
General Plasmas
M. Milosavljevic and Z. Petrovic

ISBN 1-56072-399-8

SOLUTION OF THE NON-LTE TRANSFER PROBLEM USING THE METHOD OF ITERATION FACTORS

O. Atanacković-Vukmanović[1] *and E. Simonneau*[2]
[1] Astronomical Observatory, Volgina 7, 11050 Belgrade, Yugoslavia
[2] Institut d' Astrophysique, 98bis Boulevard Arago 75014 Paris, France

1. Introduction

Since our knowledge of stars is derived from observing the radiation from their atmospheres, radiative transfer represents a fundamental problem in theoretical astrophysics and an important diagnostical tool. As is well known, stellar spectra contain a wealth of information on the physical properties of stellar atmospheres. The spectral features are formed in the interactions between the radiation field and the atmospheric gas during the radiative transfer through the outermost layers of the stars. Therefore, for an accurate interpretation of the spectra a reliable theory of radiative transfer is required.

Due to a free escape of the photons, radiation field in stellar atmospheres is highly anisotropic and deviates a lot from the equilibrium one. Moreover, in such rarefied media where the radiative processes dominate the collisional ones radiation field makes a strong influence on the excitation state of the atmospheric gas. Hence, atomic level populations are determined not only by the local thermodynamic parameters (LTE approximation) but by the radiation field as well. In other words, in a fully self-consistent (non-LTE) treatment of radiative transfer, the level populations are to be determined with the radiation field by a simultaneous solution of the transfer equations (that describe the transport of radiation) and the equations of statistical equilibrium (that describe the interactions of the radiation field and the gas).

Mathematical approach to the solution of this problem is determined by the dependence form of the atomic level populations on the radiation field. If this dependence is explicit, as in the case of the line formed by the two-level atom, it is possible to formulate radiative transfer problem in an explicit integro-differential form. The set of coupled integro-differential equations can be solved either by direct (one step) or iterative methods. On the contrary, in the case of the multi-level-atom line radiative transfer, due to a mutual coupling among different atomic levels and the corresponding radiative transitions, the aforementioned dependence cannot be cast in an explicit form. To solve such implicit non-linear problems an iterative technique must be applied.

2. Method of iteration factors

The most straightforward iteration scheme, so-called Λ iteration solves the radiative transfer (RT) and the statistical equilibrium (SE) equations in turn. However, the convergence of this simple procedure is infinitely slow when applied to the typical non-LTE conditions. This is due to numerically unfortunate treatment of the wing photons - the only active protagonists of the radiation transport at large optical depths.

This paper deals with a new, simple and fast-convergent method of iteration factors developed to solve the non-LTE line formation problem. The method revises the classical Λ iteration scheme by the use of proper quasi-invariant functions. Small modifications (and corresponding small computational efforts) of the Λ iteration lead to a rapid convergence to the exact solution.

The root of the method can be found in Feautrier's (1964) suggestion that model atmosphere computations would be made faster using the ratio of two radiation field intensity moments. Auer and Mihalas (1970) applied successfully the idea to the solution of the monochromatic (gray case) transfer problem, by introducing the variable (depth-dependent) Eddington factors (VEFs) in a simple iteration scheme. Defined as the ratios of the third to the first angular moment of the radiation field (i.e., K_ν/J_ν), the VEFs are closer to the exact values than are the corresponding current estimates of J_ν and K_ν.

However, Feautrier's idea of using good quasi-invariant properties of the homologous physical quantities ratios was not applied to a more general class of problems before the paper by Borsenberger et al. (1987). In this study on resonance line photons transfer together with the transport of excited atoms, the idea was applied to the other variables (frequencies and velocities), and two kinetic equations, coupled by their source terms, were solved using the relevant factors.

The method of iteration factors to compute the self-consistent temperature correction for a plane-parallel LTE stellar atmosphere model in the radiative equilibrium was successfully applied by Simonneau and Crivellari (1988). The same was done for the case when the convective transport is present (Crivellari and Simonneau, 1991). Fieldus et al. (1990) generalized the method to include spherically extended line-blanketed model atmospheres.

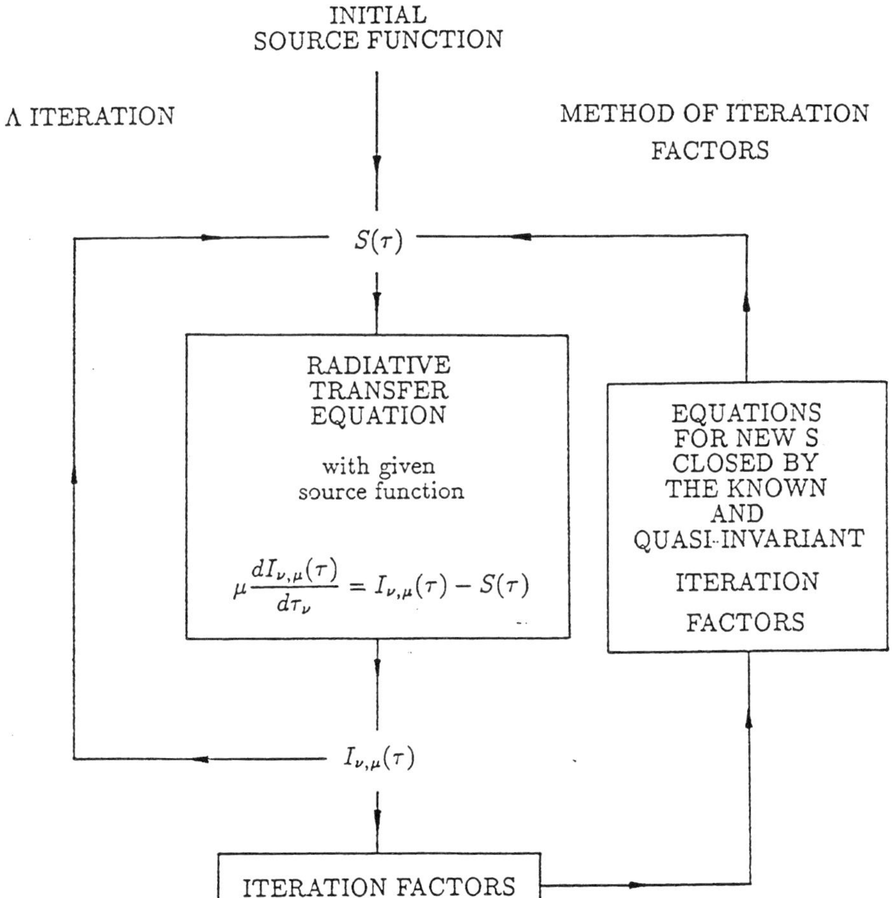

Fig.1. Flow-chart of the iteration factors method.

In the line formation problem considered in this paper the factors are defined accordingly to the physics of the problem and the form of the corresponding equations.

Computed at the beginning of each iteration step from the formal solution of the RT equation (with the given source function), the iteration factors are used as the known coefficients of the equations for the new source function (see Fig.1).

Defined as quasi-invariants of the problem, they change only little from one iteration to another and depend only slightly on the choice of the initial source function, hence becoming almost exact in the first few iterations. As the exact coefficients of the equations for the new source function they lead to the exact solution very quickly.

The major advantage of the method comes from the fact that it does not require any matrix algorithm for the solution of the RT equation.

The form of the equations for getting the new values of the source function as well as the form of the iteration factors is determined by the problem considered. In the explicit, two-level-atom like cases, these equations are the moments of the RT equation closed by the factors defined as the ratios of the corresponding intensity moments. In the implicit, multi-level-atom like cases, new source functions have to be obtained from the modified statistical equilibrium equations, "linearized" by the use of some closure relationships playing the role of iteration factors.

3. Two-level-atom line transfer problem

In the first place let us consider the most simple case of a line formed by a two-level atom in a plane-parallel, semi-infinite and constant property medium. The consideration of this simple instance is of two-fold importance. Firstly, its solution represents the solution of a whole class of similar explicit problems and secondly, we can test our method against its known exact solution.

We can start from the explicit form of the two-level-atom line source function

$$S(\tau) = \varepsilon B(\tau) + (1-\varepsilon) J_\varphi(\tau) \quad . \tag{1}$$

Here, ε is the standard non-LTE parameter, $B(\tau)$ is the Planck function and $J_\varphi(\tau)$ is the scattering integral

$$J_\varphi = \frac{1}{2} \int_{-\infty}^{\infty} \varphi_x dx \int_{-1}^{1} I_{x,\mu} d\mu \tag{2}$$

that couples the radiation field intensities at each frequency x (x being frequency displacement from the line center in Doppler units) and in each direction μ. φ_x is the normalized profile function determined by different broadening mechanisms.

The explicit form of the source function (1) implies that an improved source function can be obtained from the moments of the RT equation.

If we perform the integration of the RT equations:

$$\mu \frac{d}{d\tau} I_{x,\mu}(\tau) = \varphi_x [I_{x,\mu}(\tau) - S(\tau)] \tag{3}$$

first over angles (by applying the operators $(1/2) \int d\mu$ and $(1/2) \int \mu d\mu$, respectively) and then over frequencies (applying $\int \varphi_x^2 dx$ to the first equation and $\int \varphi_x dx$ to the second one) we get the system of two first-order differential moment equations:

$$\begin{aligned} \frac{dH_{\varphi^2}}{d\tau} &= J_{\varphi^3} - \varphi^3 S \\ \frac{dK_{\varphi}}{d\tau} &= H_{\varphi^2} \ . \end{aligned} \tag{4}$$

In order to close the system we need two additional relations among four unknown intensity moments.

As the first relation we can take the ratio:

$$F = \frac{K_\varphi}{J_\varphi} \tag{5}$$

that will be reffered to as the generalized Eddington factor similar to the variable Eddington factor in the monochromatic transfer problem. This factor takes into account the anisotropy of the radiation field and as the ratio of two similar physical quantities it is expected to be a good quasi-invariant of the problem.

The second closure relation must involve the remaining intensity moments. Since they have been obtained by frequency quadratures, weighted by different powers of the profile φ_x, the relation required will describe the frequency distribution of the intensity over the line, taking into account the given opacity profile. Thus an efficient iterative correction towards the exact solution, throughout the whole medium and especially at great optical depth, will necessarily require the description of the radiation field distribution over the line frequencies as good as possible. Our aim is to bring all these information into a single *scalar* closure relation.

Among several closure relations that can be introduced at different sophistication degrees, the most general relation which takes into account all the physics of radiative transfer (anisotropy of the radiation field via separate treatment of the inward and outward directed radiation streams, redistribution of the radiation over the line frequencies and the most active part of the transport given by the wing photons) is given by:

$$J_{\varphi^3} = f_J J_\varphi + f_H H_{\varphi^2} + f_S S \ . \tag{6}$$

Its coefficients can be easily derived from what is the best numerical simulation of the transfer process - a formal solution of the RT equation. Describing the evolution of the specific intensity at each frequency x along each direction μ, the formal solution of Eq.(3) can be written for the two streams of the radiation field - the outgoing $I^+_{x\mu}(\tau)$ and the incoming $I^-_{x\mu}(\tau)$ one, defined on the respective μ- intervals, [0,1] and [-1,0], in the form:

$$I^+_{x\mu}(\tau) = \int_\tau^\infty S(t)\exp\left(-(t-\tau)\varphi_x/\mu\right)\frac{dt\varphi_x}{\mu} \ , \ 0 \le \mu \le 1 \tag{7a}$$

$$I^-_{x\mu}(\tau) = -\int_0^\tau S(t)\exp\left((\tau-t)\varphi_x/\mu\right)\frac{dt\varphi_x}{\mu} \ , \ -1 \le \mu \le 0 \ . \tag{7b}$$

All the moments appearing in the system (4) can be easily computed in terms of the above intensities. Integrating Eqs. (7a) and (7b) over angles and frequencies, we can write a relation between any pair of "two-stream" moments and the relevant "full" moment:

$$Q_{\varphi^n} = \frac{1}{2}(Q^+_{\varphi^n} + Q^-_{\varphi^n}) \ .$$

If we now rewrite Eqs. (7a) and (7b) in the form:

$$I^+_{x\mu}(\tau) = S(\tau) + \int_\tau^\infty \left(\frac{dS}{dt}\right) e^{-(t-\tau)\varphi_x/\mu} dt \tag{8a}$$

$$I^-_{x\mu}(\tau) = S(\tau)(1 - e^{-\tau\varphi_x/\mu}) - \int_0^\tau \left(\frac{dS}{dt}\right) e^{-(\tau-t)\varphi_x/\mu} dt \ , \tag{8b}$$

we see in the above expressions the first term on the right-hand side represents the contribution of the local source function to the radiation field at any given depth point, whereas the integrals involving the first- (and higher-) order derivatives account for the non-local character of the transfer problem. In both cases the local ("passive" in radiation transport) terms consist of the source function, multiplied by a coefficient that introduces the main scale length of the transfer problem, namely the exponential decay due to the escape of the photons through the boundary surface.

Since the basic idea is to isolate the "passive" photons, we can introduce the non-local ("active-transfer") terms in the following way:

$$\tilde{I}^{+}_{x\mu}(\tau) = I^{+}_{x\mu}(\tau) - S(\tau) \tag{9a}$$

$$\tilde{I}^{-}_{x\mu}(\tau) = I^{-}_{x\mu}(\tau) - S(\tau)(1 - e^{-\tau\varphi_x/\mu}) \ . \tag{9b}$$

The elimination of the *known* "passive" term of the radiation field is already possible at the beginning of each iteration, when the formal solution is performed.

As we wish to iterate only on the ratios of the non-local ("active-transfer") contributions of the intensity moments, we are led to define the iteration factors as follows:

$$\alpha^{\pm} = \frac{\tilde{J}^{\pm}_{\varphi^3}}{\tilde{J}^{\pm}_{\varphi}} \qquad \theta^{\pm} = \frac{\tilde{H}^{\pm}_{\varphi^2}}{\tilde{J}^{\pm}_{\varphi}} \ . \tag{10}$$

Performing the integration of expressions (9) over angles and frequencies, we can rewrite the factors (10) as:

$$\alpha^{+} \equiv \frac{J^{+}_{\varphi^3} - S\varphi^3}{J^{+}_{\varphi} - S} \qquad \alpha^{-} \equiv \frac{J^{-}_{\varphi^3} - S(\varphi^3 - M_{03})}{J^{-}_{\varphi} - S(1 - M_{01})}$$

$$\theta^{+} \equiv \frac{H^{+}_{\varphi^2} - S\varphi^2/2}{J^{+}_{\varphi} - S} \qquad \theta^{-} \equiv \frac{H^{-}_{\varphi^2} - S(\varphi^2/2 - M_{11})}{J^{-}_{\varphi} - S(1 - M_{01})} \ . \tag{11}$$

Here, the $M_{mn}(\tau)$s are the frequency integrated integro - exponential functions:

$$M_{mn}(\tau) \equiv \int \varphi^n_x dx \int_0^1 d\mu \mu^m e^{-\tau\varphi_x/\mu} = \int \varphi^n_x E_{m+2}(\varphi_x, \tau) dx. \tag{12}$$

The factors $\alpha^{\pm}$ and $\theta^{\pm}$ account separately for the behaviour of the non-local parts of the radiation field, both outgoing and incoming. The $\alpha^{\pm}$ factor accounts for the frequency redistribution between the line core and the wings. The factor $\theta^{\pm}$ represents the redistribution over directions and, to some extent, also over frequencies. Both are considered as "form factors".

Known the source function, the factors are to be directly computed according to their definitions (11) from Eqs. (7a) and (7b) at the first step of each iteration. Once the $\alpha^{\pm}$s and $\theta^{\pm}$s are computed at each optical depth, their definition can be used to obtain, after some algebra, the coefficients of the second closure relationship for the system (4):

$$J_{\varphi^3} - \varphi^3 S = f_J(J_{\varphi} - S) + f_H H_{\varphi^2} + f_S S \ . \tag{13}$$

The coefficients f_J, f_H and f_S are simply related to the iteration factors by the following:

$$f_J = \frac{\alpha^+\theta^- + \alpha^-\theta^+}{\theta^+ + \theta^-}$$

$$f_H = \frac{\alpha^+ - \alpha^-}{\theta^+ + \theta^-}$$

$$f_S = \frac{1}{2}(f_J M_{01} - M_{03} - f_H M_{12}) \,. \tag{14}$$

In the frame of the "two-stream" model, the coefficient f_H is a measure of the anisotropy of the radiation field. The anisotropy is explicitly taken into account by the separate treatment of the outward and inward intensity moments. The coefficient f_J plays the rôle of a diffusion coefficient. At great optical depths f_J tends to φ^3. The coefficient f_S contains the frequency integrated integro-exponential functions reproducing the kinetic behaviour of the transfer process. The main scale length to describe the radiation field is introduced explicitly in the definition of the iteration factors, so that they carry on only second-order properties from one iteration to another. The correct behaviour of the solution at great optical depths is provided by the corresponding differences of the intensity moments and the local source function. Therefore, a high quality procedure may be expected.

The solutions confirm the predictions about the good quality of the method (Atana cković-Vukmanović and Simonneau, 1993). Hereinafter we give the solutions obtained assuming pure Doppler broadening. The use of the Voigt profile function does not lead to any substantial difference in the convergence properties of the method.

The behaviour of the four iteration factors $\alpha^{\pm}$ and $\theta^{\pm}$ under the extreme non-LTE conditions ($\varepsilon = 10^{-8}$) is presented in Fig. 2. The iterative procedure is started with an initial estimate obtained from a simple analytical consideration on a two-stream equilibrium model. Even with such initial values the convergence is achieved within a small number of iterations.

The source function for the case $\varepsilon = 10^{-8}$ is given in Fig.3. in the run of iterations. It clearly appears that even for this extreme non-LTE case, the source function approaches its exact value already in the third iteration. The error on the solution is less than 1%.

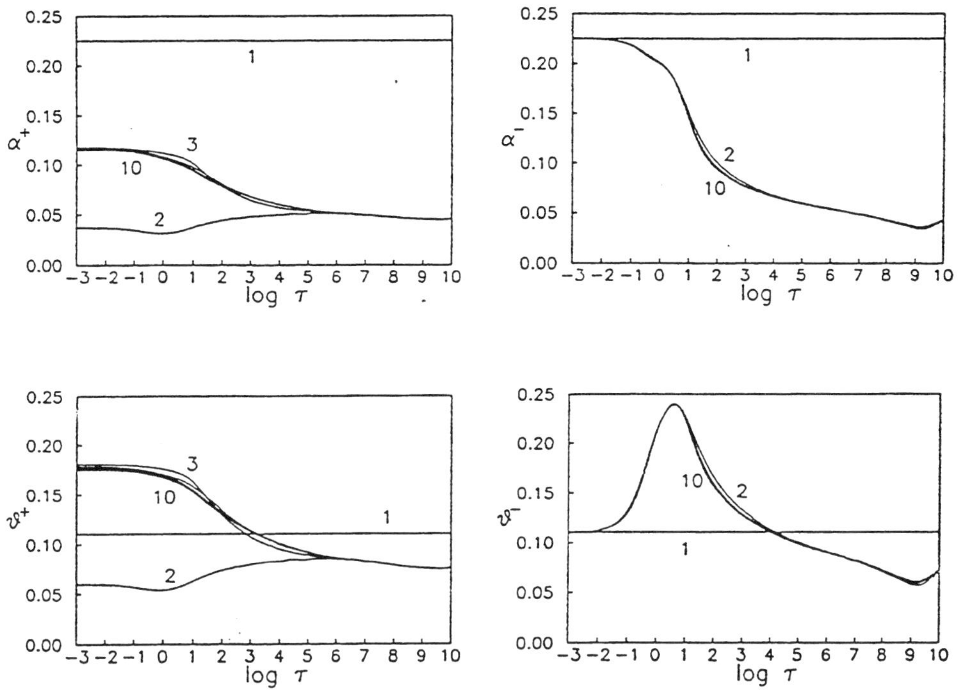

Fig.2. The variation of the iteration factors $\alpha^{\pm}(\tau)$ and $\theta^{\pm}(\tau)$ for the case $\varepsilon = 10^{-8}$ and $B = 1$ in the labelled number of iterations.

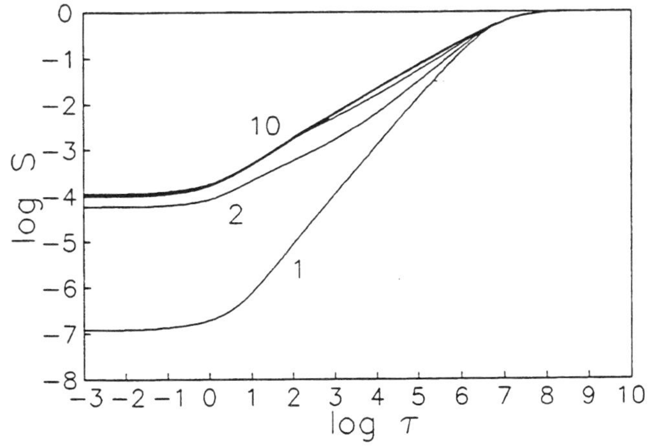

Fig.3. The convergence of $S(\tau)$ for the case $\varepsilon = 10^{-8}$ and $B = 1$ in the labelled number of iterations.

4. Multi-level-atom line transfer problem

The mutual coupling of particular radiative transitions within a multilevel atom implies an *implicit* dependence of any line source function (ratio between any two level populations) on the corresponding radiation field. Roughly speaking, three approaches have been developed so far to solve this non- linear problem. The first one is the complete linearization method of Auer and Mihalas (1969), where the linearization of all equations over the relevant variables is made prior to their simultaneous solution. The second group involves the so-called accelerated lambda iteration (ALI) methods where the radiative transfer is approximated by a more simple Λ operator and treated in conjuction with the full set of statistical equilibrium equations. This approach also needs linearization of either the both transfer and rate equations (Scharmer and Carlsson, 1985) or the rate equations alone (Werner and Husfeld, 1985). The third approach completely avoids linearization by using the "preconditioning" of the starting equations. In this iterative procedure proposed by Rybicki and Hummer (1991), in analogy to the core saturation assumption (Rybicki, 1972), approximate lambda operators (ALO) are introduced directly into the multilevel equations to make them linear from the outset. The band size of the ALO matrix corresponds to the assumed degree of non-localness of the transfer process. The procedure is extremely simple and quite direct in application when the operator is a local (diagonal) one, but somewhat optional if a non-local (tri- or higher-order diagonal matrix) operator is used. The method of this kind is much more advantageous in comparison with the methods that need linearization because of a greater simplicity and far smaller number of numerical operations.

The iteration factors method solves the above problem by the use of some quasi-invariant scalar closure relationships in a revised Λ iterative procedure. In respect to the previously mentioned methods its advantage is not only in avoiding linearization but also in no need for any matrix structure.

We shall consider again the time-independent radiative transfer equation (3) for the specific intensities $I_{x,\mu}(\tau)$ of the radiation field in a static, plane-parallel and one-dimensional medium.

The line source function $S(\tau)$, defined as the ratio of the emission to the absorption coefficient for a particular transition ij, is under the assumption of complete redistribution given by:

$$S_{ij}(\tau) = \frac{n_j(\tau)A_{ji}}{n_i(\tau)B_{ij} - n_j(\tau)B_{ji}} \quad . \tag{15}$$

Level populations are given by the statistical equilibrium equations describing the balance between all the processes (radiative + collisional) that populate and those that depopulate each atomic level considered. For each level j of an $NL-$ level atom, placed between lower levels i $(i < j)$ and upper levels k $(k > j)$, an SE equation can be written in the form:

$$n_j\{\Sigma_{i<j}(B_{ji}J_\varphi^{ji} + A_{ji} + C_{ji}) + \Sigma_{k>j}(B_{jk}J_\varphi^{jk} + C_{jk})\}$$

$$-\Sigma_{i<j}n_i(B_{ij}J_\varphi^{ij} + C_{ij}) - \Sigma_{k>j}n_k(B_{kj}J_\varphi^{kj} + A_{kj} + C_{kj}) = 0 \quad . \tag{16}$$

Here Cs are the collisional excitation or de-excitation rates - the known functions of τ. For the lowest level ($j = 1$) we have $B_{ij} = C_{ij} = 0$ and, similarly, $A_{kj} = B_{kj} = C_{kj} = 0$ for the highest one ($j = NL$).

For a particular transition $i \to j$ we have that $J_\varphi^{ij} = J_\varphi^{ji}$ as the consequence of the complete redistribution assumption.

The system (16) of $NL - 1$ equations has to be solved together with the particles conservation equation:

$$\Sigma_{i=1}^{NL} n_i = n_{total}$$

to give NL level populations $n(\tau)$. However, since J_φs depend on the level populations (via Eqs.(3) and (15)), the system (16) is a non-linear one.

Equations (3) and (16) maintain their original linear form if they are solved in turn by the classical Λ iteration. However, as is well-known, this scheme often leads to an incorrect solution at great thermalization depths.

In order to solve the above equations keeping their original linear form, we can revise the Λ iteration scheme by the use of some additional linear relations among the unknown functions.

In order to solve the problem, we propose the following iterative procedure $n_i \to S_{ij} \to J_\varphi{}^{ij} \to \Phi_{ij} \to n_i \to S_{ij}...$ where an additional linear closure relationship Φ_{ij} between the radiation field and the level populations is introduced to be solved in conjuction with the SE equations. The consistency of this relation with the problem equations will be provided by the iterative evaluation of its coefficients. In order to achieve a high convergence rate, these coefficients should be good quasi-invariants.

Following the expression (15) for the source function, we can rewrite Eqs.(16) into a more convenient form:

$$\Sigma_{k>j}(n_jB_{jk} - n_kB_{kj})J_\varphi^{jk} - \Sigma_{i<j}(n_iB_{ij} - n_jB_{ji})J_\varphi^{ij} + n_j\Sigma_{i<j}A_{ji}$$

$$-\Sigma_{k>j}n_kA_{kj} + n_j(\Sigma_{i<j}C_{ji} + \Sigma_{k>j}C_{jk}) - \Sigma_{i<j}n_iC_{ij} - \Sigma_{k>j}n_kC_{kj} = 0 \ . \quad (17)$$

According to Eqs. (15) and (17), the required closure relationship describing the non-local coupling of the radiation field with SE equations can be most directly written for each transition as:

$$J_\varphi^{ij} = \beta_{ij}\frac{n_jA_{ji}}{n_iB_{ij} - n_jB_{ji}} \ . \quad (18)$$

Substitution of (18) into (17) eliminates the unknown radiation fields J_φs and makes the SE equations linear in level populations $n_i(\tau)$.

The coefficients β_{ij} in (18) are to be computed at the beginning of each iteration from the "old" (obtained in the previous iteration) values of the source function S_{ij} and the corresponding radiation fields J_φ^{ij} (formal solutions of Eq.(3)) according to:

$$\beta_{ij} = \frac{J_\varphi^{ij}}{S_{ij}} \ . \quad (19)$$

These "iteration factors" can be used then as the known coefficients of the modified SE equations obtained by substitution of the closure relation (18) in Eqs. (17). Solving SE equations we have new set of level populations $n(\tau)$, i.e. new source functions $S(\tau)$ and the modified optical depth scale τ to start the next iteration. The formal solution of Eq.(3) gives new integrated mean intensities $J_\varphi(\tau)$ and, hence, new iteration factors $\beta(\tau)$.

The introduction of the closure (18) represents the basic difference of this scheme with respect to that of the Λ iteration. Since βs are the ratios of two homologous quantities we expect them to achieve their exact values rapidly, yielding a better convergence compared to the classical Λ iteration. Albeit mathematically correct, the above procedure failed to converge. Not surprisingly enough if we have the following physical consideration in mind. The mean intensity of the radiation field at some depth point $J_\varphi(\tau)$ depends not only upon the local source function $S(\tau)$ but also on the source function at all other points via its first- (and higher-) order derivatives as the protagonists of the transfer process. Such intensities enter the SE equations (which couple all level populations locally) providing that each point is coupled to all others. On the other hand, the closure relation (18) does not take the effect of the other layers into account, i.e. the important non-local aspect of the transfer phenomenon is neglected.

The non-local coupling of the radiation field with SE equations can be realized by the following closure relation:

$$J_\varphi = \alpha + \beta S = \alpha + \beta \frac{n_j A_{ji}}{n_i B_{ij} - n_j B_{ji}} \quad . \tag{20}$$

In the relation (20) the coefficient α accounts for the contributions to the mean intensity at layer τ coming from the other layers, whereas β represents the contribution of the radiative processes taking place in the layer τ. The relation (20) is reduced to (18) when the term α is not singled out. The substitution of (20) into SE equations (17) leads to the linear system that can be solved easily for the unknown level populations.

In order to derive the coefficients $\alpha(\tau)$ and $\beta(\tau)$ of the closure (20) at each depth point, we proceed from the formal solution of Eq.(3) as before. Once a functional representation for $S(\tau)$ is selected, the coefficients $\alpha(\tau)$ and $\beta(\tau)$ are derived by a separate treatment of the RT equation for the "in-going" and "out-going" radiation field ("two-stream" model), according to the two-point boundary nature of the transfer process. Computed with the distribution of $S(\tau)$ obtained in the previous iteration step, the coefficients enter the SE equations modified by the use of the closure relation, to give a new set of level populations, i.e. a new source function.

The rate of convergence can be improved by an alternative use of the "two-stream" model within the iteration scheme. In a separate treatment of the "inward" and "outward" intensities, we can treat the coefficients of the "inward" closure relation as the ones that will be computed, like the iteration factors, from the old solution at the beginning of each iteration step. In order to have the "full" coefficients $\alpha(\tau)$ and $\beta(\tau)$ of Eq.(20) we need the coefficients of the corresponding "outward" relation as well. Proceeding from the lower boundary condition we determine them simultaneously with the new source function during the back-substitution when sweeping back to the surface. The computation of the "inward" coefficients from the old source function and the determination of the new source function simultaneously with the "outward" coefficients is to be performed in turn until the required criterion for the convergence is satisfied.

Although the proposed scheme is quite general and can be applied to an arbitrary high number of atomic levels, we have tested its properties against the well-known solution of the two-level-atom line transfer problem in a constant-property medium. Clearly, we did not make the use of the fact that the source function in this particular instance depends on the radiation field explicitly.

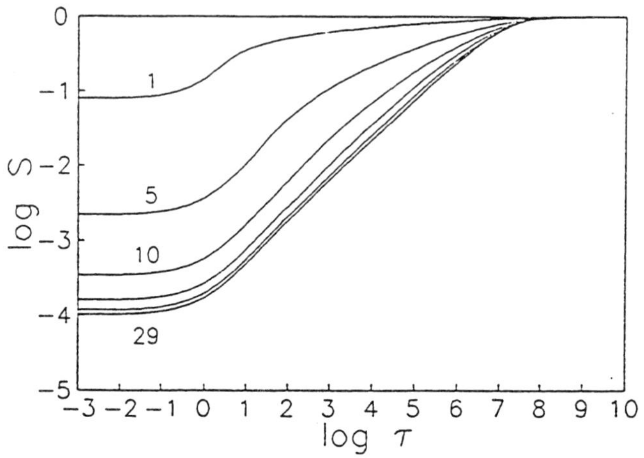

Fig.4. The convergence of $S(\tau)$ for the case $\varepsilon = 10^{-8}$ and $B = 1$ in the labelled number of iterations

When applied to the extreme non-LTE case ($\varepsilon = 10^{-8}$) the abovecited procedure gives the exact solution (with the error less than 1%) in 29 iterations (see Fig.4).

5. Conclusions

This paper deals with the computational properties of a new fast-convergent iterative method developed for the solution of the non-LTE line formation problem. The method represents a very simple iterative scheme where the computation of certain quasi-invariant iteration factors from the formal solution of RT equation and their use for obtaining new corrected source function is performed in turn. A small additional computational effort with respect to the Λ iteration scheme leads to a very fast convergence to the exact solution. No matrix structure is needed and, hence, no large computing memory and time is required. Since the use of the iteration factors enables us to replace a large set of radiative transfer equations by only one equation with the factors as the mean coefficients, this method seems promising in applications to the problems where the radiative transfer is coupled with other physical phenomena.

References

Atanacković -Vukmanović,O. and Simonneau,E.: 1993, to be published in *JQSRT*.

Auer,L.H. and Mihalas,D.: 1969, *Astrophys. J.* **158**, 641.

Auer, L.H. and Mihalas, D.: 1970, *Mon. Not. R. Astron. Soc.* **149**, 65.

Borsenberger, J., Oxenius, J. and Simonneau, E.: 1987, *JQSRT* **37**, 331.

Crivellari, L. and Simonneau, E.: 1991, *Astrophys. J.* **367**, 612.

Feautrier, P.: 1964, in *Proceedings of the First Harvard- Smithsonian Conference on Stellar Atmospheres*, SAO Spec. Rep. No. 167, Cambridge, Mass., p.108.

Fieldus, M.S., Lester, J.B. and Rogers, C.: 1990, *Astron. Astrophys.* **230**, 371.

Rybicki, G.B.: 1972, in *Line Formation in the Presence of Magnetic Fields*, R.G.Athay, L.L.House and G.Newkirk,Jr. (Eds.), High Altitude Observatory, Boulder, CO p.145.

Rybicki,G.B. and Hummer,D.G.: 1991, *Astron. Astrophys.* 245, 171.

Scharmer,G.B. and Carlsson,M.: 1985, *J. Computational Physics* 59, 56.

Simonneau, E. and Crivellari, L.: 1988, *Astrophys. J.* **330**, 415.

Werner,K. and Husfeld,D.: 1985, *Astron. Astrophys.* 148, 417.

Low Temperature and
General Plasmas
M. Milosavljevic and Z. Petrovic

ISBN 1-56072-399-8

GYRORESONANT SCATTERING OF RADIATION BELT ELECTRONS BY OBLIQUE WHISTLER WAVES

Jasna D., U.S. Ianan and T.F. Bell

Abstract. Radiation belts consist of high energy (ten of eV to tens of MeV) electrons and ions trapped by the Earth's magnetic field with particle source and loss processes in a dynamic equilibrium. When scattered in pitch angle, the energetic electrons precipitate into the upper atmosphere where they cause significant disturbances (produce heating, secondary ionization and conductivity enhancement, emit light or X-rays). Interactions with electromagnetic waves are an important means of loss of these particles from the radiation belts. The Earth's magnetoplasma supports a large variety of natural and man-made electromagnetic waves amongst which are whistler-mode waves. Whistler mode waves are typically in the very low frequency range (hundreds of Hz to tens of kHz), below the electron gyrofrequency so that they can resonantly interact with radiation belt electrons resulting in their precipitation out of the radiation belts and into the atmosphere. A computer simulation model is developed to study the interaction between relativistic charged particles and whistler mode waves traveling obliquely with respect to the ambient magnetic field. We use this test particle simulation model based on gyro-averaged equations of motion to study the influence of oblique Magnetospherically Reflected (MR) whistlers on the near-loss-cone distribution function of radiation belt electrons. We find that MR whistlers originating in lightning can resonantly interact with radiation belt electrons over a broad range of L-shells and precipitate higher energy electrons from lower L-shells. Electrons in the energy range of 1-2.6 MeV are precipitated from L=2, whereas from L=4 the precipitated electron energy range is 150-220 keV. The precipitated differential electron flux, due to this interaction, is higher for higher L-shells, and the maximum value ranges from $\Phi_{Eprec}(1.11\,\mathrm{MeV}) = 5.2\times10^{-4}$ el/cm^2-s-keV at L=2 to $\Phi_{Eprec}(173\,\mathrm{keV}) = 4.6\times10^{-1}$ el/cm^2-s-keV at L=4. The lifetimes of radiation belt electrons in a given magnetic flux tube around the L-shell on which the interaction takes place are found to be of the order of several days, comparable to lifetimes corresponding to electron loss induced by hiss, which was heretofore assumed to be the dominant loss mechanism [1]. The minimum electron lifetimes vary from 2.47 days for E=1.11 MeV electrons at L=2 to 4.64 days for E=173 keV electrons at L=4.

Magnetosphere

The Earth's ***magnetopshere*** is the space surrounding the Earth, filled with relatively low density ionized gas, permeated by the Earth's magnetic field. Its boundaries separate plasmas which are under the influence of interplanetary and Earth's magnetic fields, and are defined by the interaction of the supersonic solar wind with the geomagnetic field [2]. A schematic picture of the magnetosphere is shown in Figure 1. In the sunward direction the Earth's dipole-like internal magnetic filed, whose axis is inclined at ~11° with respect to the rotational axis, is compressed by the solar wind. A collisionless bow shock is formed where the solar wind, due to its interaction with the magnetosphere, becomes thermalized and subsonic. Further downstream is the first magnetosphere boundary – the ***magnetopause***. A rough

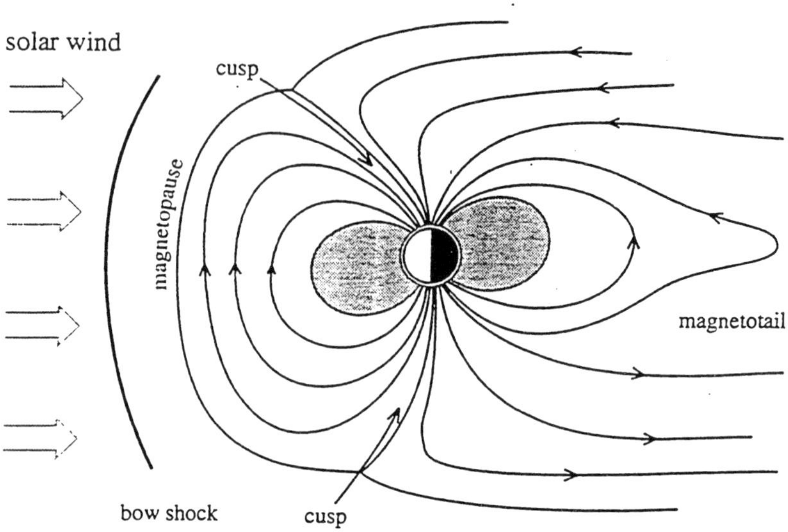

Figure 1. Magnetosphere. The Earth's dipole like magnetic field lines are distorted by the solar wind, forming a bow shock where the supersonic solar wind is thermalized and becomes subsonic. The magnetopause separates Earth's and interplanetary magnetic fields and plasmas.

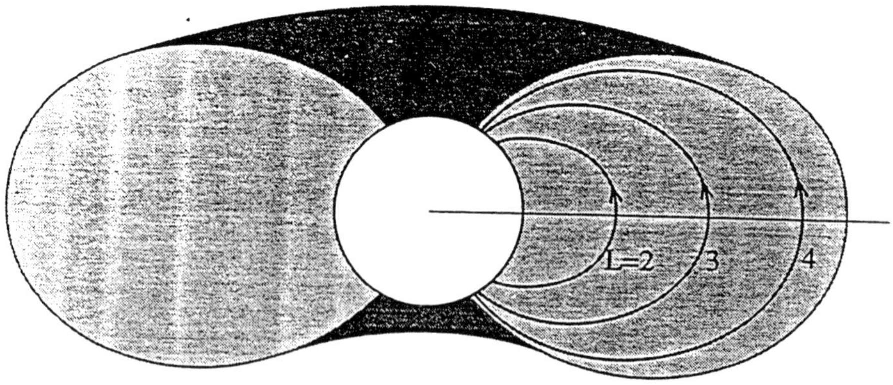

Figure 2. Inner magnetosphere. The Earth's dipole like magnetic field lines in the inner magnetosphere are not significantly distorted by the solar wind. Magnetic field lines crossing the equatorial geomagnetic plane at the same distance from the center of the Earth form an L-shell. The L-shell number represents its equatorial radial distance in Earth radii. The shaded surface represents L=4.5, whereas L=2,3 and 4 are each represented only by a single corresponding magnetic field line.

estimate of its position, determined by balancing the solar wind pressure and the Earth's magnetic field pressure, places the magnetopause at about ten Earth radii under geomagnetically quiet conditions. In the anti-sunward direction, magnetic field lines are dragged away from the Earth to form the ***magnetotail*** which represents the extension of the magnetosphere beyond the moon to several hundred Earth radii.

The part of the magnetopshere where the Earth's magnetic field lines are not significantly distorted by the solar wind and where the field can be approximated by a centered dipole will be referred to as the ***inner magnetosphere***, as shown in Figure 2.

Magnetic field lines crossing the geomagnetic equatorial plane at the same distance from the center of the Earth form a surface that is defined as an ***L-shell***. The ambient relatively low density and low energy (cold) ionized gas in the inner magnetosphere is created through ionization of neutrals by solar ultra-violet radiation at the daytime atmosphere – magnetopshere boundary [3], and it consists of electrons and positive ions (H^+, He^+ and O^+) in diffusive equilibrium [4]. The particle density in this background cold plasma ranges from a few thousands to several particles per cubic centimeter [5]. The typical variation of the electron concentration at the equatorial plane with radial distance from the Earth in Earth radii (or L-value by above definition) is shown in Figure 3. An abrupt boundary at which the electron concentration drops by roughly two orders of magnitude is known as the ***plasmapause*** [6,7]. Its position (equatorial distance from the Earth's center) varies significantly with geomagnetic activity, from two Earth radii following major magnetic substorms up to seven Earth radii after long (several days) quit periods [7]. Electrons and ions in this background 'cold' plasma have low energies, typically in the 0.1 eV – 1 eV range.

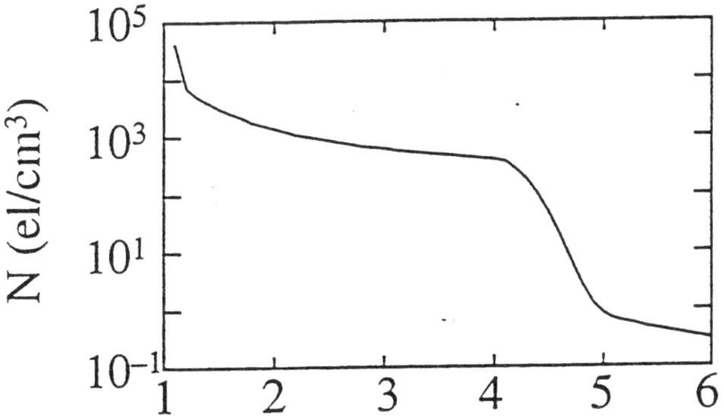

Figure 3. Equatorial electron density. Typical equatorial ambient cold electron density as a function of L-shell, with the plasmapause being at L~4.5.

As a magnetized plasma, the magnetosphere constitutes a dispersive, anisotropic, and inhomogeneous medium that supports a large variety of natural and man-made electromagnetic waves over a very broad frequency range, ranging from fractions of a Hz to a few MHz. Amongst the waves supported by the magnetosphere is a special class of extremely low frequency (ELF) and very low frequency (VLF) waves (tens of Hz to tens of kHz) called whistler-mode waves, due to the fact that a particular kind of whistler-mode wave, injected into the magnetosphere by lightning discharges and known as a 'whistler', produces a whistling sound when detected in the conjugate hemisphere with the receiver connected to an audio speaker [8]. ELF/VLF waves can propagate through the magnetopshere in either ducted or non-ducted modes as shown in Figure 4. Ducted waves travel within background density enhancements called

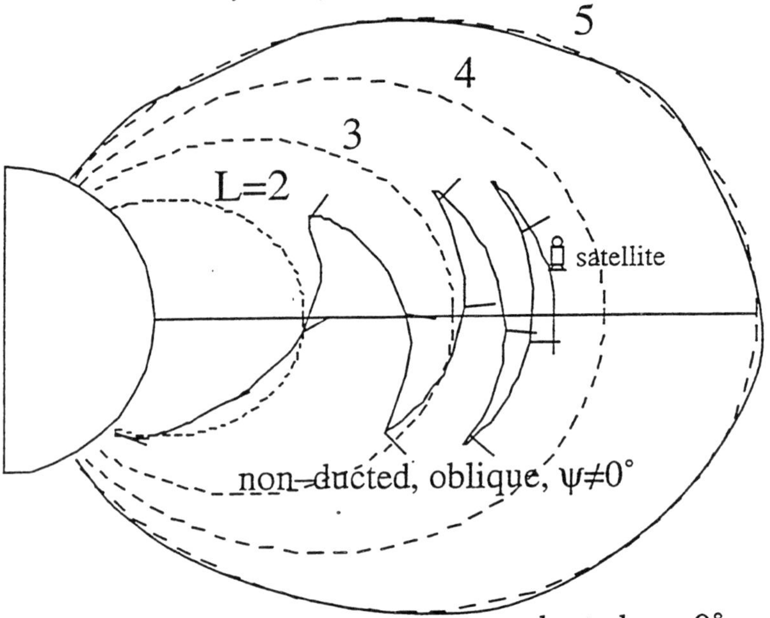

Figure 4. Ducted and non-ducted whistler-mode waves. Ducted waves travel within background plasma density enhancements called 'ducts'; their wave normal angle is approximately parallel to the magnetic field line, they penetrate through the sharp lower ionospheric boundary and are detected on the ground. Non-ducted waves propagate with their wave normal at a substantial angle with respect to the magnetic field line; they are typically reflected within the magnetosphere before they reach the ionosphere, and they can typically be detected only *in situ*.

ducts, formed around particular magnetic field lines [8]. As the waves propagate in the duct, much like light traveling in a graded optical fiber, the wave vector remains approximately parallel (within ±30^c) to the magnetic filed line [8]. Because of their near normal incidence to the sharp lower ionospheric boundary, ducted waves can penetrate through the narrow transmission cone of the lower ionosphere to reach the Earth's surface where they can be detected by ground based receivers. Non-ducted waves propagate obliquely with their wave vectors at a substantial angle with respect tot he magnetic field line, and they are either reflected within the magnetosphere before they reach the ionosphere (the reflections occur where the wave frequency is equal to the local lower hybrid resonance frequency [9] or are specularly reflected at the sharp lower ionospheric boundary due to their highly oblique incidence. Non-ducted waves can typically be detected only *in situ*, on spacecraft or rockets. Whistler mode waves can be incoherent or coherent signals. Incoherent, broadband noise-like signals, generally with well defined upper and lower cutoff frequencies, are referred to as hiss [10, 11, 12], whereas examples of narrowband coherent VLF waves are natural whistlers generated by lightning, discrete 'chorus' emissions, harmonic radiation from large scale power grids, signals from ground based transmitters and triggered emissions [8, 13, 14, 15]. Whistler mode waves are of special importance because their frequencies and propagation velocities are low enough for resonant interactions with radiation belt electrons. Examples of whistler mode waves recorded on the Dynamic Explorer-1 (DE-1) satellite are shown in Figure 5.

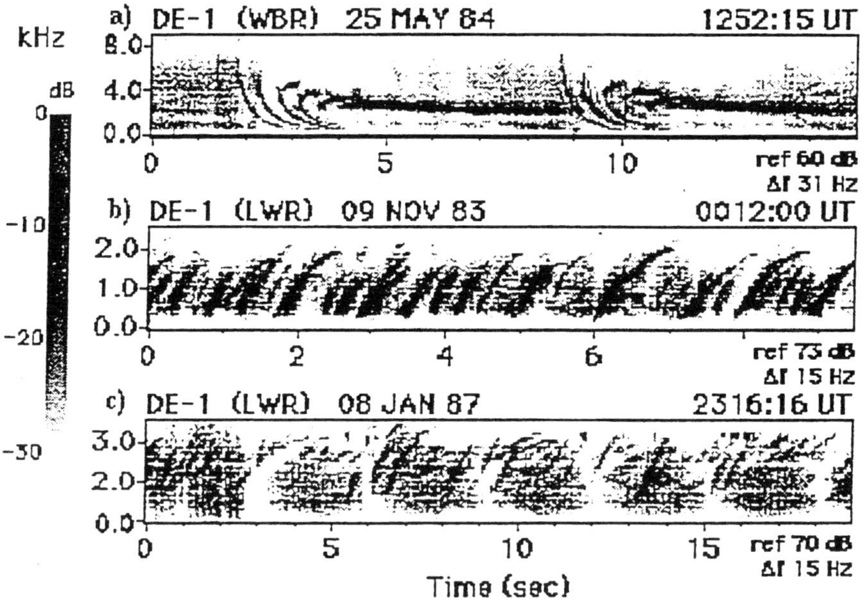

Figure 5. Whistler mode waves. Frequency vs time spectra of electromagnetic whistler-mode signals observed on the DE-1 satellite. Note that darker shades represent stronger signals. a) 25 May 84, two multihop whistlers with well distinguished first four hops. b) 9 Nov. 83, chorus with strong and clear rising tones. c) 8 Jan. 87, chorus with more complex structure.

Radiation Belts

In addition to the cold plasma, the inner magnetosphere is also populated by suprathermal (tens to hundreds of eV) and energetic (few keV to tens of MeV) electrons and ions, which are trapped by the Earth's magnetic field and which constitute the ***radiation belts*** [16, 17, 2]. These particles execute helical motions around the magnetic field lines, reflect back and forth between the two mirror points in opposite hemispheres and drift around the Earth due to magnetic field gradient and curvature drifts [17]. Trapping of the particles by the Earth's magnetic filed constrains their motion so that the complete trajectory of a given particle along a given magnetic field line can be defined by only two parameters, for example by $\upsilon_{z_{eq}}$ and α_{eq} (equatorial velocity in the direction of the magnetic field and equatorial pitch angle respectively, where pitch angle is the angle between the particle's velocity vector and the magnetic field), as illustrated in Figure 6. Along a given magnetic field line, the mirror points are determined uniquely by the equatorial pitch angle; they are independent of the particle's total velocity (i.e., energy). The scattering of the particles by external forces (such as those due to resonant whistler-mode waves) may lead to changes in their equatorial pitch angle and displacement of their corresponding mirror points along the field line. If its mirror point is lowered into the increasingly denser atmosphere the particle interacts with the atmospheric constituents and can be lost from the radiation belts and be 'precipitated' into the atmosphere. The critical equatorial pitch angle which corresponds to the mirror point defined with respect to an 'effective' upper atmospheric boundary is called the ***loss cone angle***. All particles that have equatorial pitch angles smaller than the loss cone angle mirror below this boundary and are therefore considered to be lost from the radiation belts.

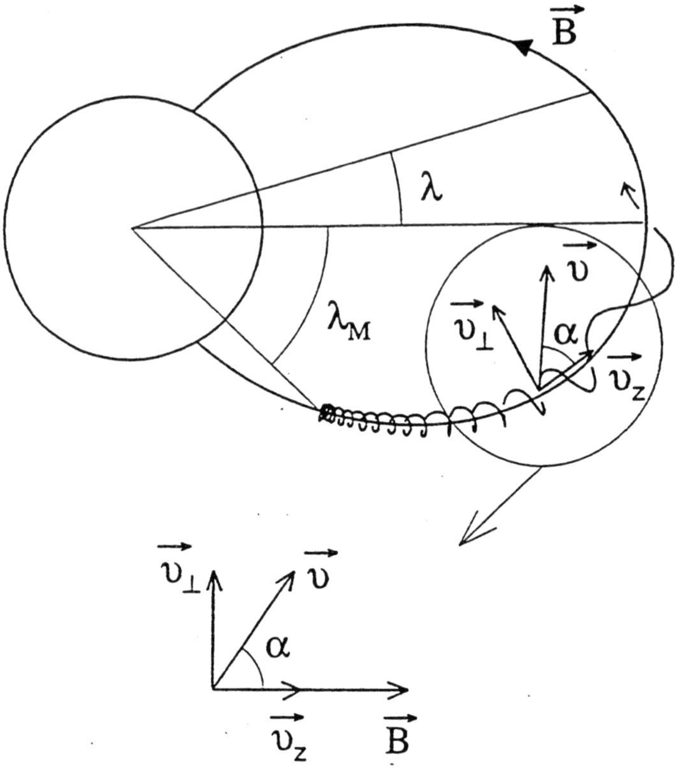

Figure 6. The trajectory of a radiation belt electron. Radiation belt particles (electrons and ions) execute complex helical motions consisting of a gyration around the magnetic field line, reflection between the two mirror points along the field line, and drift around the Earth across the magnetic field lines. The particle's trajectory and mirror points (λ_M) are defined by only two parameters, for example by equatorial velocity in the magnetic field direction $\upsilon_z(\lambda=0^c)$, , and equatorial pitch angle, $\alpha(\lambda=0^c)$.

The determination of the loss rates of radiation belt electrons as well as the quantitative evaluation of the contributions to this loss of different mechanisms are important both in terms of our understanding of the dynamics of the radiation belts and also because the precipitated energetic particles cause significant disturbances in the upper atmosphere such as heating, ionization, conductivity enhancements, emission of light and X-rays [18]. Although not fully understood, the source and loss processes of radiation belt particles are in a dynamical equilibrium. The solar wind is believed to be one source of energetic particles which enter the magnetosphere through the cusp or magnetotail and drift across the magnetic field lines to form the radiation belts via radial diffusion. Since the boundary between the atmosphere and the magnetosphere is gradual rather than abrupt, physical and chemical processes in the intermediate region, the ionosphere, are also important as sources and losses of radiation belt particles [19].

Resonant Interactions Between Radiation Belts and Waves

Past investigations of resonant whistler-mode wave-particle interactions have recognized their importance in the loss of radiation belt electrons (e.g., [20]). Experimental observations have shown that ionospheric effects of precipitated electrons, such as sub-ionospheric VLF perturbations, X-ray emissions, or photoemissions can often be correlated with natural ducted VLF waves (whistlers, chorus emissions, noise bursts, etc.) [21, 22, 23, 24, 25, 26, 27, 28, 29]. In particular, the ionospheric signatures of the scattering of energetic electrons out of the radiation belts in cyclotron resonant interactions with lightning-generated ducted (i.e., parallel propagating) whistlers have been both theoretically studied in detail (e.g. [30]) and are now commonly observed [31, 32, and references therein]. As a measure of the effectiveness of this interaction, precipitated energetic electron fluxes have been theoretically estimated and compared with experimental data [33, 34, 35].

However, the bulk of the wave energy injected into the magnetosphere by lightning discharges propagates in the non-ducted mode. Although non-ducted waves may be less efficient than ducted waves in pitch angle scattering and precipitation of electrons in a single encounter cyclotron resonant interaction (due to elliptical polarization of non-ducted waves versus circular polarization of ducted waves), their overall effect on radiation belt electron precipitation is yet to be ascertained.

Magnetospherically reflected (MR) whistlers

Non-ducted (oblique) whistler mode waves originating in lightning discharges are often found to undergo multiple reflections between hemispheres. These reflections occur at points where the wave frequency matches f_{LHR} the local lower hybrid resonance frequency [9].

Figure 7 shows examples of MR whistlers observed on the DE-1 satellite. The format is one of a typical frequency vs time spectrogram with the density of the shading representing the intensity of the recorded signal. Two whistler events originating in two successive lightning flashes (in this case ~ 8 s apart) are shown, each initiating a sequence of MR components. The first few (four) hops of both whistlers show very pronounced dispersion after which the hops begin to merge into each other to form a long enduring relatively narrowband signal. The duration of the narrowband part of the signal for this example is about 2 s. Although the total duration of each MR event is ~ 10 s for the cases shown, MR whistlers are known to last for as long as 100 s [36].

Both ducted and MR (oblique) whistlers are generated by lightning discharges. It should be noted that non-ducted whistlers do not require the presence of any specific density structures (e.g., ducts) for their propagation. They generally occupy larger regions of the magnetosphere [37]. On this basis, although the statistics of the occurrence rates of non-ducted whistlers are not well documented, it may safely be assumed that they occur at least as often as ducted whistlers. The ducted whistler rate depends on the time of the year and time of the day and varies from 0.3 whistlers per minute for an average summer day to 22 whistlers per minute for an average winter night [38]. For an extremely active day, the ducted whistler rate can be as high as 195 whistlers per minute, but year-round average is 6 whistlers per minute.

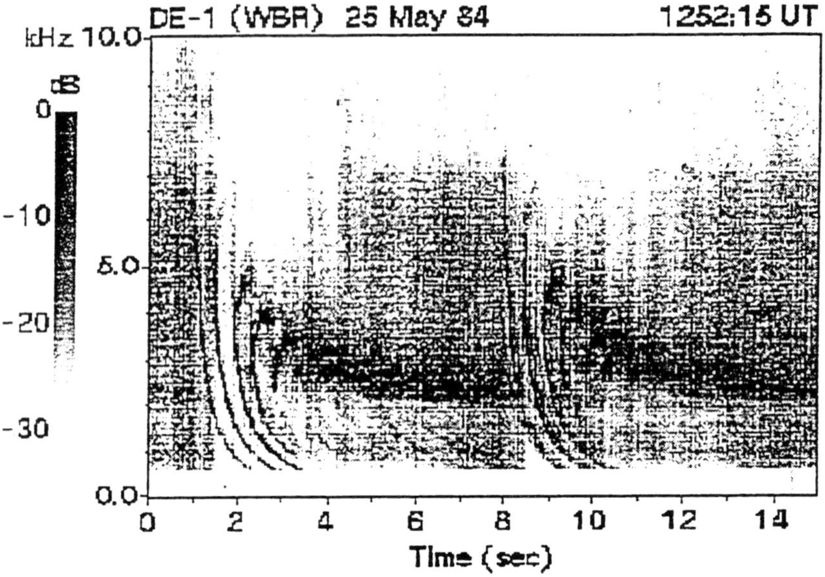

Figure 7. MR whistlers observed on the DE-1 satellite. The examples are given in the form of a typical frequency vs time spectrogram with the shading representing the intensity of the recorded signal. Two whistler events originating in two successive lightning flashes, ~ 8 s apart, ate shown, each initiating a sequence of MR components. After very pronounced dispersion of the first four hops of both whistlers, the hops begin to merge into each other to form a relatively narrow band signal lasting for about 2 s.

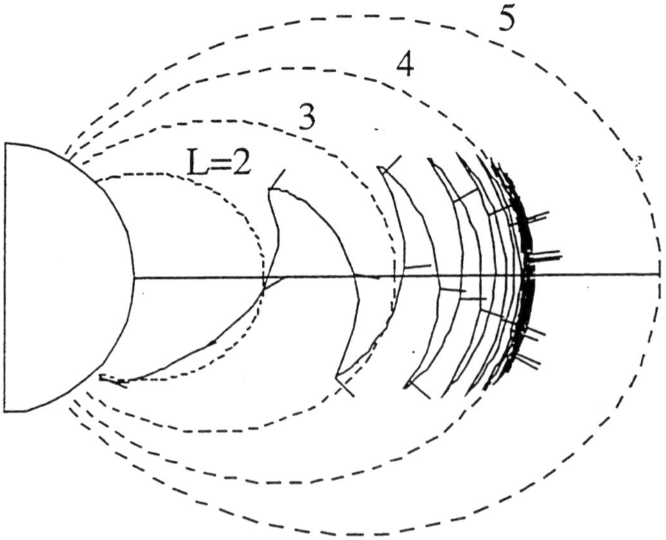

Figure 8. Sample raypath for an MR signal. The raypath for a f=0.33 kHz wave injected at 400 km altitude, vertically upwards from the southern hemisphere at L=2, propagating to higher L-shells by reflecting back and forth between hemispheres and eventually settling down at L~4.

The magnetospheric reflection of oblique whistler mode waves can be effectively studied by raytracing. Here we utilize the Stanford VLF raytracing code (e.g., see [39]). The example in Figure 8 shows a f=0.33 kHz wave injected vertically upwards from the southern hemisphere at L=2. The ray travels vertically upwards reflecting back and forth between the hemispheres as expected and eventually settles down at $L \cong 4$. More extensive raytracing studies [36] show that the L-shell at which the whistler wave settles down depends strongly on the wave frequency, but is relatively independent of the injection latitude or initial wave propagation direction. The frequency of the whistler wave vs L-shell of settlement as obtained from raytracing is plotted in Figure 9.

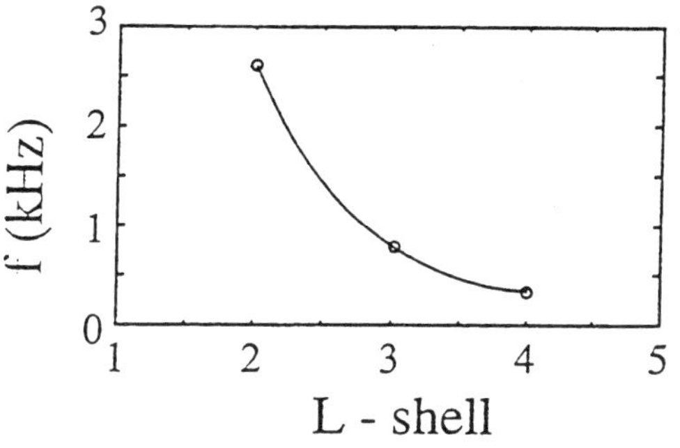

Figure 9. The frequency of the MR whistler wave vs L-shell of settlement. This result is derived on the basis of simulated raypaths as shown on Figure 8.

To summarize, MR whistlers are oblique whistler mode waves injected into the magnetosphere by lightning, they can endure in the magnetosphere for as long as 100 s each, and after several (5-6) reflections the raypaths settle down into a multiply reflecting pattern at an L-shell uniquely corresponding ($f \approx f_{LHR}$) to the signal frequency. Thus, at any given L-shell, we can expect accumulation of oblique whistler mode wave energy at the corresponding frequency. The wave normal angle of the multiply reflecting waves slowly varies with latitude along the field line in a particular manner as dictated by raytracing. As such; these multiply reflecting waves, continuously present along the given L-shell constitute a narrowband wave which can resonantly interact with the radiation belt electrons. It should be noted that the wave distribution along the field line consists of a superposition of multiply reflecting waves. In the following, we use a test particle simulation to represent the interactions of a distribution of electrons with these continuously present nearly monochromatic MR whistlers, the wave normal angle of which vary along the field line as described by raytracing. Our assumption of a monochromatic wave with specified phase variation amounts to considering one of the many multiply reflecting wave components, the superposition of which may generate standing wave patterns and slow variations in wave phase.

Electron velocity space

In general, electron velocity space is three dimensional. For the special case of electron motion in the Earth's magnetosphere, important features of the unperturbed particle motion can be described by only two velocity space coordinates, for example the equatorial pitch angle α_{eq} and the component of the particle velocity along the

Earth's magnetic field line $\upsilon_{z_{eq}}$. We represent the full distribution function by a large number of individual test particles and infer the modification of the distribution from the simulated trajectories of each of the test particles in the presence of the wave. Once a new particle distribution is determined, measurable quantities such as differential (per unit particle energy) and total precipitated particle fluxes and radiation belt electron lifetimes, are easily derived.

Figure 10 shows a single test electron scattering in $\upsilon_{z_{eq}} - \alpha_{eq}$ velocity space. Note that in the absence of the wave, the trajectory of a test electron moving in the Earth's magnetic field would be a single point in $\upsilon_{z_{eq}} - \alpha_{eq}$ space. Due to the interaction with the wave, the electron trajectory in velocity space evolves along a curve with initial and final points corresponding to the initial and final velocity space coordinates. Since the wave forces acting on the particle are functions of the initial electron phase η (the third velocity coordinate that determines the direction of the electron velocity perpendicular to the Earth's magnetic field), electron trajectories and their final points in velocity space depend on the initial η [40, 41]. In view of this dependence of scattering on η, each cell in velocity space needs to be represented by a distribution of several test electrons with different initial phases. We choose 12 test electrons equally spaced in phase to represent a given cell based on the fact that the final scattering i.e. the root-mean-square value of the final pitch angle change at the end of the resonant interaction, namely $\overline{\Delta\alpha_{eq_f}} = \sqrt{\langle \Delta\alpha_{eq}^2(\lambda_f) \rangle}$, (where averaging ($\langle\ \rangle$) is over the number of test electrons representing each cell) does not change significantly with further increase of the number of test electrons representing each cell.

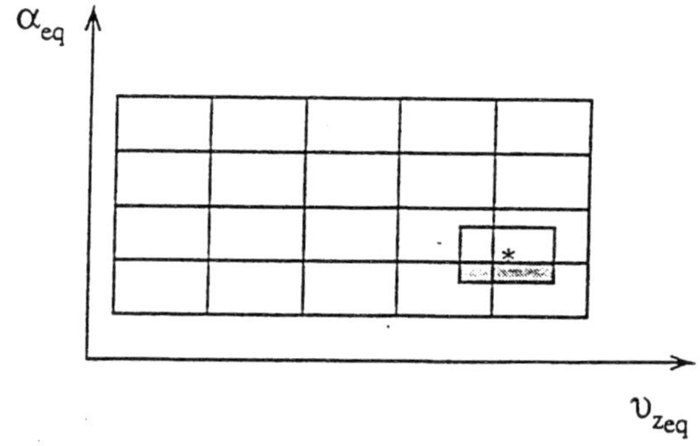

Figure 10. A single test electron scattering in velocity space. The schematic shows the initial and final points of the test electron trajectory in the velocity space. Electron population from the initial cell represented by the test electron is scattered into the four adjoining cells as represented by the cell corresponding to final point on the electron trajectory. The number of electrons scattered into any of the four adjoining phase cells is proportional to the area of the final cell that overlaps the corresponding cell.

The modified electron distribution after a one-pass interaction with the wave is obtained using the scattering of all test electrons weighed by the initial distribution function (electron density per velocity space volume) and its dependence on the velocity space coordinates [i.e., $f(\upsilon_{z_{eq}}, \alpha_{eq})$].

Initial conditions and assumptions

In the inner magnetosphere ($L \leq 6$), we use the diffusive equilibrium background 'cold' plasma density model [4, 42] and centered dipole field as a model of the Earth's magnetic field.

We describe the initial electron distribution function in a manner similar to that of [43] and [30]. We assume that radiation belt electrons are uniformly distributed in the electron pitch angle space between α_{lc} and $\pi/2$ and with an energy dependence of the form $f_E(E) \propto E^{-3}$ based on the experimental data [44]. Since the energetic electron population is commonly described and measured in terms of the differential energy spectrum $\Phi_{E_{diff}}(E)$, the proportionality constant is chosen such that the corresponding differential energy spectrum at E=1 keV is $\Phi_{E_{diff}}(1\text{keV}) = 10^8 \text{ el}/\text{cm}^2 \text{ s keV ster}$.

One of the fundamental assumptions of our test particle formulation is the fact that we neglect the effects on the wave of the distribution of energetic electrons. This means that we are assuming either that the currents stimulated in the particle population do not lead to significant damping or amplification of the wave or that this effect has been included in the model chosen for the wave structure. This approximation was also adopted in the test particle model studies of ducted wave-particle interactions [45, 30]. In the context of this assumption, it is only necessary to define the wave characteristics as an input to the test particle code. For non-ducted obliquely propagating waves, taking the wave as a predefined structure and neglecting the effects of the energetic particles on the wave is probably an even better approximation, since growth and emission triggering by non-ducted waves is observed much less often [46].

In our modeling, then, we assume the wave to be monochromatic (single frequency), and propagating at a given angle ψ with respect to the magnetic field. With the wave frequency ω and wave normal angle ψ specified, all properties of the wave are defined in terms of the local cold plasma parameters. The properties (dispersion relation, polarization of the wave fields, group velocity) of the electromagnetic waves that are supported by the cold, infinite, homogeneous, collisionless plasma in the presence of the external homogeneous, static magnetic field are derived from Maxwell's equations [47]. The magnetosphere is not a homogeneous medium since both the plasma density and magnetic field (and therefore plasma and cyclotron frequencies of the *k*th species, ω_{pk} and ω_{Hk} respectively) vary in space. Fortunately, these spatial variations are generally small over the distance of the order of the wavelength, so that at any given point, the wave propagation can be represented using the slowly varying approximation (WKB) in which the wave is assumed to have the same characteristics as those of a wave traveling in a homogeneous medium having the same refractive index. The Poynting flux (power density) of the slowly varying wave was also assumed to be constant during its interaction with the radiation belt electrons, although the wave normal angle ψ was allowed to slowly vary with latitude as dictated by raytracing. The assumed constant Poynting flux value was used to calculate the wave field components at different points along the field line.

The precipitated differential electron energy spectrum caused by the electron-wave interaction $\Phi_{E_{prec}}$ is used to infer the electron lifetimes in a given magnetic flux tube subject to certain assumptions as discussed below. We note in this context that the precipitated differential energy spectrum may change as a function of time both because of temporal drifts, or because of variations in the parameters of the wave interacting with the electrons. For the purpose of estimating electron lifetimes $\tau(E)$, we assume that the precipitated differential energy spectrum remains constant in time, i.e., $\Phi_{E_{prec}}(E,t) \cong \Phi_{E_{prec}}(E)$. We define N_f as the number of energetic electrons with a given energy per unit energy in a flux tube, per unit area of a tube cross section, at the equator so that the SI unit of the variable N_f is el/m^2-J. We assume that N_f decays exponentially in time as a result of the resonant scattering due to the electron-wave interaction $N_f(E,t) = N_f(E)\exp(-t/\tau(E))$ while $\partial N_f(E,t)/\partial t = -\Phi_{E_{prec}}(E,t) = -\Phi_{E_{prec}}(E)$ so that the lifetime of the electrons in a given tube is $\tau(E) = N_f(E)/\Phi_{E_{prec}}(E)$.

Note that the electron lifetimes defined above are derived for the electron population confined to a narrow flux tube around the field line where the interaction takes place. The extent to which this loss process contributes to the electron lifetimes on a global scale depends on the extent of magnetospheric regions illuminated by the waves, since similar scattering processes would be expected in all regions where the oblique waves are present. Experimental data indicates that magnetospheric regions illuminated by whistlers from a given thunderstorm center may extend to many tens of degrees in longitude around that of the storm center [37].

Electron precipitation by MR whistlers

We now investigate the interaction between radiation belt electrons and MR whistlers in the equatorial region at three typical L-shells. The Poynting flux of the whistler waves was assumed to be S=113 pW/m^2 corresponding to the intensity of a f=6.82 kHz ducted wave ($\psi = 0°$) with $B_w = 5$ pT in the equatorial plane at L=4. The frequencies of the waves that settle down at the chosen L-shells and the dependence of their wave normal angle ψ on the geomagnetic latitude are determined using the Stanford raytracing simulation code. The $\psi(\lambda)$ used for simulation represents best-fits to the numerical results obtained from raytracing. The variation of wave normal angle along the field line $\psi(\lambda)$ is slow enough so that the wave can be considered monochromatic in the WKB sense.

We now study the resonant interaction between these monochromatic MR signals and radiation belt electrons represented by the distribution function as defined earlier, using the test particle simulation model.

Figure 11 shows the near loss cone electron distribution function resulting from a one-pass interaction with a 0.33 kHz oblique whistler wave at L=4. Local maxima and minima in electron scattering occur due to the constructive and destructive interference between the two resonant encounters (on both sides of the equator) of the particles with the wave. Maximum penetration into the loss cone (~ 5×10^{-4} deg) occurs for $\upsilon_{z_{eq}} = 198.4\times10^6$ m/s.

The distribution functions similar to that shown in Figure 11 and which result from a one-pass interactions with the wave are also computed for L=2 and 3 and are subsequently used to calculate the precipitated differential electron spectra as shown in Figure 12 for all three L-shells. In general, MR whistler components which settle down on higher L-shells resonantly interact with and scatter lower energy electrons. For

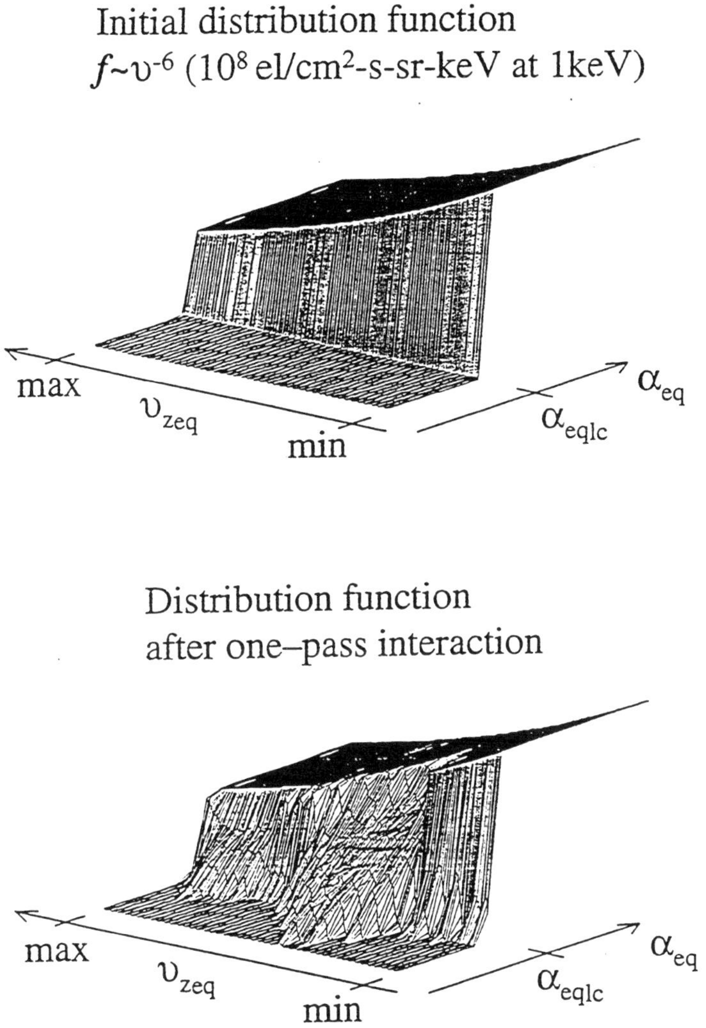

Figure 11. The initial and final near-loss-cone electron distributions for a one-pass interaction with 0.33 kHz whistler wave at L=4. Local maxima and minima in electron scattering occur due to the constructive and destructive interference between the two resonant encounters of the distribution with the wave.

example, a 0.33 kHz whistler wave at L=4 scatters electrons in the 150-220 keV range, whereas the 2.6 kHz MR whistler component that settles down at L=2 scatters electrons in the 1-2 MeV energy range. Precipitated differential electron flux is in general higher for lower energy electrons that are scattered on higher L-shells. Maximum value of the precipitated differential electron flux at L=4 is $\Phi_{E_{prec}}(173\ \text{keV}) = 4.6 \times 10^{-1}$ el/cm^2-s-keV. Corresponding total precipitated energy fluxes are $Q(L=2) = 0.48 \times 10^{-9}$ J/m^2-s, $Q(L=3) = 2.66 \times 10^{-9}$ J/m^2-s, $Q(L=4) = 3.14 \times 10^{-9}$ J/m^2-s and are generally higher for higher L-shells.

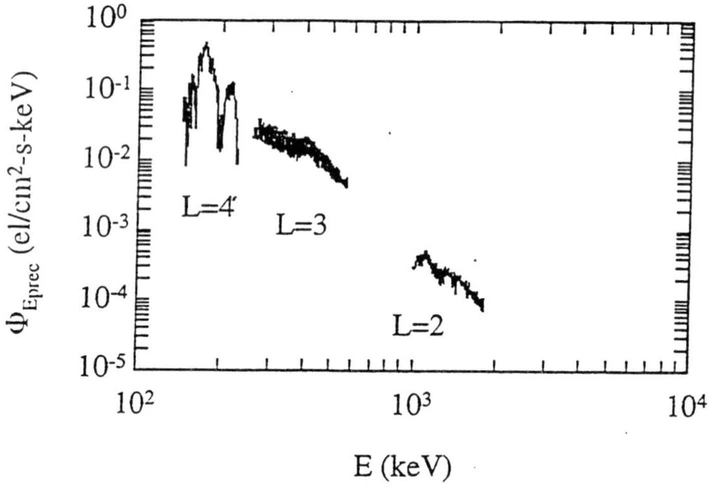

Figure 12. Precipitated differential electron spectra. For all three cases considered, precipitated differential electron spectra is calculated from the distribution function (such as that in Figure 11) after a one-pass interaction with the corresponding wave.

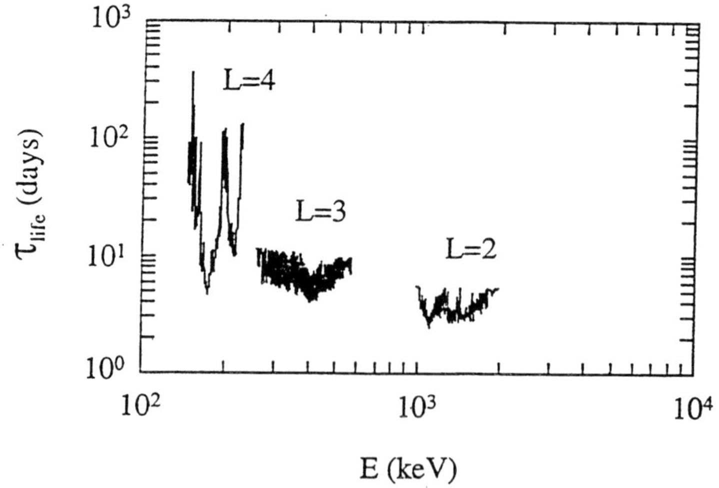

Figure 13. Electron lifetimes. Precipitated differential electron spectra are used to calculate electron lifetimes as discussed in the text.

The electron lifetimes corresponding to the precipitated fluxes of Figure 12 and calculated using the definitions and assumptions described earlier are given as a function of electron energy in Figure 13. The minimum lifetimes of the electrons with the corresponding energy that are scattered by MR whistlers are of the order of several days and do not differ significantly for interactions on different L-shells. The minimum electron lifetimes vary from 2.47 days for E=1.11 MeV electrons at L=2 to 4.64 days for E=173 keV electrons at L=4.

In Figure 14 we compare our lifetime estimates with those due to cyclotron resonant scattering by whistler mode hiss following large magnetic storms [1]. Interaction between radiation belt electrons and hiss causes precipitation and loss from L=3 and 4 of electrons in a broader energy range than that due to the resonant electron-MR whistler interaction. At L=3 and 4 electron lifetimes due to hiss induced losses are smaller than the electron lifetimes due to resonant electron-MR whistler interaction. However, a L=2 MR whistlers can contribute more than hiss to the electron loss from the radiation belts. Note that the results given by [1] are calculated based on the assumption of the continuous presence of a wide band oblique whistler mode signal throughout the entire plasmasphere. In order to be able to compare the two, we also assumed the continuous presence of coherent, narrowband oblique whistler mode signal throughout the entire plasmasphere. However, our results can be scaled to more realistic values by simple multiplication once the ratios for spatial and temporal presence of the MR whistlers throughout the plasmasphere are available.

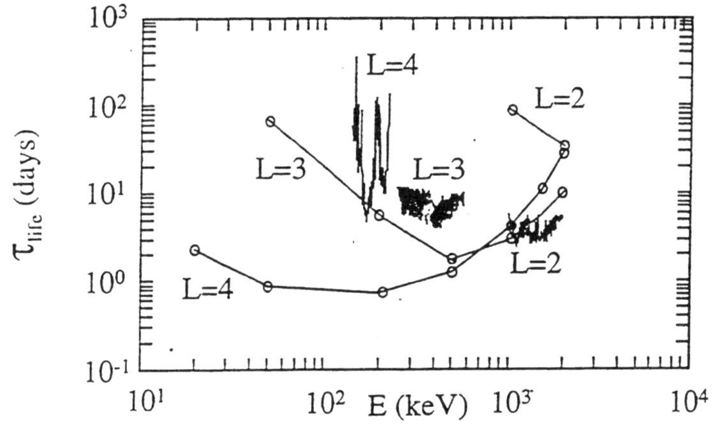

Figure 14. Electron lifetimes compared with previous work. We compare our lifetime estimates (_____) with those due to cyclotron scattering by whistler mode hiss following large magnetic storms [1] (__o__).

Conclusions

We have carried out test particle simulations of the cyclotron resonant scattering of radiation belt electrons in a single resonant encounter (one pass) with obliquely propagating monochromatic whistler-mode waves in the magnetosphere. Our results indicate that, for typical parameters as used in this paper, these interactions result in significant perturbations of the trapped particle distributions and lead to precipitation of particle fluxes into the ionosphere. Although our individual simulations were limited to one-pass interactions with monochromatic oblique waves, when viewed as a whole, they imply that wave energy injected into the magnetosphere by lightning and propagating obliquely in the non-ducted mode may substantially contribute to the establishment of radiation belt equilibrium.

In this context, we have shown that oblique MR waves which settle down on given L-shells and exist for up to 100 s can significantly affect the lifetimes of the energetic electrons. In the case of the latter, it should be noted that non-ducted whistlers injected into the magnetosphere typically have components over a wide range of frequencies, so that each frequency component would 'settle' on a different L-shell and the scattering (as computed) can be expected to occur simultaneously over a broad range of L-shells during the 10-100 s following each lightning discharge.

The implications of our results should be evaluated in the context of recent theoretical and experimental findings which clearly indicate that ducted whistler waves originating in lightning regularly precipitate energetic radiation belt electrons. The ionospheric signatures of the scattering of energetic electrons out of the radiation belts in cyclotron resonant interactions with lightning-generated ducted (i.e., parallel propagating) whistlers has been both theoretically studied in detail [48, 45, 49, 50, 30] and commonly observed [31, 32, and references therein]. As a measure of the effectiveness of this interaction, precipitated energetic electron fluxes have been theoretically estimated and compared with experimental data [33, 34]. Although ducted whistlers often reflect from the lower ionospheric boundary and can bounce back and forth along the duct multiple times, most experimental evidence of ducted whistler induced precipitation involves single encounter interactions with a discrete one-hop whistler. In terms of the contribution of ducted-whistler-induced precipitation to radiation belt losses, the rate of occurrence of lightning and whistlers is thus a determining factor. For an extreme day ducted whistler rate can be as high as 195 whistlers per minute, but year-round average is 6 whistlers per minute [38]. However, the bulk of the magnetospheric wave energy from lightning discharges propagates in the non-ducted mode. Non-ducted whistlers do not require the presence of any specific density structures (e.g., ducts) for their propagation and they generally occupy larger regions of the magnetosphere [37]. On this basis, although the statistics of the occurrence rates of non-ducted whistlers are not well documented, it may safely be assumed that they occur at least as often as ducted whistlers.

Interaction of radiation belt electrons with MR whistlers at a later stage of their propagation (after the settlement of ray paths on a certain L-shell, when the wave normal angle is high, close to 90°) endures for as long as the waves exist (~ 100 s) so that it is appropriate to discuss the consequences of the losses in terms of radiation belt electron lifetimes. Interactions with MR whistlers causes precipitation of higher energy electrons from lower L-shells. Electrons in the energy range of 1-2.6 MeV are precipitated from L=2, whereas from L=4 precipitated electron energy range is 150-220 keV. The precipitated differential electron flux, due to this interaction, is higher for higher L-shells, and the maximum value is ranging from $\Phi_{Eprec}(1.11\,\text{MeV}) = 5.2\times10^{-4}$ el/cm^2-s-keV at L=2 to $\Phi_{Eprec}(173\,\text{keV}) = 4.6\times10^{-1}$ el/cm^2-s-keV at L=4. The lifetimes of the radiation belt electrons in a tube around the L-shell on which the interaction takes place range depending on electron energy from several days to ~ 100 days and are comparable with lifetimes corresponding to electron loss induced by hiss [1]. The minimum electron lifetimes vary from 2.47 days for E=1.11 MeV electrons at L=2 to 4.64 days for E=173 keV electrons at L=4.

References

[1] Lyons, L. R., R. M. Thorne, and C. F. Kennel, *J. Geophys. Res.*, 1972, **77**, 3455.

[2] Akasofu, S. I., and Chapman, Sc'lar Terrestrial Physics, Oxford at the Clarendon Press, 1972.

[3] Banks, P. M., and G. Kockarts, Aeronomy (Part B), Academic Press, New York and London, 1973.

[4] Angerami, J. J., and J. O. Thomas, *J. Geophys. Res.*, 1964, **69**, 4537.

[5] Park, C. G., D. L. Carpenter, and D. B. Wiggin, *J. Geophys. Res.*, 1978, **83**, 3137.

[6] Carpenter, D. L., *J. Geophys. Res.*, 1963, **68**, 1675.

[7] Carpenter, D. L., *J. Geophys. Res.*, 1966, **71**, 693.

[8] Helliwell, R. A., Whistlers and Related Ionospheric Phenomena, Stanford University Press, Stanford, Calif., 1965.

[9] Edgar, B. C., *J. Geophys. Res.*, 1976, **81**, 205.

[10] Gurnett, D. A., and L. A. Frank, *J. Geophys. Res.*, 1972, **77**, 172.

[11] Thorne, R. M., E. J. Smith, R. K. Burton and R. E. Holzer, *J. Geophys. Res.*, 1973, **78**, 1581.

[12] Laaspare, T. and R. A. Hoffman, *J. Geophys. Res.*, 1976, **81**, 524.

[13] Helliwell, R. A., and J. P. Katsufrakis, *J. Geophys. Res.*, 1974, **79**, 2511.

[14] Helliwell, R. A., J. P. Katsufrakis, T. F. Bell and R. Raghuram, *J. Geophys. Res.*, 1975, **80**, 4249.

[15] Park, C. G., Proc. International Symp. on Solar-Terrestrial Physics, Boulder, Colo., D. Williams (ed.), June 1976.

[16] Hess, W. N., The Radiation Belt and the Magnetosphere, Blaisdell Publishing Co., Mass., 1968.

[17] Roederer, J. G., Dynamics of Geomagnetically Trapped Particles, Springer, New York, 1970.

[18] Rees, M. H., Physics and Chemistry of the Upper Atmosphere, Cambridge University Press, Cambridge, 1989.

[19] Walt, M., W. M. MacDonald, and W. E. Francis, Physics of the Magnetosphere, edited by R. L. Carvillano, J. F. McClay and H. R. Radoski, pp. 534-555, D. Reidel, Dordrecht, Holland, 1968.

[20] Dungey, J. W., *Planet. Space Sci.*, 1963, **11**, 591.

[21] Rosenberg, T. J., R. A. Helliwell, and J. P. Katsufrakis, *J. Geophys. Res.*, 1971, **76**, 8445.

[22] Helliwell, R. A., J. P. Katsufrakis, and M. L. Trimpi, *J. Geophys. Res.*, 1973, **78**, 4679.

[23] Foster, J. C., and T. J. Rosenberg, *J. Geophys. Res.*, 1976, **81**, 2183.

[24] Lohrey, B., and A. B. Kaiser, *J. Geophys. Res.*, 1979, **78**, 2142.

[25] Helliwell, R. A., S. B. Mende, J. H. Doolittle, W. C. Armstrong, and D. L. Carpenter, *J. Geophys. Res.*, 1980, **85**, 3376.

[26] Mende, S. B., R. M. Arnoldy, L. J. Cahill, Jr., J. H. Doolittle, W. C. Armstrong, and A. C. Fraser-Smith, *J. Geophys. Res.*, 1980, **85**, 1194.

[27] Rosenberg, T. J., J. C. Siren, D. L. Mathews, K. Marthinsen, J. A. Holtet, A. Egeland, D. L. Carpenter, and R. A. Helliwell, *J. Geophys. Res.*, 1981, **86**, 5819.

[28] Dingle, B., and D. L. Carpenter, *J. Geophys. Res.*, 1981, **86**, 4597.

[29] Carpenter, D. L., and J. W. LaBelle, *J. Geophys. Res.*, 1982, **87**, 4427.

[30] Chang, H. C., and U. S. Inan, *J. Geophys. Res.*, 1985, **90**, 1531.

[31] Inan, U. S., F. A. Knifsend, and J. Oh, *J. Geophys. Res.*, 1990, **95**, 17217.

[32] Burgess, W. C., and U. S. Inan, *Geophys. Res. Lett.*, 1990, **17**, 259.

[33] Inan, U. S., H. C. Chang, R. A. Helliwell, W. L. Imhof, J. B. Reagan, and M. Walt, *J. Geophys. Res.*, 1985, **90**, 359.

[34] Inan, U. S., and D. L. Carpenter, *J. Geophys. Res.*, 1987, **92**, 3293.

[35] Inan, U. S., M. Walt, H. D. Voss, and W. L. Imhof, *J. Geophys. Res.*, 1989, **94**, 1379.

[36] Draganov, A. B., U. S. Inan, V. S. Sonwalkar, and T. F. Bell, *Geophys. Res. Lett.*, 1992, **19**, 233.

[37] Sonwalkar, V. S., and U. S. Inan, to be submitted to *J. Geophys. Res.*, 1993.

[38] Burgess, W. C., and U. S. Inan, *J. Geophys. Res.*, 1992, in press.

[39] Inan, U. S., and T. F. Bell, *J. Geophys. Res.*, 1977, **82**, 2819.

[40] Inan, U. S., *J. Geophys. Res.*, 1987, **92**, 127.

[41] Jasna D., U. S. Inan, and T. F. Bell, *Geophys. Res. Lett.*, 1992, **19**, 1639.

[42] Park, C. G., Methods of Determining Electron Concentrations in the Magnetosphere from Nose Whistlers, Stanford, Radioscience Laboratory, Stanford Electron. Labs., Stanford Univ., 1972.

[43] Inan, U. S., T. F. Bell and R. A. Helliwell, *J. Geophys. Res.*, 1978, **83**, 3235.

[44] Schield, M. A., and L. A. Frank, *J. Geophys. Res.*, 1970, **75**, 5401.

[45] Inan, U. S., T. F. Bell, and H. C. Chang, *J. Geophys. Res.*, 1982, **87**, 6243.

[46] Bell, T. F., U. S. Inan, and R. A. Helliwell, *J. Geophys. Res.*, 1981, **86**, 4649.

[47] Stix, T. H., The Theory of Plasma Waves, New York, McGraw-Hill, 1962.

[48] Inan, U. S., Non-Linear Gyroresonant Interactions of Energetic Particles and Coherent VLF Waves in the Magnetosphere, Stanford, Radiosci. Lab., Stanford Electron. Labs., Stanford Univ., 1977.

[49] Chang, C. H., Cyclotron Resonant Scattering of Energetic Electrons by Electromagnetic Waves in the Magnetosphere, Stanford, STARLAB, Stanford Electron. Labs., Stanford Univ., 1983.

[50] Chang, H. C., and U. S. Inan, *J. Geophys. Res.*, 1983, **88**, 10,053.

Jasna D., Institute of Physics, Maksima Gorkog 118, 11080 Beograd - Zemun, Yugoslavia.

U. S. Inan and T. F. Bell, STAR Lab, Stanford University, Stanford, CA 94305, USA.

Low Temperature and
General Plasmas
M. Milosavljevic and Z. Petrovic

ISBN 1-56072-399-8

THE ION REDISTRIBUTION IN THE DYNAMO REGION OF THE IONOSPHERE

Davorka p. Grubor

Geomagnetic Institute, 110306 Grocka, Belgrade, Yugoslavia

INTRODUCTION

Four topics considering the phenomena taking place in the middle ionosphere were studied in their mutual connection and in the connection with processes taking place in upper ionosphere and magnetosphere:

1. thermospheric general circulation model - tidal neutral winds,
2. ionospheric current system,
3. solar variation of the geomagnetic field in quiet conditions,
4. formation of thin dense layers due to ion redistribution.

Two, mostly three of these phenomena are usually treated simultaneously. In this work the effort was taken to found the connection between all four of them. At the beginning, a short reminding about Geospace regions, in the terms familiar to the plasma physicists, electron density and electron temperature, is presented .The picture of Geospace (ionosphere and magnetosphere) was created in early 1960s and generally it still holds, although the details of fine structure discovered with numerous sophisticated diagnostic methods recently developed, have to be added to it. The electron density and electron temperature height profiles as given in figure below, are plotted according to International Reference Ionosphere data, Rawer (1981).

The region situated in altitude range from 80 to 200 km is characterized by the strong interaction between the neutral particles flow and the ionized medium. The mobility of ions and so the electrical conductivity of the medium is determined by the ratio of ion-neutral collision frequency and ion gyro-frequency in the geomagnetic field. This ratio and therefore the total electrical conductivity in the altitude range 80-200 km, is the greatest in the ionosphere. The ion motion in the direction of the neutral particle flow with velocity **u** across the lines of the geomagnetic field with induction **B**, generates the eddy electric field **u** x **B**. This field causes the separation of charged particles and consequently, the formation of the *polarization field*, **E** = -∇ V. The resultant field, -∇ V + **u** x **B**, induces the ionospheric current. That is why this region is named the *dynamo region*.

For understanding the phenomena mentioned above, a good knowledge of electric fields in the Geospace is necessary. It is known that there is a strong electrodynamic coupling between ionosphere and magnetosphere. A lot of data obtained by ground based (whistlers, radar techniques), balloon-borne, rocket-borne, low orbit and high orbit satellite-borne measurements, as well as theoretical studies, enabled the modeling of the global electrodynamics of the ionosphere-magnetosphere system. The results reviewed by Fälthammar (1989), can be accepted as the current state of knowledge, though the analyses of the recent satellite data (ISEE, Dynamic Explorer, Viking) are still going on. Some of previously existing models are proofed by last satellite-borne experiments (the properties of convection field in magnetosphere, auroral acceleration processes), but new outstanding questions appeared.:

a) the way in which plasma enters the magnetosphere, both from the solar wind and from the ionosphere, and b) the properties of magnetic field-aligned electric fields and their role in the auroral particle acceleration. The explanation that the polarization field from dynamo region is mapped in the magnetosphere and that the disturbances of the magnetospheric field are transferred to the dynamo region along the geomagnetic field lines, is not clear enough. Nevertheless, the evidences of such interaction exist and some of them will be mentioned here.

GEOSPACE REGIONS

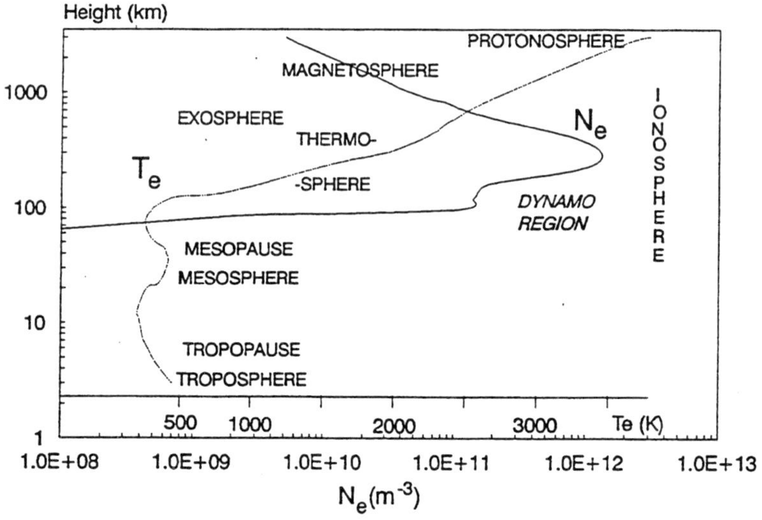

THE ION REDISTRIBUTION

The vertical distribution of electron density over the dynamo region, undergoes the regular diurnal, seasonal and solar cycle variations. In the height interval between 115 and 125 km there is a electron density enhancement during the insolation period, usually called the regular E layer. The redistribution of ions yields the disturbance in electron density profile, manifested as the formation of the thin (up to 2 km), dense layer. As it does not undergo the regular time variations like the E layer, it is named: sporadic E layer, (or Es layer). However, the statistical studies carried out in many midlatitude observatories show that there are *nonsporadic* properties of the Es layer (Whitehead, 1970; Chavdarov et al., 1975; Derblom, 1981; Grubor, 1990a; Grubor, 1991).

To understand the mechanism of ion redistribution one have to know the origin and the behavior of so called neutral winds in the ionosphere. The most complete treatment is given by the thermospheric general circulation model (TGCM), by Fesen et al., (1986). The model takes into account the following processes:

- heating of the stratosphere, mesosphere and thermosphere by insolation absorption (UV, EUV),
- ion drag effect,
- gravitational force of the Moon,
- molecular gas diffusion,
- eddy turbulence gas motion,
- Coriolis force.

All this processes together, but with different importance, cause the perturbation in temperature and composition in thermospheric plasma. The perturbations can be excited in situ or in the lower parts of the atmosphere and then propagate upwards known as the *tidal atmospheric waves*. Meridional temperature gradients cause the neutral particle flow - *tidal neutral winds*. Depending on the processes involved, the periodicity of perturbations can be expressed through diurnal, semidiurnal, terdiurnal and higher order wind modes. The equation of motion for ions in the plasma characterized by ion-neutral collisions, controlled by electric field **E** and geomagnetic field **B** is:

$$(\mathbf{E} + \mathbf{v}_i \times \mathbf{B}) - \rho B(\mathbf{v}_i - \mathbf{u}) = 0, \quad (1)$$

where **v** is the ion velocity, **u** is the velocity of neutral wind and $\rho = \nu_{in}/\omega_i$ - the ratio between ion-neutral collision frequency and ion gyro-frequency. The idea of ion redistribution as the mechanism responsible for thin, dense layer formation, dates from the early stage of ionospheric research (Dangey, 1959). However, the complex theory describing this mechanism was developed by Nygrén et al. (1984). They assumed the helical shape of the wind velocity height dependence (so-called "corkscrew" model). The main result of the theory is that the layer can be formed at the height at which a set of conditions concerning the local neutral wind pattern is fulfilled. The detailed presentation of the theoretical approach will be omitted and only the explanation of the final result will be given here: The divergence of the vertical ion velocity component, v_{iz}, at the height where the layer is to be formed, z_0, has to satisfy the relationship:

$$[dv_{iz} / dz]_{z0} = [- u(z_0) \cos I \sin\alpha_0 / (1 + \rho^2 (z_0)] \cdot \{[d\rho /dz]_{z0} - [\rho^2 (z_0) + \sin I] [d\alpha / dz]_{z0}\}. \quad (2)$$

In (2), z denotes the vertical axis, I - magnetic dip angle and α is the direction angle of wind velocity vector. Subscripts "0" refer to the height of the layer. The condition that has to be fulfilled is that the divergence of v_{iz} at certain level has to be negative, i.e. the ion motion has to converge to that level. Situation illustrated by Fig. 1, changes with time. That means that the condition (2) will be fulfilled at different heights, depending on the time of the day, following the downward phase propagation of the tidal wave. Also the whole pattern is different for different season.

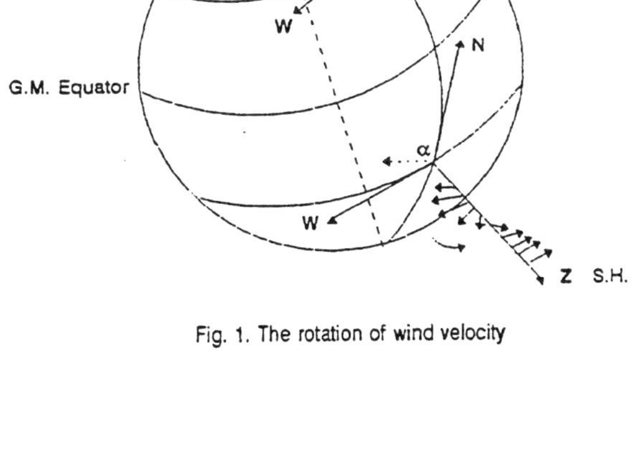

Fig. 1. The rotation of wind velocity

DETERMINATION OF THE LOCAL WIND PATTERN

The diurnal and seasonal variations of the height at which the sporadic E layer is registered by vertical sounding at Grocka observatory (44^0 38'N, 20^0 46'E), were studied, Grubor and Stupar (1988). Actually, the percentage occurrence of Es in certain height intervals was calculated. The example referring to the summer solstice is shown in the Fig. 2. Each hourly value is the average of five years hourly data, registered in the period of high solar activity of the XX solar cycle (1978-1982). As it can be seen, the layer is formed at higher altitudes in the morning hours and than descends to the so called dumping level, reaching it around the midday. The same behavior takes place in the afternoon hours. This semidiurnal pattern permanently changes to the diurnal one, starting from summer solstice to winter months, in both directions.

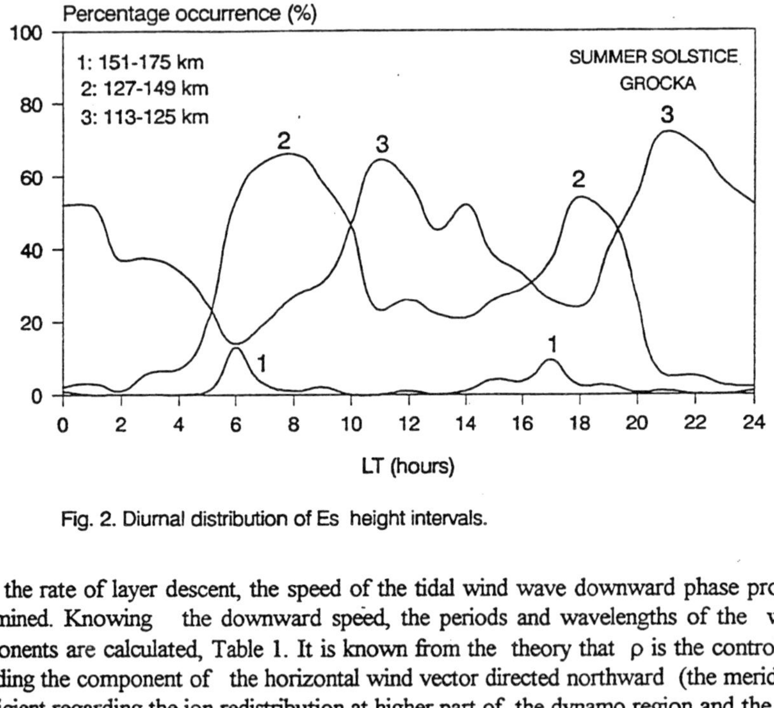

Fig. 2. Diurnal distribution of Es height intervals.

From the rate of layer descent, the speed of the tidal wind wave downward phase propagation is determined. Knowing the downward speed, the periods and wavelengths of the wind wave components are calculated, Table 1. It is known from the theory that ρ is the controlling factor, providing the component of the horizontal wind vector directed northward (the meridional - u_N) is efficient regarding the ion redistribution at higher part of the dynamo region and the component directed westward (the zonal - u_W) is efficient in the lower part of the dynamo region.

TABLE 1.The characteristics of zonal and meridional wind components at 113-125 km and 151-175 km range, respectively; τ is the period in hours and λ is the vertical wavelength in kilometers.

	113-125 km; u_W		151-175 km; u_N	
Season	τ	λ	τ	λ
Summer	9.8	64.4	10.8	71.0
Equinox	8.3	81.1	9.6	93.8
Winter	8.1	94.4	6.8	79.3

Summarizing the results of this study, the following conclusions are pointed out:

- the departure of the semidiurnal pattern for all seasons except June solstice; the pattern seems to be more terdiurnal.
- the disturbance of the descending pattern that takes place in winter months.
- the appearance of two close midday peaks in the medium height interval, 127-149 km, in winter months.
- the apparent diurnal pattern at 105-111 km height interval, and its variation with semiannual period.

These observations somewhat change the conventional standpoint about the midlatitude tidal wind pattern as predominantly semidiurnal. The information that is missing to complete the local wind pattern and that can not be determined from vertical sounding data, is the intensity of the wind velocity u(z). It is known from rocket-borne and Incoherent Radar Scatter (IRS) data that the amplitude of wind velocity, u_0, varies in the range from 50-100 m/s, depending on the solar activity, season and latitude. Assuming here that $u_0 = 100$ m/s, the variation with altitude, i.e. u(z) profile, was constructed. The elements of the local wind pattern as determined above, were used. The example is given in Fig. 3, and corresponds to height profiles of the meridional and zonal wind velocities at 0500 LT in July.

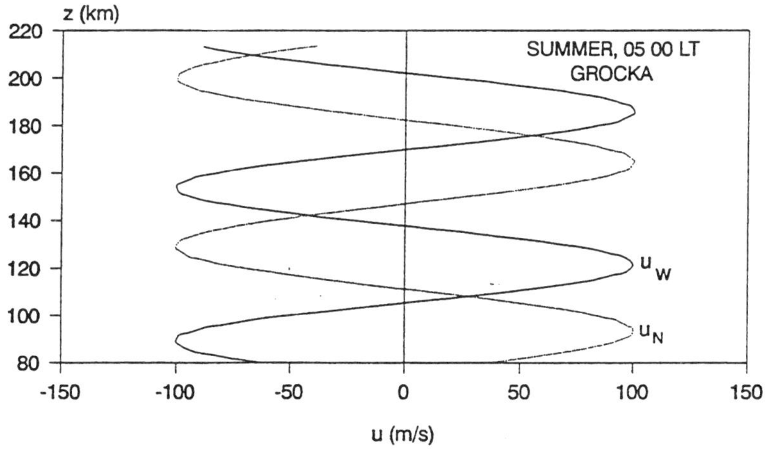

Fig. 3. Height profiles of u_N and u_W

DETERMINATION OF THE LOCAL CURRENT PATTERN

As mentioned above, the dynamo mechanism produces the current flow, and current density is given by the equation:

$$\mathbf{j} = \sigma(-\nabla V + \mathbf{u} \times \mathbf{B}), \tag{4}$$

where σ is electric conductivity tensor, V is the potential of the polarization electric field, **u** is the neutral wind velocity and **B** is geomagnetic induction. In the stationary approximation the continuity equation has to be satisfied, i.e.:

$$\nabla \mathbf{j} = \nabla\sigma(-\nabla V + \mathbf{u} \times \mathbf{B}) = 0, \qquad (5)$$

yielding the *dynamo equation* :

$$\nabla\sigma\nabla V = \nabla\sigma(\mathbf{u} \times \mathbf{B}) \qquad (6)$$

The dynamo equation is three-dimensional partial differential equation with variable coefficients, not exactly solvable in the analytical way. Introducing some approximations, convenient for ionospheric conditions, the solution, i.e. the distribution function V(r) can be found. The most successful method for solution was given by Takeda and Maeda (1980).

In the conducting layer limited in the vertical direction, as the dynamo region is, the two-dimensional approximation of the potential distribution can be adopted. According to this model (*thin shell*), the ionospheric dynamo current flows in the horizontal plane. The component of the horizontal current density directed northward (the meridional one), and directed westward (the zonal one), are given below:

$$j_N = \sigma_P (E_N - u_W B \sin I) / \sin^2 I + \sigma_H (E_W + u_N B \sin I) / \sin I \qquad (7a)$$

$$j_W = \sigma_H (E_N - u_W B \sin I) / \sin I + \sigma_P (E_W + u_N B \sin I). \qquad (7b)$$

The subscripts "H" and "P" refer to the Hall and Pedersen conductivity respectively; subscripts "N" and "W" refer to the meridional and zonal vector components, respectively. The contributions to the development of thin shell model are reviewed by Richmond (1979). It is obvious that the horizontal current density has the vertical structure, due to the vertical changes of conductivity and vertical changes of horizontal wind velocity components u_N and u_W. The polarization field is constant in the vertical direction and the change of geomagnetic induction in the range of dynamo region is negligible. Effects of dynamo mechanism registered at the Earth's surface as the variation of geomagnetic field are determined by integrated current densities:

$$J_N = \int_{z_1}^{z_2} j_N \, dz, \quad J_W = \int_{z_1}^{z_2} j_W \, dz, \qquad (8a,b)$$

where z_1 and z_2 are lower and upper limit of the dynamo region, respectively. One example of current density height profiles, obtained assuming the thin shell model and using the data from rocket measurements of the wind velocity, is given in Fig. 4, Rees (1979). Taking the vertical profiles of the electrical conductivity from Richmond et al. (1976), and using the vertical profiles of u_N and, u_W obtained as explained in section 3, the current density profiles according to 7 (a, b) were determined, Grubor (1990a). The procedure was applied for every hour of data registration and for each month. In that way the typical diurnal and seasonal changes of the current density profiles were established for Grocka location. The examples are shown in Figs. 5(a, b) and 6(a,b,c).

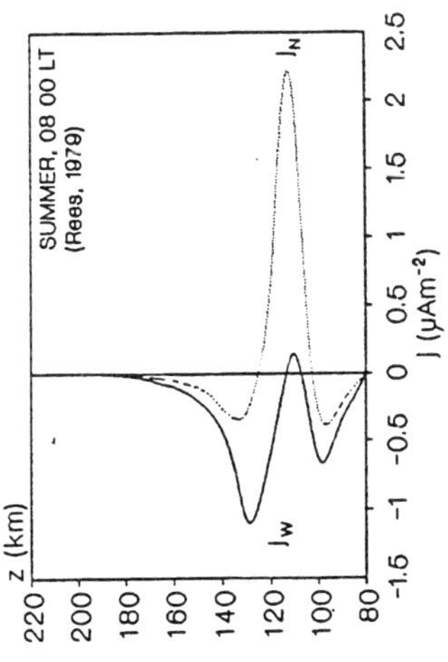

Fig. 4. Height profiles of J_N and J_W.

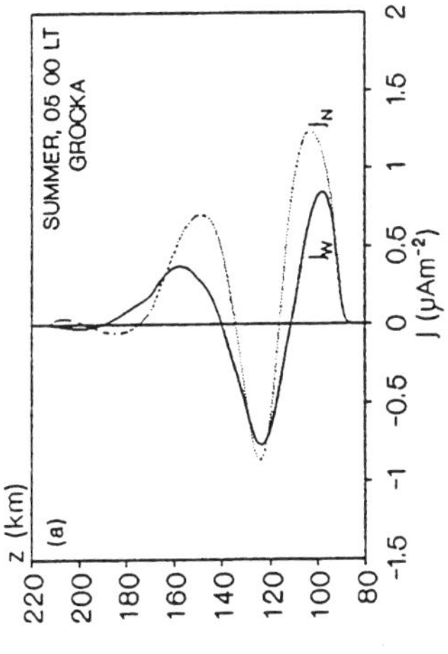

Fig. 5(a). Summer sunrise profiles.

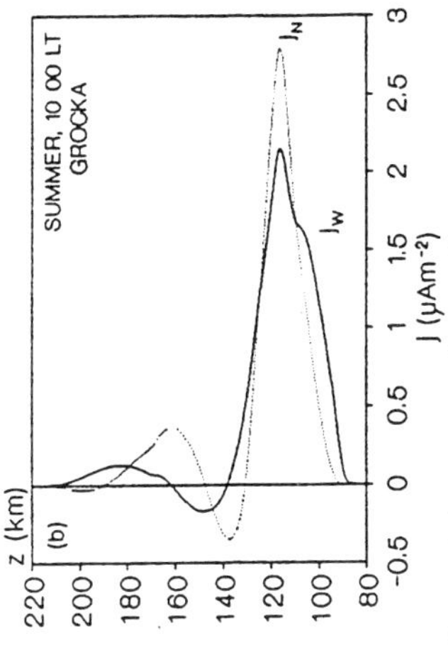

Fig. 5(b). Summer daytime profiles.

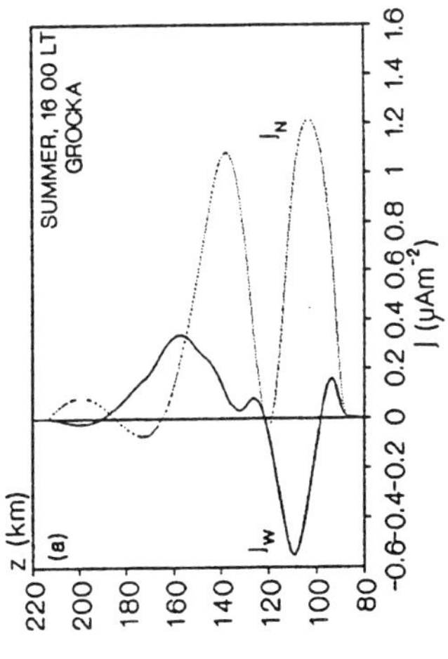

Fig. 6(a). Summer afternoon profiles

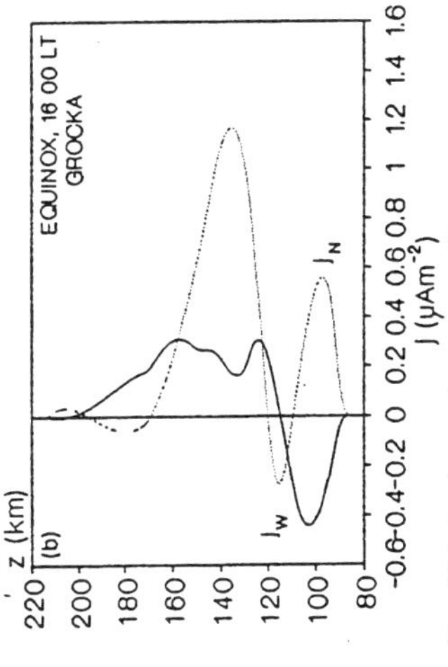

Fig. 6(b). Equinox afternoon profiles

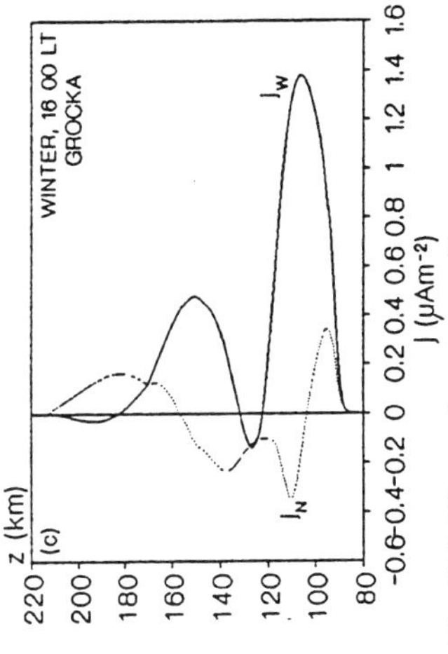

Fig. 6(c). Winter afternoon profiles

IONOSPHERIC DYNAMO INFLUENCE ON THE LOCAL GEOMAGNETIC FIELD

It is well known that the diurnal changes of geomagnetic field, so-called solar variation of geomagnetic field in quiet days or *S_q variation*, are generated by dynamo region currents, Chapman and Bartels (1940). The local S_q variation has typical diurnal and seasonal shape. The subject of this investigation is not the generation of S_q variation itself, but the *variability* of S_q variation. Namely, the day-to-day departures from the typical diurnal pattern can occur. The aim was to find out whether the day-to-day variability of S_q variation is due to the disturbances of the local dynamo current system, or finally, the local neutral wind pattern. The disturbances were deduced from the vertical sounding data, i.e. diurnal variations of sporadic E heights. Any irregularity in the typical behavior indicates that the neutral wind pattern is disturbed.

Using the Biot-Savar law applied to the ionosphere, the change of the geomagnetic induction registered at the Earth's surface, can be expressed as follows:

$$|\Delta B| = |(2\pi/10f)J| \ \ (nT), \qquad (9)$$

where J (A/km) is the density of the horizontal current, integrated over the height range of dynamo region and f~0.6 is the empirical factor taking in account the contribution of the current induced at the Earth's surface, by the dynamo current flow. The geomagnetic induction components are influenced by current density, in the following sense:

$$\Delta B_N = -J_W, \quad \Delta B_W = J_N, \qquad (10)$$

and the vertical component B_Z is not influenced by the horizontal current flow. The elements of the geomagnetic field, usually registered at geomagnetic observatories are: horizontal projection, $H = [B_N^2 + (-B_W)^2]^{1/2}$, east declination $D = \mathrm{arctg}(-B_W/B_N)$ and vertical projection $Z = -B_z$. At the midlatitudes $B_W \ll B_N \Rightarrow H \cong B_N$ and $\Delta H \cong \Delta B_N$. Therefore, instead of S_q variation the S_q (H) variation is considered. The diurnal changes of horizontal projection are registered on magnetograms as well as D and Z.

The method for detection of day-to-day variability of S_q(H), based on comparative analysis of sporadic E diurnal height distribution and diurnal H distribution, was developed, Grubor (1988). If the diurnal variation of sporadic E percentage occurrence and S_q(H) variation are plotted parallel, from month to month, it can be noticed that the shape of the S_q(H) variation is following the seasonal changes of the Es layer activity. It is not surprising that two diurnal variations are in phase: both phenomena are generated by the same mechanism - the action of the local neutral tidal winds. The following features in the diurnal changes of Es heights and S_q(H), simultaneously occur:

1 - phase disturbance,
2 - irregularities in S_q(H) shape,
3 - simultaneous phase and amplitude disturbance,
4 - isolated disturbances of S_q(H), corresponding to intense Es,
5 - winter phase pattern,
6 - S_q(H) pattern at disturbed days,
7 - the changes in S_q(H) due to solar rotation cycle (27 days).

The cases 1, 4 and 6 are discussed bellow: The example of phase disturbance is shown in Fig. 7: For two from the series of six subsequent quiet days the sets of diagrams are plotted versus the universal

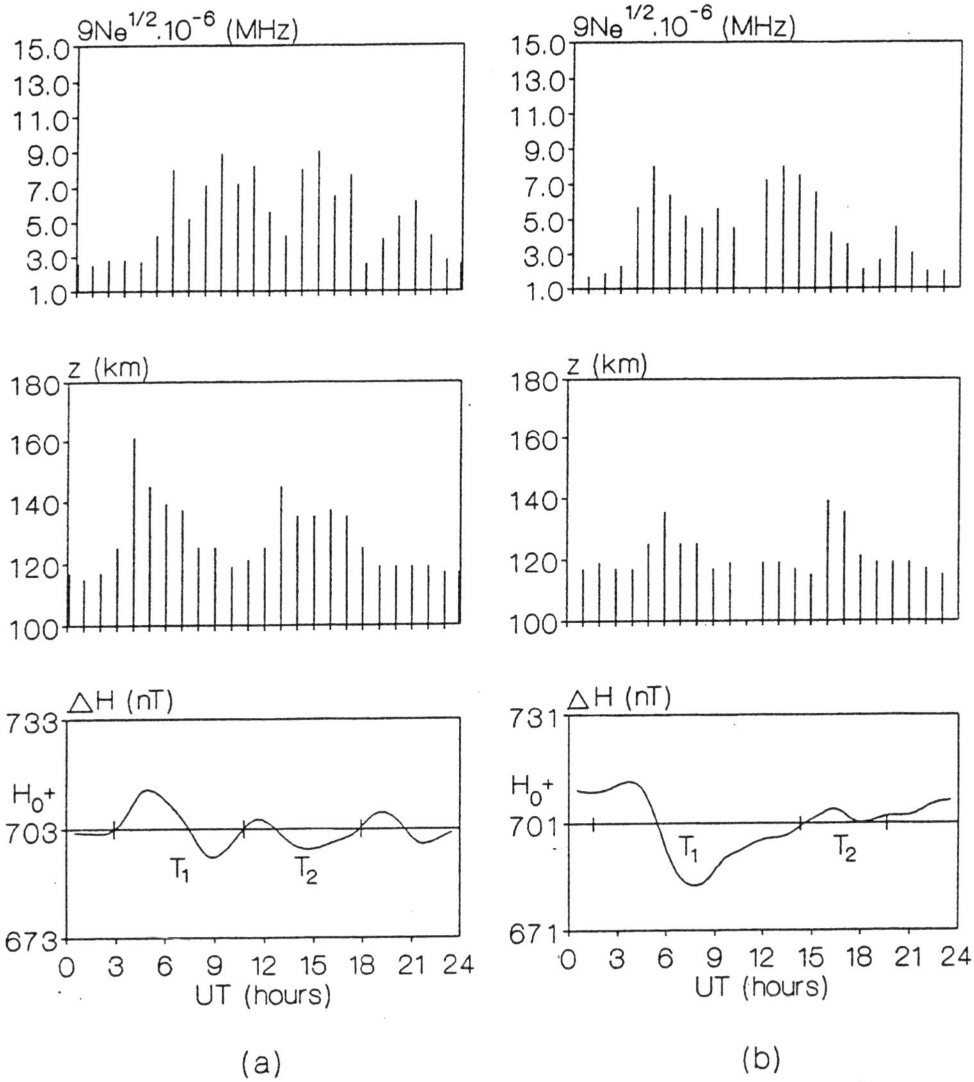

Fig. 7. (a) Normal and (b) perturbed Sq(H) phase; Ne u m^{-3}.

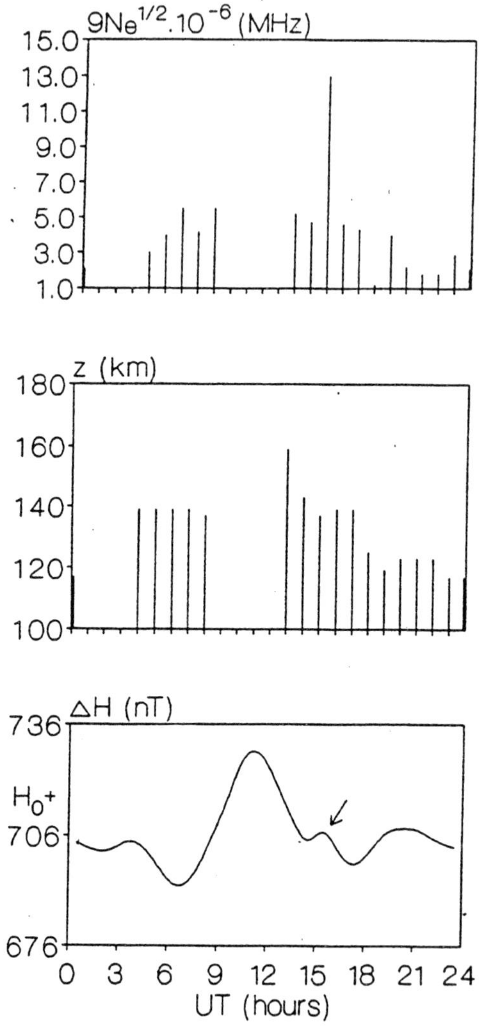

Fig. 8. Isolated disturbance in Sq(H); Ne u m^{-3}

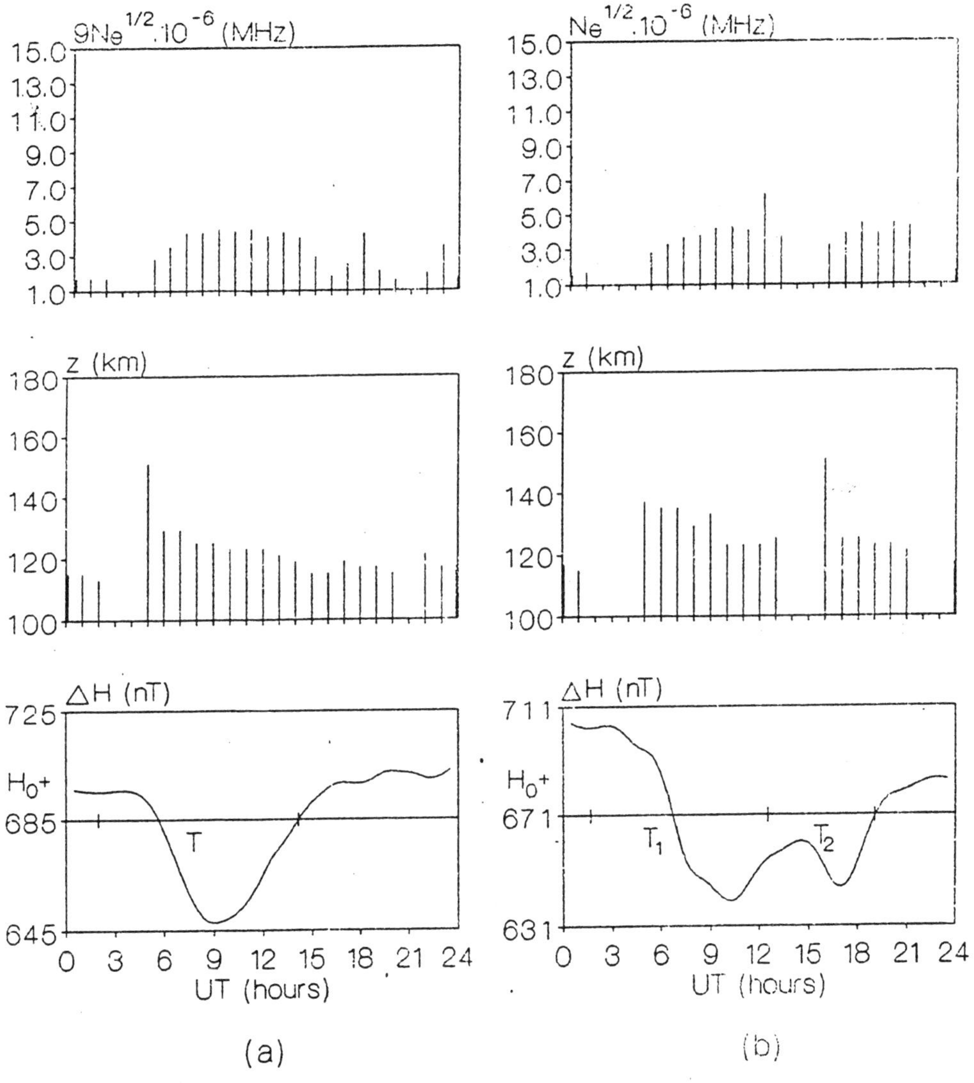

Fig. 9. (a) Quiet and (b) disturbed day; Ne u m^{-3}.

time, UT. The first set (Fig. 7a) refers to the 11 June 1981 and the second set (Fig. 7b) refers to the 13 June 1981. Each set contains the $S_q(H)$ variation (the bottom diagram), diurnal changes of Es layer heights (the medium diagram) and diurnal changes of Es blanketing frequency, fbEs, (the top diagram). S_q variation represents the diurnal oscillations of the horizontal projection, around the daily average added to the value of the permanent field H = 22000 nT.
The blanketing frequency is the maximum plasma frequency of the layer: $fbEs = 9N_e^{1/2}10^{-6}$ (MHz), if electron density, N_e, is given in (m^{-3}). As can be seen from bottom and medium diagram, the regular behavior occurs on 11 June: two periods of $S_q(H)$ variation are well defined and are coincidental with two cycles of the Es layer descent, in the course of the day. The prolongation of the "morning" cycle to the late afternoon hours in the Es height distribution and prolongation of the prenoon $S_q(H)$ period can be seen on 13 June. At the same time the shortening of the afternoon period occurs. The opposite phase disturbance (the shortening of the prenoon period), was also observed, but not shown here. The top diagram is not of interest in this case.

Another coincidental feature is the appearance of small isolated amplitude disturbances in $S_q(H)$, and appearance of very dense Es layer as can be seen from the top diagram in Fig 8. The blanketing frequencies fbEs>10 MHz, indicate the presence of the layer with electron density over 10^{12} m^{-3}, providing the ten times higher electrical conductivity than ambient one. It follows from the fact that the thin shell model of potential distribution can be applied on the Es layer, providing the essentially meridional horizontal current flow in it, Rishbeth (1983), Jalonen et al. (1984). It is shown (Grubor 1990b), that the poleward meridional current contributes to the integrated current density in dynamo region causing the enhancement of $S_q(H)$ amplitude, and the equatorward current have the opposite effect. The number of examples for both cases are found and discussed in Grubor (1990). The sudden enhancement of Es electron density can be attributed to the complex interaction between the ionosphere and magnetosphere, in which electric fields, including magnetic field-aligned fields enable the transfer of high energy particles of extraterrestrial origin into dynamo region, providing the additional ionization effects. However, the mechanism is not quite understood yet.

At the end of this expose, the comparison between the $S_q(H)$ variation on a quiet day when the phase disturbance occurs and a disturbed day without phase disturbance, is presented in Fig. 9. From the Es height variation it can be seen that the "morning" descending period is prolonged to the end of the day, (1 September 1981, Fig. 9a.), indicating the sever phase perturbation. The $S_q(H)$ variation follows this pattern, showing only one oscillation during the day. On the next day, (2 September 1981., Fig. 9b.), disturbed by magnetic substorm, the wind pattern is undisturbed, as can be deduced from two clear descending periods in Es height diurnal variations. In spite of rather disturbed H values, the $S_q(H)$ variation shows two oscillations which coincide with descending periods of Es heights. This example contribute to the opinion that the day-to-day variability of the $S_q(H)$ variation is due to the variability of the local dynamo region winds pattern (Greener and Schlapp, 1979; Briggs, 1894, rather than to the magnetic disturbances of extraterrestrial origin (Butcher, 1987).

CONCLUSION

The relationship between phenomena due to the ionospheric dynamo is approved by analyses of the data registered at Grocka observatories. The method for calculation of the neutral wind and current density height profiles on the basis of sporadic E data is established and applied to the estimation of the local ionospheric wind and current pattern. Several types of disturbances in the solar variation of the geomagnetic field in quiet conditions are observed and explained from the aspect of the ion redistribution taking place in the dynamo region of the ionosphere. All examples considered in comparative analysis of the ionospheric and geomagnetic data, suggest the causality between the day-to-day variability of the S_q(H) variation and the local dynamo region winds pattern.

REFERENCES:

Briggs, B. H., The variability of ionospheric dynamo currents, J. Atmos. Terr. Phys., 46, 419, 1984.

Butcher, E. C., Currents associated with abnormal quiet days in S_q (H), Geophys. J. R. Astr. Soc., 88, 111, 1987.

Chapman, S. and Bartels, J., Geomagnetism, Oxford Univ. Press (Clarendon), London, 1940.

Chavdarov, S. S., Chasovitin, Yu. K., Chernisheva, S. P. i Sheftel, V. M., Midlatitude ionospheric sporadic E layer, Nauka, Moskva, 1975.

Derblom, H., Nonsporadic properties of sporadic E layer, UIO-SR-81-03, Uppsala, 1981.

Dungey, J. W., Effects of a magnetic field on turbulence in an ionized gas, J. Geophys. Res., 64, 2188, 1959.

Fesen, C. G., Dickinson, R. E., and Roble R. G., Simulation of the thermospheric tides at equinox with the National Center for Atmospheric Research Thermospheric General Circulation Model, J. Geophys. Res., 91, 4471, 1986.

Fälthammar, C.-G., Electric fields in the magnetosphere -a review, Planet. Space Sci., 37, No 8, 1989.

Greener, J. G. and Schlapp, D. M., A study of day-to-day variability of S_q over Europe, J. Atmos. Terr. Phys., 41, 217, 1979.

Grubor, D. P., Observations of the day-to-day variability of the S_q (H) variation and the ion redistribution in the dynamo region, Planet. Space Sci., 36, No 11, 1165, 1988.

Grubor, D. P., Ph. D. thesis, Faculty of Electrical Engineering, University of Belgrade, 1990a.

Grubor, D. P., Possible effects of the ion redistribution in the dynamo region caused by sequential sporadic E layers, Adv. Space Res. 10, No 10, 113, 1990b.

Grubor, D., A method of including sporadic E in the ionospheric N_e (h) models, Adv. Space Res. 11, No 10, 113, 1991.

Grubor, D. P. and Stupar, P. Dj. Wind pattern variations as indicated by the effect of wind shear mechanism in the lower thermosphere, Physica Scripta, 37, 512, 1988.

Jalonen, L., Nygrén, T. and Turunen, T., On the current-carrying properties of mid-latitude type sporadic E layers, J. Atmos. Terr. Phys., 46, 383, 1984.

Nygrén, T., Jalonen, L., Oksman, J. and Turunen, T., The role of electric field and neutral wind direction in the formation of sporadic E layers, J. Atmos. Terr. Pays., 46, 373, 1984.

Rawer, K., International Reference Ionosphere - IRI 79, Report UAG-82, WDC-A, Boulder/Co., 1981.

Rees, D., Mid-latitude winds and electric fields in the lower thermosphere and their relationship with the global S_q ionospheric current system, J. Geomag. Geoelectr., 31, 267, 1979.

Richmond, A. D., Ionospheric wind dynamo theory: A review, J. Geomag. Geoelectr., 31, 287,1979.

Rishbeth, H., Further studies of directional E-layer currents, Planet. Space Sci., 31, 1177, 1983.

Takeda, M. and Maeda, H., Three-dimensional structure of ionospheric currents 1. Currents caused by diurnal tidal winds, J. Geophys. Res., 85, 6895, 1980.

Whitehead, J. D., Production and Prediction of sporadic E, Rev. Geophys. Space Phys., 8, 65, 1971.

Low Temperature and
General Plasmas
M. Milosavljevic and Z. Petrovic

ISBN 1-56072-399-8

SOLUTONS IN STRONG LANGMUIR TURBULENCE

Lj.R. Hadžievski and M.M. Škorić

Vinča Institute of Nuclear Sciences, P.O.B. 522, 11001 Belgrade, Yugoslavia

Introduction

Both for laboratory and astrophysical plasmas with a large energy content, it is typical that plasma reaches a level where nonlinear wave effects overcome the dispersive ones. In such physical environment solitons are readily formed, evolving rapidly from an arbitrary plasma state; thus determining basic features of an emerging strong plasma turbulence. Generally taken, two different global turbulence scenarios are possible, depending on whether solitons remain stable entities or alternatively being unstable, they exhibit a unique nonlinear phenomenon of a soliton collapse. It is important to stress the fundamental difference between these two turbulent descriptions, in particular, concerning the macroscopic consequences, such as absorption rates, wave energy spectra and heated particle distributions.

The simplest example of strong plasma turbulence and currently with the most developed theory is the process of Strong Langmuir Turbulence (SLT), where the main interacting modes are high frequency Langmuir and low frequency ion sound modes.

Much of the current effort in the theory of Strong Langmuir Turbulence (SLT) comes from its importance in understanding the problem of intense electromagnetic radiation of subcritical density as in laser plasma interactions (Ref. 1) and high frequency (HF) modification of the ionosphere (Ref. 2). In particular, Zakharov's model3 (Ref. 3) of a weakly magnetized SLT is frequently invoked in connection with HF heating experiments of the ionospheric F layer plasmas confined by the geomagnetic field (Ref. 2). Important additional features of SLT are efficient nonlinear heating (acceleration) mechanism of plasma particles, as well as nonlinear generation of electromagnetic field radiation at fundamental, second and higher harmonics of plasma frequency. Furthermore, nonlinear generation of zero–frequency magnetic fields by SLT has been proposed (Ref. 4). A variety of SLT applications in space and astrophysical plasmas have been recently attemped; for example: nonthermal planetary radio emission, solar phenomena, such as radio bursts (Type III) and flares and emission from pulsar magnetosphere.

On the other hand, the SLT theory is of more general interest, in particular the questions of existence, formation and stability of Langmuir solitons and their subsequent collapse toward a singularity point in a finite time. Solitons and nonstationary, soliton-like wave structures (cavitons) are expected to play a role of elementary turbulent cells in the global SLT scenario. Therefore the problem of Langmuir soliton stability and collapse is of particular importance.

The soliton stability problems according their dimensionality are naturally devided into two groups. First, there is a problem of soliton stability with respect to perturbations preserving the original soliton dimensionality. Such is the problem of Langmuir soliton stability with respect to pure one-dimensional perturbations. It was shown (Ref. 3) that in this case Langmuir solitons are always stable. To the second group belong the problems of soliton stability with respect to symmetry breaking (increasing the original soliton dimensionality) perturbations. Such is the problem of Langmuir soliton stability with respect to transverse perturbations along the direction of the original soliton homogenity. In this case Langmuir solitons in a plasma without (Ref 3.) and with (Ref. 5) an external magnetic field under the transverse perturbations with wave numbers in the range between $k=0$ and $k=k_c$ (cut-off) are always unstable.

Recent studies of the instability of weakly magnetized Langmuir solitons in the framework of the vectorial type Zakharov equations instead of using the simpler scalar model represented by the magnetized nonlinear Schroedinger (NLS) equation (Ref. 4), have revealed the detailed growth rate spectral structure in a small amplitude limit (Ref. 5), pointing to an inadequacy of the standard analytical treatments to accurately predict the instability spectra. More recently, the instability of weakly magnetized large amplitude Langmuir solitons has been numerically treated solving the corresponding linear eigen value problem (Ref. 6).

In this progress report we present numerical results of the transverse instability of weakly magnetized Langmuir solitons in the framework of Zakharov's model of SLT. We use the more realistic vectorial form of the basic equations accounting the full ion dynamics (Ref. 5, 6, 7), valid for both small and large amplitude Langmuir solitons. Results on a linear regime (an early instability stage with exponential growth) are based on two independent numerical methods. The first one is direct two-dimensional simulation of the model equations, where linear growth rate coeficients are calculated from the early time evolution of the soliton electric field. The second one is a numerical solution of the corresponding eigen value problem (Ref. 6).

Dinamical equations

Zakharov's model of a weakly magnetized SLT is represented by the system of time averaged nonlinear dynamical equations [3] describing a nonlinear coupling between the Langmuir wave potential anvelope (ψ) and the ion density variation (n) from its mean value (n_o). In convenient dimensionless units

$$t \to \frac{3}{2}\mu\, \omega_{pe}^{-1}\, t\,, \qquad r \to \mu^{1/2} r_{De}\, r,$$

$$\psi \to \frac{T}{e}\mu (12)^{1/2} \psi\,, \quad n \to \frac{4}{3}\mu n_o n\,, \quad \sigma = \frac{3}{4}\mu \left(\frac{\omega_{ce}}{\omega_{pe}}\right)^2,$$

the system reads

$$\nabla^2 (i\psi_t + \nabla^2 \psi) - \sigma \nabla_{\perp}^2 \psi - \nabla (n \nabla \psi) = 0\,, \tag{1a}$$

$$n_{tt} - \nabla^2 n = \nabla^2 |\nabla \psi|^2\,, \tag{1b}$$

where ω_{ce} and ω_{pe} are the electron cyclotron and and electron plasma frequency, respectively, μ is the ion to electron mass ratio, r_{De} is the Debye radius and T is the electron temperature in energetic units. The system (1) is justified for $\sigma << \mu$ corresponding to the physical condition of a weak magnetic field (in x-direction) $\omega_{pe} >> \omega_{ce}$, while for $\sigma=0$ it reduces to the original set of curl-free Zakharov equations (Ref. 3). The vectorial form of (1) readely simplifies to a scalar model by replacing $(-\nabla\psi)$ by scalar electric field $E(r)$ (Ref. 4). In the small amplitude (static) limit $n_{tt} << \nabla^2 n$ the set of equations (1) simplifies to a single NLS type equation. However, for large Langmuir fields the retention of the full ion inertia (1b) is essential (Ref. 5, 6).

A stationary and spatialy localized analitycal solutions of the system (1) in a form of four–parametric family of planar (1D) solitons, for the external magnetic field along x-axixs is given as

$$\psi_{sx} = \frac{\partial}{\partial x}\psi_s = \lambda\sqrt{2(1-\beta^2)}\, \mathrm{sech}\, \lambda(x-\beta t - x_o)\, \exp i\left[\left(\lambda^2 - \frac{\beta^2}{4}\right)t + \frac{\beta x}{2} + \alpha_o\right], \tag{2a}$$

$$n_s = -\frac{|\psi_{sx}|^2}{(1-\beta^2)}, \tag{2b}$$

where λ stands for the soliton strength parameter, β – for the soliton velocity, while x_o and α_o are initial soliton position and phase, respectively.

The character of these solutions depends significantly on the ratio between soliton velocity β and soliton strength parameter λ.

When

$$\frac{\beta}{\lambda} >> 1 \quad (\beta << 1)$$

the solution is an oscilating planar soliton with a quasimonochromatic filling and passing to the envelope it is possible to make additional simplification of the model equation (1) to obtain generalized nonlinear Schroedinger equation (Ref. 3).

In the oposite case

$$\frac{\beta}{\lambda} << 1 \quad (\beta << 1)$$

the solution is a simple soliton with monotonically varing electric field which in the limit $\beta = 0$ takes a form of "standing soliton"

$$\psi_s = \sqrt{2}\ \mathrm{arctan}[\ \sinh(\lambda x)\exp(i\lambda^2 t)\,. \tag{3}$$

Eigenvalue problem

In order to investigate stability of the standing soliton (3) with respect to small transverse perturbations, the solution of the system (1) will be assumed in a form

$$\psi = \psi_s + (f+ig)\exp(i\lambda^2 t)\,,\ n = n_s + u\,, \tag{4}$$

where f and g are real functions and represent the electric field potential of the transverse perturbations, while u represents the transverse perturbations of the ion density. We assume exponentially growing perturbations with a growth rate γ, harmonic in a transverse direction with a wave number $\vec{k}_\perp$

$$f, g, u \approx \exp(\gamma t + i\vec{k}_\perp \vec{r})$$

Linearizing Eq. (1) on the background of the soliton solution (3), with scaling

$$x \to \frac{x}{\lambda}\,,\quad \sigma \to \lambda^2\sigma\,,\quad u \to \lambda^2 u\,,\quad k_\perp \to \lambda^2 k\,,\quad \gamma \to \lambda^2 p$$

the following eigenvalue problem is obtained (Ref. 6)

$$\left(p^2\lambda^2 + k^2 - \frac{\partial^2}{\partial x^2}\right)u = 2\sqrt{2}\,\mathrm{sech}(x)\left(\frac{\partial^2}{\partial x^2} - k^2\right)\frac{\partial}{\partial x}f\,,$$

$$p\left(\frac{\partial^2}{\partial x^2} - k^2\right)f + \left[\frac{\partial^4}{\partial x^4} - \left(1+2k^2\right)\frac{\partial^2}{\partial x^2} + k^2(\sigma + k^2 + 1) - 2k^2\mathrm{sech}^2(x) + 2\frac{\partial}{\partial x}\mathrm{sech}^2(x)\frac{\partial}{\partial x}\right]g = 0,$$

$$-p\left(\frac{\partial^2}{\partial x^2}-k^2\right) g+\left[\frac{\partial^4}{\partial x^4}-\left(1+2k^2\right)\frac{\partial^2}{\partial x^2}+k^2(\sigma+k^2+1)-2k^2\mathrm{sech}^2(x)+2\frac{\partial}{\partial x}\mathrm{sech}^2(x)\frac{\partial}{\partial x}\right] f$$
$$-\sqrt{2}\frac{\partial}{\partial x}\mathrm{sech}(x)u = 0, \qquad (5)$$

The increment p (growth rate) of instability is given by the eigenvalues of the system (5) corresponding to the spatially localized eigenfunctions f, g and u.

Under the assumption $p^2\lambda^2 << k^2$, the system (5) turns into the corresponding eigenvalue problem in a static limit (Ref. 5)

It is evident that odd and even (with respect to x) solutions of (5) can be treated independently. Even modes correspond to marginally stable soliton deformations (Ref. 5) while odd modes are unstable with the eigenvalues p (normalized growth rate) as a function of k, σ and λ.

In the literature there exist two standard analytical methods of solutions of $p(k, \sigma, \lambda)$ in the long-wavelength limit. The first one is a perturbation method based on the local proximity of the eigenfunctions for the neutrally stable perturbations (5) (Ref. 3). The second one is variational method based on the trial functions with different number of varying parameters (Ref. 5). However, these methods although give correct results for some cases of the soliton stability problems (Ref. 3, 8) particulary for scalar dynamical equations (e.g. NLS equation) does not offer a correct answer in the case of the Langmuir soliton stability within the framework of the more complex vectorial type of the dynamical equations (Ref. 5, 6, 7).

Numerical treatment

a) Direct simulation method

Direct numerical simulation of the complete nonlinear set of equations (1) in two dimensions (2D) is based on the spectral Fourier method with respect to spatial coordinates with an explicit time integration scheme (Ref. 5).

The onset and the time evolution of the instability was followed by imposing initial conditions in a form of a standing planar 1D soliton (3), perturbed in a transverse direction

$$\psi_x(t=0) = \psi_{sx}(t=0)\,(1+2\varepsilon\cos k\,y)\,, \qquad (6)$$

where for the initial ion density we chose the adiabatic matching $n(t=0) = -|\psi(t=0)|^2$ in order to shorten plasma transients. We have used periodic boundary conditions (L_x, L_y), a numerical gridof 64x64 points (checked upon 128x128) and perturbation level $\varepsilon=0.01$ with regularly checking the conserved integrals of the system (1), plasmon number (N)

$$N = \int |\nabla\psi|^2 d\vec{r}, \qquad (7)$$

and Hamiltonian (H).

$$H = \int (\mid \nabla^2 \psi \mid^2 + \sigma \mid \nabla_n \psi \mid^2 + n \mid \nabla \psi \mid^2 + \frac{\mid \nabla \phi \mid^2}{2} + \frac{n^2}{2}) d\vec{r} \qquad (8)$$

In order to obtain a detailed spectral structure of the soliton instability we have performed runs with different values for k, σ and λ. The corresponding growth rates are calculated from 2D simulation results in the early time evolution of the maximum Langmuir energy by numerical fitting to a single exponential growth.

b) Shooting method

Another independent numerical approach to obtain the growth rate structure $p(k, \sigma, \lambda)$ is a numerical solution of the eigenvalue problem (5). It is necessary to calculate a set of eigenfuctions f, g and u which vanish at infinity with the corresponding eigenvalue p, for different values of the parameters k, σ and λ.

The fact that the solutions with a different parity are independent enables us to solve the system (5) in the interval [0, ∞). For the odd type (unstable) of the electric potential perturbations, we are looking for the solution of (5) in the interval [0, ∞) with the following boundary conditions at $x=0$:

$$f_{xxx\, x=0} = g_{xxx\, x=0} = f_{xx\, x=0} = g_{xx\, x=0} = u_{x\, x=0} = 0\,, \qquad (9)$$

while the conditions $f(0) = g(0) = 0$ based on the linearity of (5) are automatically satisfied.

On the other hand, for $x>>1$ the following asymptotic solutions of (5) are valid:

$$u_a = C_0 \exp-(p^2\lambda^2+k^2)^{1/2}x$$

$$\delta\psi_a = f_a + ig_a = (C_1 + xC_2)\exp-kx \quad (\text{for } \sigma = 0)$$

$$\delta\psi_a = f_a + ig_a = C_1 \exp \alpha x + C_2 \exp -\alpha^* x \quad (\text{for } \sigma \neq 0)\,, \qquad (10)$$

where α and α^* are the complex conjugate roots of the characteristic equation of the system (5) when $x \rightarrow 0$ with the largest negative real parts. The values of C_1 and C_2 are arbitrary complex constants, while C_0 is arbitrary real constant. By making use of (10) to calculate the values of the functions f, g and u and their derivatives at the right-hand side of the interval [0, R], $R>>1$, it is possible to solve the system (5) as a Cauchi problem and to calculate values of the functions f, g and u and their derivatives at $x = 0$. As a next step, using the boundary conditions (9) at $x = 0$, we define an auxiliary function $F(p, C_0, C_1, C_2)$ in the following way:

$$F(p, C_0, C_1, C_2) = \left[f_{xx}^2(0)+g_{xx}^2(0)+f_{xxx}^2(0)+g_{xxx}^2(0)+u_x^2\right]^{1/2}. \tag{11}$$

The function F depends on six independent parameters. Based on the linearity of (5) it is posible to fix one of them. By varying the remaining five parameters, we look for the minimum of the function F ("shooting method"). If $F=0$, obtained functions $f(x)$, $g(x)$ and $f(x)$ appear to be eigenfunctions and $p(k,\sigma,\lambda)$ an eigenvalue of the system (5) with boundary conditions (9) and (10).

Our numerical algorithm is based on two standard numerical methods. We have used the fourth-order Runge-Kutta method with an adaptive step to solve the system (5) as a Cauchy problem in a backward direction (from $x=R$ to $x=0$) and the Newton-Raphson method to perform the minimization of the function $F(p, C_0, C_1, C_2)$. The numerical procedure was started with arbitrary initial values of the parameters p^o, C_0^o, C_1^o and C_2^o and it was terminated when the value of F become smaller then 10^{-4}.

It has been demonstrated numerically that for all considered sets of the parameters k, σ and λ inside the instability region, a single minimum of F exist with the corresponding nonzero eigenfunctions $f(x)$, $g(x)$ and $f(x)$ and the eigenvalue $p(k, \sigma, \lambda)$ being independent on the initial conditions p^o, C_0^o, C_1^o and C_2^o. In the case when the values of the parameters k, σ and λ lay outside the instability region, the calculated eigenfunctions were identically equal zero for all values of x.

As an additional check we have applied our numerical algorithm to the well known problem of a soliton stability in the framework of cubic NLS equation. An approximative analitycal solution based on the variational method (Ref. 8), which is in good agriment with the numerical results in Ref. 11, is given in the form

$$\gamma = 4\frac{8}{\pi^2}\left(1-\frac{k^2}{3}\right) k^2 , \tag{12}$$

with exact values of cut-off wavenumbers ($k_{c1} = 0, k_{c2}^2 = 3$) and corresponding eigenfuctions

$$f(x) = 0\,, \quad g(x) = \text{sech}(x)\,, \ \text{for } k = k_{c_1}\,,$$

$$g(x) = \text{sech}(x)\,, \ g(x) = 0\,, \quad \text{for } k = k_{c_2}\,, \tag{13}$$

In Fig. 1. we present a comparison between above analytical results (solid lines) and our numerical results (symbols). A good agriment is evident.

a)

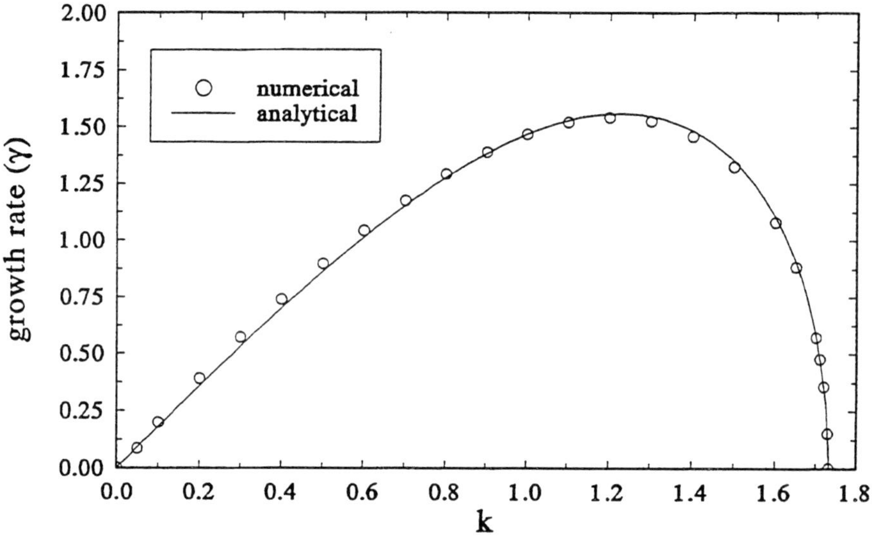

b)

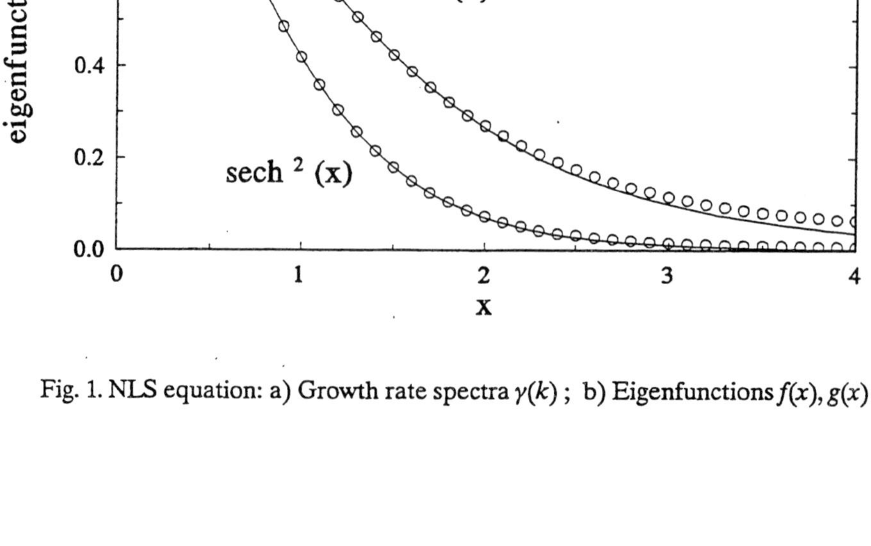

Fig. 1. NLS equation: a) Growth rate spectra $\gamma(k)$; b) Eigenfunctions $f(x), g(x)$

Results and disscussion

In order to find detailed growth rate straucture for the Langmuir soliton instability in a weak external magnetic field we have performed runs with both numerical methods for different values of the parameters k, σ and λ. In Fig. 2. we present comparison between linear eigen value results (lines) obtained numerically by "shooting" method (Ref. 6) and results based on direct 2D simulations (symbols) (Ref. 7).

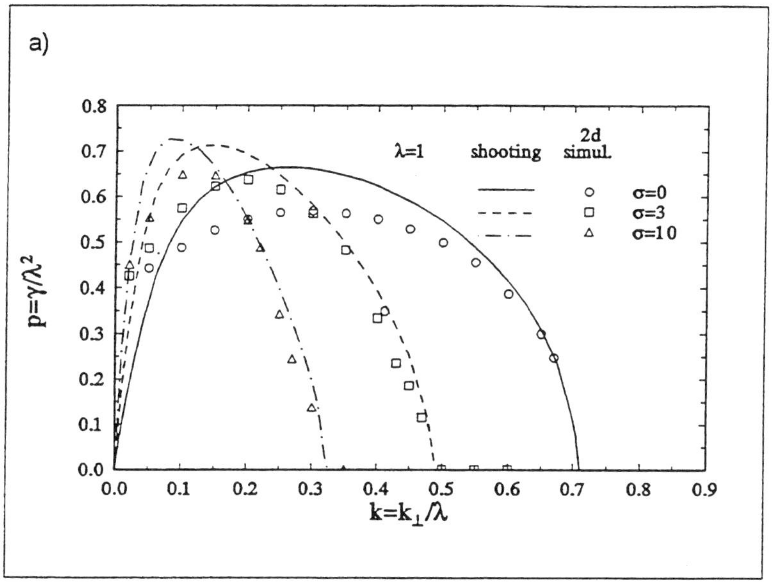

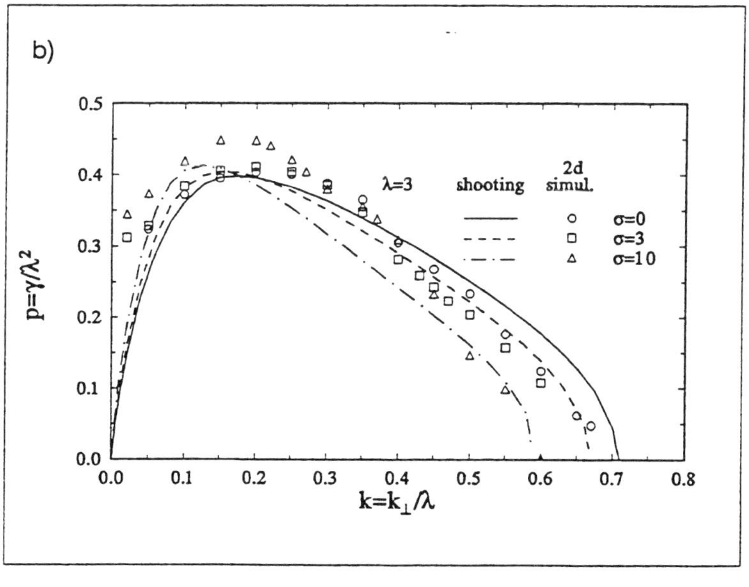

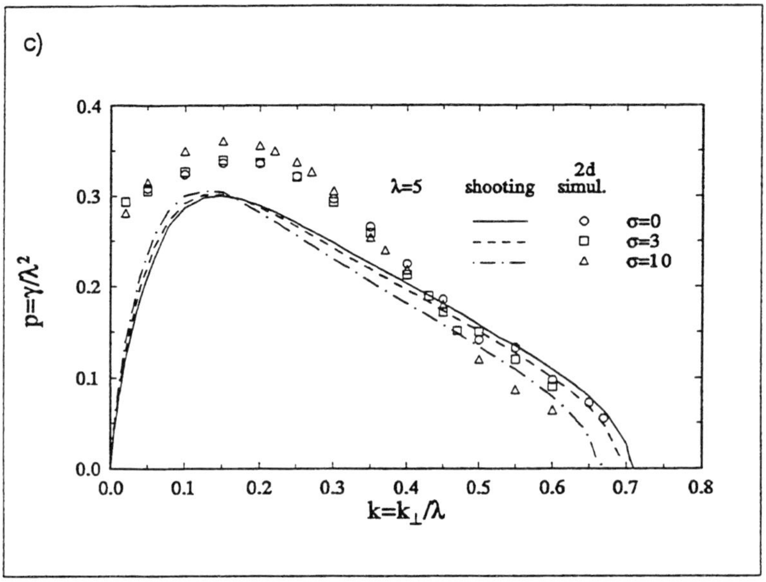

Fig. 2. Soliton instability growth rate $p(k)$ versus transverse perturbation wave number (k) for different values of the magnetic field (σ=0,3,10) and three values of the soliton strength parameter (λ): (a) λ=1, (b) λ=3 and (c) λ=5.

For all values of the parameters σ and λ spectral curves $p(k)$, ($p = {}^{\gamma}\!/\lambda^2$, $k = k_{\perp}/\lambda$) exhibit a similar structure; a single maximum between $k = 0$ and $k = k_c$ cutoff wave numbers. It is evident that, for a fixed λ, the magnetic field increase leads to a continuos shift of the maximum and cut-off wave number to the long-wavelength region, while the maximum growth rate weakly increases with s.

For larger magnetic fields the spectral dependance $\gamma(k)$ on the magnetic field strength appears to be universal (Ref. 5)

$$\gamma(k,\sigma) = \gamma(k\sqrt{\sigma}) \tag{14}$$

As the instability cutoff wave number k_c depends only on the normalized magnetic field σ_n ($\sigma_n = {}^{\sigma}\!/\lambda^2$), one can expect that for the universal dependance (14) the quantity $k_c\sigma_n^{1/2}$ has to be conserved. In Fig. 3. we show a dependence of the instability cut off wave number k_c (left y-axes) and the quantity $k_c\sigma_n^{1/2}$ (right y-axes) on the magnetic field strength σ_n revealing for large values of σ_n the generic universal dependence on $k_c\sigma_n^{1/2}$ [6].

Suppression of the magnetic field effect due to an increasing of the soliton strength is observed.

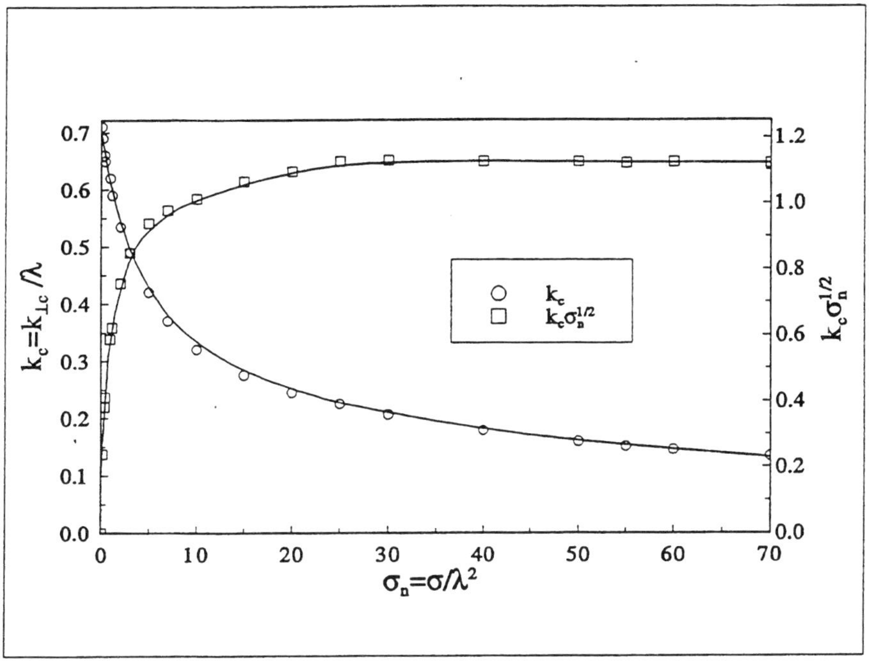

Fig. 3. Dependence of the instability cut off wave number k_c (left y-axes) and the quantity $k_c\sigma_n^{1/2}$ (right y-axes) on the magnetic field strength σ_n.

It is important to stress that the numerically calculated growth rate spectral dependance differs qualitatively from the analytical formulas obtained by the perturbation method (Ref. 12)

$$\gamma = 2k^2\{[\ 12-7\zeta(3)]\ -\ 7\sigma\zeta(3)\}\ , \tag{15}$$

and by the variational method (Ref. 5)

$$\gamma = 2\frac{[\ 12-7(\lambda_0^2+\sigma)]\ +7\zeta(3)k^4}{[\ 1+\frac{7}{2}\zeta(3)k^2\lambda_0^{-2}]^{\ 2}}\ , \tag{16}$$

where

$$\zeta(3) = \sum_n n^{-x}$$

is Reimann's zeta function

From the expression (15) it seems that for sufficiently large values of σ

$$\sigma = \frac{12 - 7\zeta(3)}{7\zeta(3)}$$

the magnetic field stabilizes the instability. On the other hand, from the expression (16) one can obtain that for sufficiently large magnetic fields islands (regions) of stability around

$$\sigma = \frac{12 - 7\zeta(3)}{7\zeta(3)} \lambda_0^2$$

should exist. However, numerical results show that the magnetic field increase does not stabilize the soliton instability and does not produce an island of stability; i.e., instability exists in the entire interval between $k = 0$ and $k = k_c$.

Further, we investigate the maxumum growth rate dependance on soliton strenght parameter for both linear eigen value (Ref. 6) and direct simulation. In Fig. 4. (in log-log scales) the growth rate maximum (γ_{max}) versus the soliton strength (λ) covering the range from small (static limit) to large (supersonic) Langmuir soliton amplitudes in the Zakharov model of SLT is given. The power law like $\gamma_{max} \sim \lambda^{2\alpha}$ dependance recovers analytically predicted values $\alpha=1$ in the static limit region ($\lambda<1$) and $\alpha<1$ for the hydrodynamic region ($\lambda>1$).

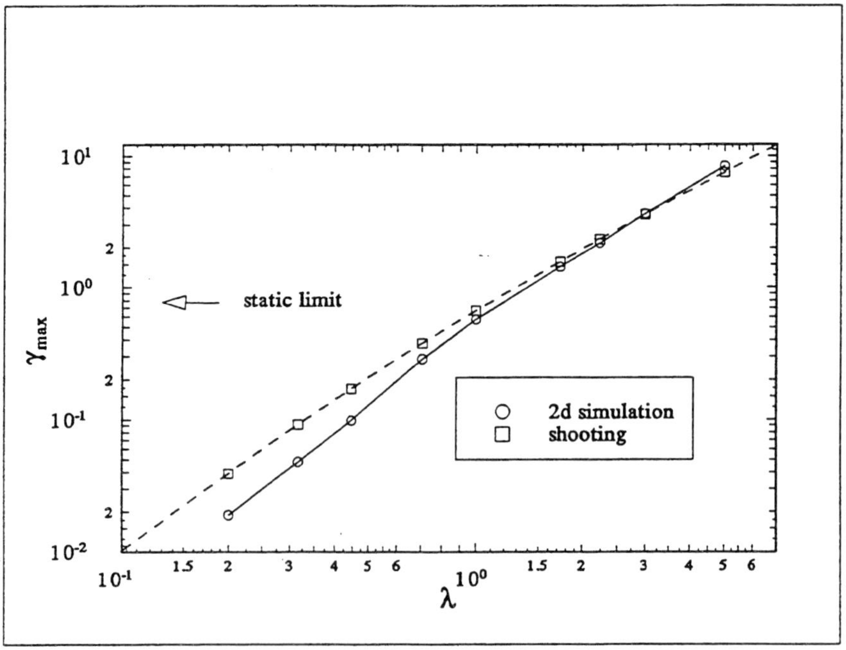

Fig. 4. The growth rate maximum (γ_{max}) versus the soliton strength (λ) covering the range from small (static limit) to large (supersonic) Langmuir soliton amplitudes in the Zakharov model of SLT.

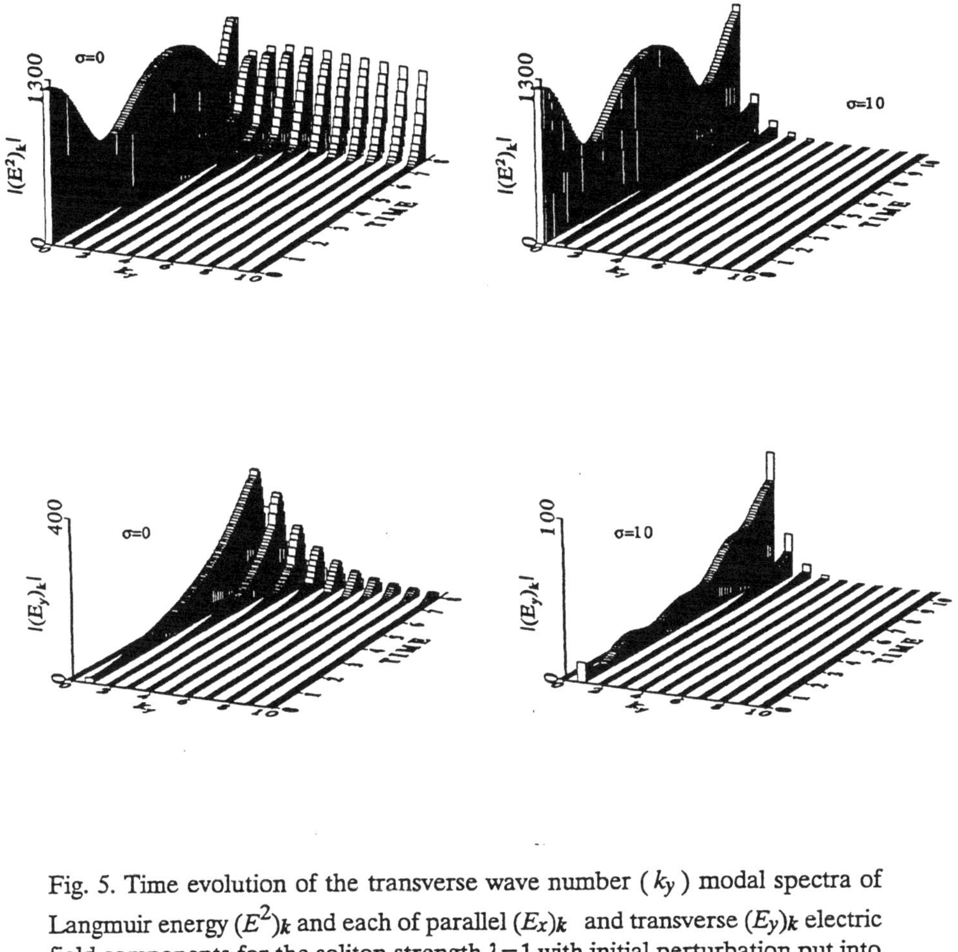

Fig. 5. Time evolution of the transverse wave number (k_y) modal spectra of Langmuir energy $(E^2)_k$ and each of parallel $(E_x)_k$ and transverse $(E_y)_k$ electric field components for the soliton strength $\lambda=1$ with initial perturbation put into the first harmonic $k_y=k_\perp=0.25$ for zero ($\sigma=0$) and weak ($\sigma=10$) magnetic field, where k_y is in the units of $k_\perp$ .Only the first 11 of 32 available harmonics for $k_y>0$ are plotted due to the symmetry k_y ky space.

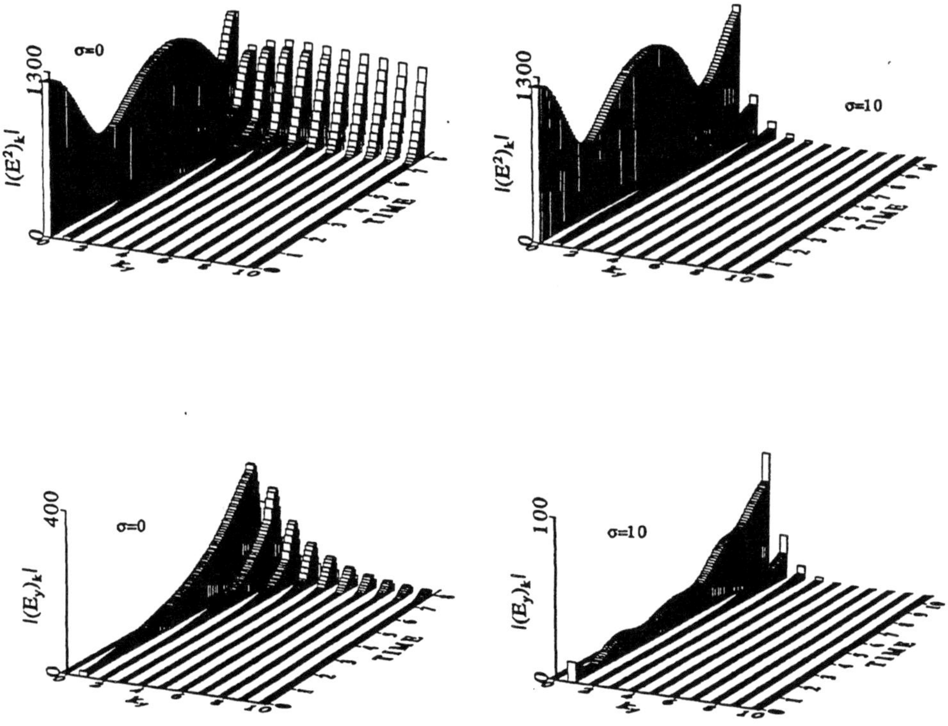

Fig. 6. Time evolution of the transverse modal spectra of Langmuir energy electric field components as described in Fig. 4., except for the increased soliton strength $\lambda=5$.

Finally, we expose the time resolved spectral structure of Langmuir energy together with electric field components in order to enlighten the onset, instability development and highly nonlinear collapse stage (Ref. 11). In Fig. 5. we present the time evolution of the transverse wave number (k_y) modal spectra of Langmuir energy $(E^2)_k$ and each of parallel $(E_x)_k$ and transverse $(E_y)_k$ electric field components for the soliton strength $\lambda=1$ with initial perturbation put into the first harmonic $k_y=k_\perp=0.25$ for zero ($\sigma=0$) and weak ($\sigma=10$) magnetic field. Periodic oscillations of the zero harmonic mode $k_y=0$ correspond to "eigen frequency" ($\sim\lambda^2$) of the standing Langmuir soliton (3). The growth of higher harmonics is readily observed in an early "linear" stage.

Moreover, in Fig. 6. we give the time evolution of the transverse modal spectra of Langmuir energy electric field components as described in Fig. 5., except for the increased soliton strength $\lambda=5$. As compared to small amplitude case (Fig. 5.) pronounced effect of nonlinear frequency shift changing the period of the zero mode oscillations, together with spectral broadening of growing higher modes is readily observed. However, increasing magnetic fields (σ) narrows region of unstable modes. The stage of Langmuir soliton collapse is followed by the rapid growth of higher order harmonics as a result of an intense spatial localization, still for increasing σ smaller number of harmonics participate in an collapse process.

References

1. A.M. Rubenchik and V.E. Zakharov, in Handbook of Plasma Physics, vol. 3 Physics of Laser Plasmas, eds. A.M. Rubenchik and S. Withowski, p. 335 (1991) Elsevier Science

2. D.F. DuBois, H.A. Rose and D. Russel, Phys. Rev. Lett. 66,1970 (1991)

3. E.A. Kuznetsov, V.E. Zakharov and A.M. Rubenchik, Phys. Rep. 142, 103 (1986)

4. M. Kono, M.M. Škorić and D. ter Haar, J. Plasma Phys. 26, 123, (1981)

5. Lj.R. Hadžievski, M.M. Škorić, A.M. Rubenchik, E.G. Shapiro and S.K. Turitsin, Phys. Rev. A 42, 3561 (1990)

6. Lj.R. Hadžievski and M.M. Škorić, Phys. Fluids B 3, 2452 (1991)

7. Lj.R. Hadžievski and M.M. Škorić, Phys. Fluids B 5, 2076 (1993)

8. K. Rypdal and J.J. Rasmussen, Phys. Scr. 40, 192 (1989)

9. L. Berge, P. Donssean, G. Pelletier and D. Pesme, Phys. Rev. A 42, 4952 (1990)

10. N.R. Pereira, R.N. Sudan and J. Denavit, Phys. Fluids 20, 936 (1977)

11. Lj.R. Hadžievski and M.M. Škorić, ESA SP-351, 281 (1993)

Low Temperature and
General Plasmas
M. Milosavljevic and Z. Petrovic

ISBN 1-56072-399-8

NONLINEAR SATURATION OF STIMULATED RAMAN SCATTERING IN A PLASMA

M.S. Jovanović[1,2] *and M.M. Škorić*[2]

[1] Department of Physics, University of Niš, P.O. Box 91, 18001 Niš, Yugoslavia

[2] Vinča Institute of Nuclear Sciences, P.O. Box 522, 11001 Belgrade, Yugoslavia

1. INTRODUCTION

Propagation of intense laser light in a plasma produces numerous nonlinear processes [1,2]. The so-called laser plasma instabilities [1,3] couple between plasma modes, provided a laser–supported electromagnetic wave (EMW) appears as an incident (pump) wave, that seeds its energy to other waves. With regard of the principle that the three waves, taking part in the nonlinear three–wave interaction, can not belong to the same mode, a limited number of different types of electromagnetic decay are feasible: stimulated Raman scattering, stimulated Brillouin scattering, two–plasmon decay and ion–acoustic decay. In addition to these processes the laser light can produce density modulations which lead to either self-focusing or filamentation. All the above processes occur in the underdense plasma of laser fusion target, ie. in the area where plasma density is lower than the critical density value: $n_{cr} = \varepsilon_0 m_e \omega_0^2/e^2$, (the word "critical" coming from the fact that plasma frequency ω_{pe} equals the laser frequency ω_0 at the critical surface).

We put our interest here into stimulated Raman scattering (SRS) [1,2–4] process which couples an incident EMW to another EMW, referred to as a scattered wave, and an electron plasma wave (EPW), assuming the latter two waves grow from the noise level in plasma. Perfect matching conditions are assumed here:

$$\omega_0 = \omega_1 + \omega_\ell, \quad \mathrm{k}_0 = \mathrm{k}_1 + \mathrm{k}_\ell, \tag{1}$$

where indices '0', '1' and 'ℓ' relate to the pump, scattered EMW and EPW (Langmuir wave), respectively. Consequently, the process is possible for $\omega_0 \geq 2\omega_{pe}$ which implies $n_0 \leq \frac{1}{4} n_{\mathrm{cr}}$.

Raman backscattering instability is of a particular interest for two reasons: it has the maximum growth rate $\gamma_0 = \frac{1}{4} k_\ell v_{osc} \sqrt{\omega_0/\omega_1}$, where

$$v_{osc} = \frac{eE_0}{m_e\omega_0}, \quad \omega_1 = \omega_0 - \omega_{pe}, \quad k_\ell = \frac{\omega_0}{c}\left(\sqrt{1-\frac{\omega_{\mathrm{pe}}^2}{\omega_0^2}} + \sqrt{1-2\frac{\omega_{\mathrm{pe}}}{\omega_0}}\right),$$

and the backward SRS carries the laser energy out of plasma and, thus, reduces the efficiency of energy deposition into the fusion target.

SRS growth is a fast process in regard to other instabilities. Therefore, laser *ns* pulses, which are usually applied today, are long enough for the scattered EMW to attain the level of the pump. At that time linear theory ceases to describe the system correctly, and nonlinear theory comes indispensable for further prediction of the wave evolution.

According to the nonlinear theory, the pump EMW is also subject to temporal and spatial variations, so the terms coupling among different modes will appear in the corresponding equations. It has recently become clear that SRS is often absolute in non-uniform plasmas [5] and the problem of the nonlinear saturation has come to a spotlight of SRS investigations. Recent workers have applied rather complex models, incorporating SRS–SBS coupling [6–8] or a parametric decay of SRS-driven EPW, based on Zakharov–Maxwell equations [9,10]. Their results support particular features of SRS noted by experimental workers [3,11]. Still, we estimate that the results of similar importance can be derived from a much simpler generic model, such as the one described here [12,13].

We consider a nonlinear saturation of SRS as a paradigm of a nonlinear three-wave interaction (3WI) in a spatially extended system [14,15] as it is explained in the following. Our model of nonlinear saturated SRS includes the pump-depletion effect as well as the generic nonlinear frequency shift (NLFS) [16,17], induced by a large amplitude EPW, that could be of a ponderomotive [18] and/or a relativistic origin [19,20]. This nonlinearity, when large enough, significantly affects the SRS evolution and causes a transition of the system from a regular to a strongly nonlinear incoherent regime, or spatio-temporal chaos [21,22].

2. BASIC MODEL EQUATIONS

In a hydrodynamic approach, chosen here to be applied, one starts from Maxwell and fluid equations for electron component of the plasma (since ions are immobile, serving to provide plasma electroneutrality):

$$\nabla \cdot \mathbf{E} = -e\delta n_e, \quad \nabla \times \mathbf{E} = -\frac{\partial \mathbf{B}}{\partial t}, \quad \nabla \times \mathbf{B} = -e\left(n_0 + \delta n_e\right) + \frac{\partial \mathbf{E}}{\partial t},$$

$$\frac{\partial \delta n_e}{\partial t} + \nabla \cdot \left[\left(n_0 + \delta n_e\right)\mathbf{v}_e\right] = 0, \tag{2}$$

$$\frac{\partial \mathbf{v}_e}{\partial t} + \left(\mathbf{v}_e \cdot \nabla\right)\mathbf{v}_e = -\frac{e}{m_e}\left(\mathbf{E} + \mathbf{v}_e \times \mathbf{B}\right) - \frac{3v_{th}^2}{\left(n_0 + \delta n_e\right)}\nabla \delta n_e - \nu_e \mathbf{v}_e.$$

We assume a plasma slab of a uniform initial density profile and of a finite spreading L. Spatial uniformity is certainly an idealization, however, it is evident that the corona density in fusion plasmas is a weak function of space coordinates, thus the characteristic length scale of the density variation is much greater than the laser light wavelength, and the uniformity assumption appears justified. Laser beam irradiates plasma perpendicularly from $x < 0$ and enters the plasma slab with an initial amplitude $\mathcal{E}_0$. Scattered EMW and EPW grow from thermal noise levels (ϵ_1 for EMW and ϵ_ℓ for EPW). The EPW suffers the linear damping characterized by its rate ν_e. Under the assumption that the Raman instability growth rate γ_0 is lower than all characteristic frequencies in the problem ($\gamma_0 < \omega_0, \omega_1, \omega_{pe}$), it is convenient to make use of slowly-varying envelope approximation. Thus, if $\mathbf{E}_0(x,t)$ and $\mathbf{E}_1(x,t)$ are electric field vectors associated with the pump and scattered EMW's, and $\delta n_e(x,t)$ – plasma density variation in Raman-driven EPW, we may write in the one-dimensional case:

$$E_{0,1}(x,t) = \mathcal{E}_{0,1}(x,t)\exp(i\omega_{0,1}t \pm ik_{0,1}x), \quad \delta n_e(x,t) = \rho(x,t)\exp(i\omega_\ell t - ik_\ell x), \quad (3)$$

where $\mathcal{E}_0$, $\mathcal{E}_1$ and ρ are slowly-varying wave amplitudes, whose second derivatives in x and t are assumed negligibly small.

For practical purposes the case of backward SRS is of particular interest. Therefore, combining Maxwell and fluid equations, and assuming resonant matching (1) the following system describing spatio-temporal evolution of weakly-coupled waves in stimulated Raman backscattering, is obtained [4,13]:

$$\frac{\partial a_0}{\partial \tau} + V_0\frac{\partial a_0}{\partial \xi} = -a_1 b, \quad \frac{\partial a_1}{\partial \tau} - V_1\frac{\partial a_1}{\partial \xi} = a_0 b^*, \quad \frac{\partial b}{\partial \tau} + V_\ell\frac{\partial b}{\partial \xi} + \Gamma b = a_0 a_1^*, \quad (4a,b,c)$$

where $\xi = x/L$, $\tau = \omega_0 t$, and

$$a_{0,1}(\xi,\tau) = \frac{k_\ell v_{osc}}{4\omega_1}\sqrt{\frac{\omega_{pe}}{\omega_{0,1}}}\frac{\mathcal{E}_{0,1}(x,t)}{E_0}, \quad b(\xi,\tau) = -i\frac{\omega_p^2}{4\omega_0\sqrt{\omega_0\omega_1}}\frac{\rho(x,t)}{n_0}, \quad (5)$$

v_{osc} designating the electron quiver velocity in the pump EMW electric field ($v_{osc} = eE_0/m_e\omega_0$); Γ – normalized EPW damping rate ($\Gamma = \nu_e/2\omega_0$), while (normalized) group velocities are: $V_0 = k_0c^2/\omega_0^2 L$, $V_1 = k_1c^2/\omega_0\omega_1 L$, and $V_\ell = 3k_\ell v_{th}^2/\omega_0\omega_\ell L$.

The most important information on the Raman backscattering is its reflectivity, designating a fraction of incident laser intensity reflected through the SRS process:

$$R = \frac{V_0\,|a_1(0)|^2}{V_1\,|a_0(0)|^2}. \quad (6)$$

We note that the expression (6) normalizes the maximum reflectivity value to unit in the steady state.

The equations (4) require the appropriate initial and boundary conditions. We have chosen the ones involving the nonzero source assumption [23,24]: $\mathcal{E}_0(x,0) = 0$ (except $\mathcal{E}_0(0,0) = E_0$), $\mathcal{E}_0(0,t) = E_0$, $\mathcal{E}_1(x,0) = \mathcal{E}_1(L,t) = \epsilon_1 E_0$, and $\delta n_e(x,0) = \delta_e(0,t) = \epsilon_\ell n_0$.

3. STABILITY OF THE STATIONARY RAMAN BACKSCATTERING SATURATED BY PUMP DEPLETION

In analysis of the system of equations (4), the first task was to study a possibility of establishing of steady state characterized by stationary wave amplitudes. We assume the pump depletion mechanism leading to a saturation of SRS and its transition to a steady state (in which all temporal derivatives are neglected). According to the system of equations (4), the three wave envelopes in such a state are described by the following system:

$$V_0\frac{dA_0}{d\xi} = -A_1B, \quad V_1\frac{dA_1}{d\xi} = -A_0B^*, \quad V_\ell\frac{dB}{d\xi} + \Gamma B = A_0A_1^*, \tag{7}$$

where $A_0(\xi)$, $A_1(\xi)$ and $B(\xi)$ are time-independent spatial envelopes of the pump, backscattered and plasma waves, obeying the boundary conditions indicating that the daughter waves raise from the nonzero thermal noise level:

$$A_0(0) = \frac{k_\ell v_{osc}}{4\omega_1}\sqrt{\frac{\omega_{pe}}{\omega_0}}, \quad A_1(1) = \frac{k_\ell v_{osc}}{4\omega_1}\sqrt{\frac{\omega_{pe}}{\omega_1}}\epsilon_1, \quad B(0) = \frac{\omega_{pe}^2}{4\omega_0\sqrt{\omega_0\omega_1}}\epsilon_\ell. \tag{8}$$

In the case of zero EPW damping ($\nu_e = 0$), the wave amplitudes are expressed through the spatially periodic Jacobi elliptic functions [13]:

$$A_0(\xi) = \mathrm{dn}\left(\frac{1}{\sqrt{V_1V_\ell}}\xi, \sqrt{\frac{R}{V_0V_1}}\right), \qquad A_1(\xi) = \sqrt{R}\,\mathrm{cn}\left(\frac{1}{\sqrt{V_1V_\ell}}\xi, \sqrt{\frac{R}{V_0V_1}}\right),$$
$$B(\xi) = \sqrt{\frac{V_1}{V_\ell}R}\,\mathrm{sn}\left(\frac{1}{\sqrt{V_1V_\ell}}\xi, \sqrt{\frac{R}{V_0V_1}}\right). \tag{9}$$

Stationary solutions for the backscattered EMW $A_1(\xi)$ are presented in Fig. 1 for the laser and plasma parameters used in our computations ($L = 20\lambda_0$, $n_0 = 0.1n_{cr}$, $\kappa T_e = 2\,keV$, $v_{osc} = 0.1c$, $\nu_e = 0$, $\epsilon_1 = 0.01$, $\epsilon_\ell = 0$).

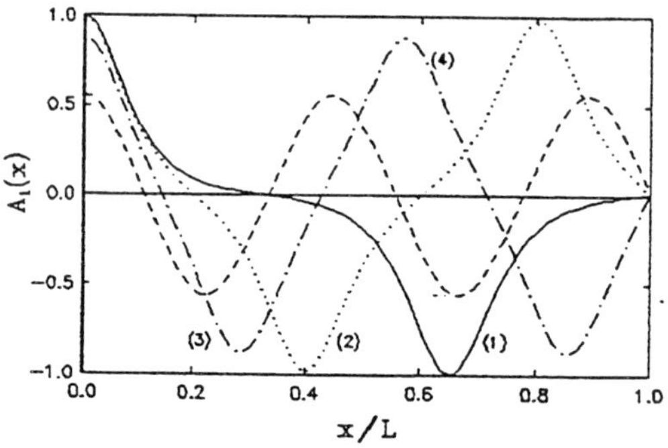

FIGURE 1: Spatial dependence of the stationary backscattered electromagnetic wave amplitudes, driven by SRS, for the case of zero wave damping, and for laser and plasma parameters: $L = 20\lambda_0$, $n_0 = 0.1n_{cr}$, $\kappa T_e = 2\,keV$, $v_{osc} = 0.1c$, $\nu_e = 0$, $\epsilon_1 = 0.01$, $\epsilon_\ell = 0$. *Four backscattering modes correspond to the reflectivity (R) values: (1) 0.999 (two nodes on the x-axis), (2) 0.967 (two nodes), (3) 0.777 (four nodes) and (4) 0.313 (four nodes).*

It is seen from Fig. 1 that the system of equations (7) for steady–state wave amplitudes has a multiple spectrum of eigensolutions. This spectrum is discrete and contains a finite number of scattering modes. Thus, for the chosen parameters there exist four possible values of R (0.313, 0.777, 0.967, 0.999) that match the boundary conditions (8). All the modes are spatially periodic, with various numbers of nodes on the x–axis (4, 4, 2 and 2, respectively), and are fitted to EMW noise 0.01 on the rear boundary ($x = L$).

Whether any of the eigenmodes will be established in simulations is a question of a particular interest, whose answer can be obtained by means of linear stability analysis [25]. This method investigates a behaviour of small perturbations added to the calculated stationary solutions:

$$a_{0,1}(\xi,\tau) = A_{0,1}(\xi) + \delta a_{0,1}(\xi,\tau), \quad b(\xi,\tau) = B(\xi) + \delta b(\xi,\tau),$$

which are assumed to show an exponential growth (or decrease):

$$\delta a_0,\ \delta a_1,\ \delta b \sim \exp(\gamma\tau),$$

where the sign of the eigenvalue γ decides whether the solution is stationary indeed ($\gamma < 0$) or is subject to a decay ($\gamma > 0$). The four periodic modes plotted above appear all with positive γ's, and consequently unstable, which is to be confirmed by a direct simulation.

The Raman process exhibits a high sensitivity to the plasma length. Thus, if the length increases, the above modes still remain active, but their reflectivities tend to unit and their nonlinear instabilities decrease. In addition, with increasing L the conditions are created for appearance of new periodic modes, with more and more nodes on x–axis (see Fig. 2). Moreover, all of them are unstable, as shown in Fig. 3.

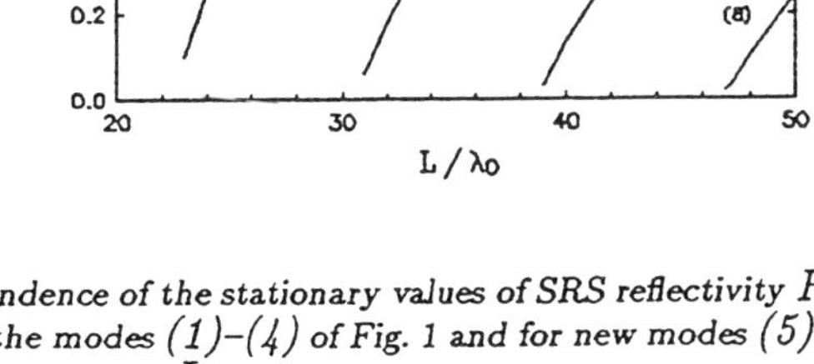

FIGURE 2: Dependence of the stationary values of SRS reflectivity R on the plasma length L/λ_0 for the modes (1)–(4) of Fig. 1 and for new modes (5)–(8) appearing successively with increasing L.

However, the direct simulation of the system of equations (7) for the no-damping case gives the time evolution of the reflectivity R, demonstrated in Fig. 4, which does not show any instability; moreover, the reflectivity tends asymptotically to the state with R close to unity. This state, by inspection, appears aperiodic in

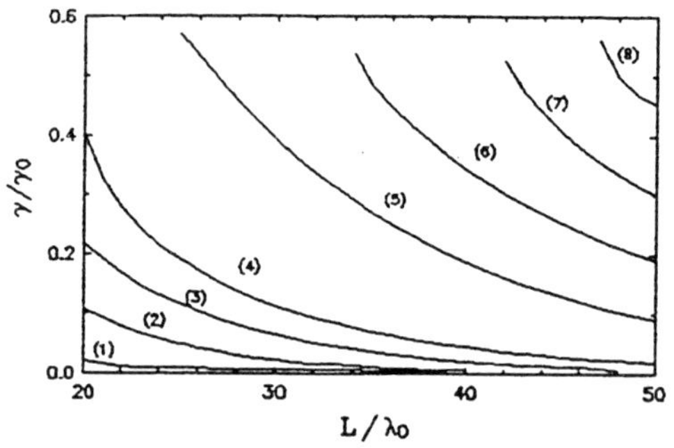

FIGURE 3: Dependence of the instability of stationary SRS states on the plasma length L/λ_0 for the modes of Fig. 2.

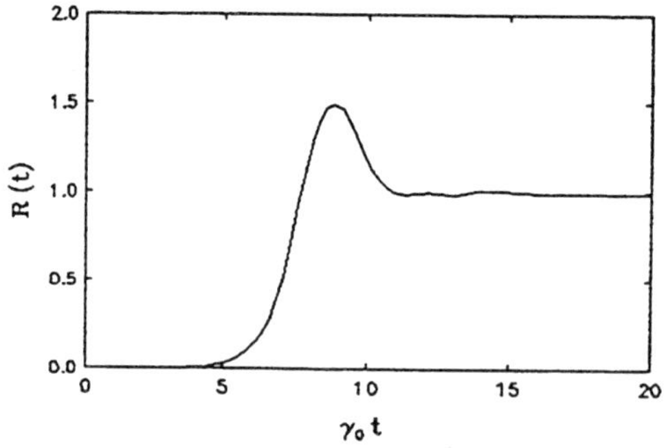

FIGURE 4: Results of direct numerical simulation of the equations describing SRS process: time evolution of SRS reflectivity $R(t)$ in the plasma with zero EPW damping for the parameters of Fig. 1.

space and is quasistationary in nature, because it does not result from the stationary amplitude analysis.

In realistic plasmas the waves are always more or less damped. Namely, if electron–ion collisions are rare, the collisionless (Landau) damping is inevitably present. For this reason, an effect of the EPW damping on the nonstationarity of the backscattering process was studied. With nonzero damping term ($\Gamma > 0$) the system of differential equations (7) for steady–state wave amplitudes ceases to be analytically solvable. Thus, the system (7) was treated numerically, by using shooting method [26]. The wave envelopes are determined in the whole interaction region. Variations of the stationary reflectivity values for the above mentioned SRS backscattering modes *(1)–(4)* by various values of the EPW damping rate ν_e/ω_{pe}

are presented in Fig. 5.

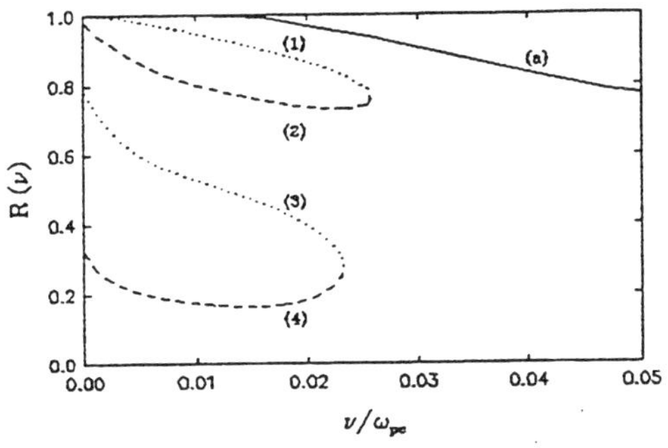

FIGURE 5: Stationary SRS reflectivity R vs. EPW damping rate for the periodic modes (1)–(4) of Fig. 1, and for the aperiodic mode (a) appearing for EPW damping values $\nu_e > 0.009\omega_{pe}$.

The curve *(a)* relates to a spatially aperiodic mode which appears when the EPW damping exceeds a certain critical value (approximately 0.009 in our case). For greater values of EPW damping rate (greater than 0.025 in our case) the aperiodic mode is only present.

The shooting method has been used again to investigate the stability of the scattering modes presented in Fig. 5. Stationary SRS modes periodic in space *(1)–(4)* appear, again, linearly unstable to small perturbations, in the whole region of their existence. The growth rate of this instability is comparable to Raman instability linear growth rate γ_0. In that case the stationary state may be not established at all, or be disintegrated after its establishment. For EPW damping $\nu_e/\omega_{pe} > 0.009$ an aperiodic scattering mode appears as an exact solution of the system (7). For greater values of the damping ($\nu_e/\omega_{pe} > 0.009$) there exist no spatially periodic stationary solutions and the aperiodic one is only possible. Applying the former method, the aperiodic mode turned out stable. For instance, for the damping $\nu_e/\omega_{pe} = 0.1$ we obtained the SRS saturated state instability $\gamma/\gamma_0 = -1.583$. This indicates that SRS for such a set of parameters is stable. Thus, for the damping values usually met in laser plasmas, the present analysis, which does not take into account the other nonlinear effects on the EPW, such as nonlinear frequency shift, nonlinear damping, strong coupling, influence of the SBS etc., predicts that the SRS process evolves to a final steady state. Similar conclusions were obtained for the case of SBS in a bounded plasma [27,28].

In conclusion of this section, it was found that stimulated Raman backscattering is steady in a finite plasma with moderate linear damping of electron plasma waves. In distinction, in the case of sufficiently small damping effect, below some critical value, the stimulated backscattering process becomes intrinsically nonstationary, relaxing only asymptotically in time to a quasistationary state with high reflectivity ($R \to 1$). The above is possibly relevant to the nonstationary regime characteristic of pulsations, observed in the backscattering process [29]. The (non)stationarity is attached to the character of the steady–state envelope: spatially periodic solutions are unstable and vice versa. The same conclusions are clearly applicable to stimulated Brillouin backscattering and, potentially, other parametric decay processes in a bounded homogeneous plasma.

4. EFFECT OF THE ELECTRON PLASMA WAVE NONLINEARITY ON THE STIMULATED RAMAN BACKSCATTERING

In contemporary laser fusion experiments [5,11], SRS has displayed rich and exciting physical properties, not predicted by the nonlinear theory based merely on the pump depletion effect, such as: anomalously low reflectivity, spectral broadening, gaps in spectra, spiky burst-like structures and transition to incoherent (irreproducible) dynamics. All these point to a rich and complex physical nature of nonlinear saturated SRS regimes in plasmas, which is due to a strong nonlinearity of EPW playing a critical role in a nonlinear saturation of SRS. Recent workers presumably accounted for EPW and ion–acoustic wave coupling [7,8,30] through the quenching of SRS by SBS [6], or by applying rather complex models incorporating a parametric decay of SRS–driven EPW and based on Zakharov–Maxwell equations [9,10]. The results presented in publications referred to above, support particular features noted by experimental workers. Still, the results of similar importance can be derived from a much simpler generic model, such as the one we offer here.

We concentrate in this section on the nonlinear saturation of SRS as a paradigm of nonlinear three–wave interaction (3WI) in a spatially extended system [14,15]. The nonlinear 3WI model of saturated SRS we exploit here includes the generic nonlinear frequency shift (NLFS) [16,17], induced by a large amplitude EPW. Namely, high intensity laser beams induce large amplitude plasma waves. Thus, a relativistic correction to the electron mass makes it necessary to take into account the nonlinear term in the equation (4c) [18]. On the other hand, the corresponding nonlinearity could be also of a ponderomotive origin, because large amplitude plasma waves experience coupling to ion modes through the ponderomotive force effect [19,20]. The both effects produce the nonlinear relativistic detuning term in the EPW equation (4c), yielding a generic cubic nonlinearity, which affects considerably the SRS evolution in the large amplitude EPW regime.

Thus, we add the NLFS term in the equation describing the spatio–temporal evolution of the EPW in the Raman backscattering process, so that the equation (4c) takes the form:

$$\frac{\partial b}{\partial \tau} + V_\ell \frac{\partial b}{\partial \xi} + \Gamma b - i\alpha |b|^2 b = a_0 a_1^*. \tag{10}$$

This ubiquitous type of partial differential equation arises in many different contexts, because this model accounts for the lowest order nonlinear effects of some generic interaction. For the spatially uniform case ($\frac{\partial}{\partial \xi} \to 0$), it was shown elsewhere to exhibit a low–dimensional chaos [31] under restricted conditions. A spatially extended model of 3WI, we consider here, is more difficult to investigate. The recent results [22] have revealed a rich physical behavior of saturated regimes corresponding to low–dimensional as well as spatio–temporal chaos (STC). The latter term refers to a chaotic behavior of coherent structures, or spatial patterns, and, therefore, featuring the coexistence of coherence and chaos [21,22], unlike the more familiar low–dimensional chaos, where spatial degrees of freedom are not involved [32].

A series of numerical simulations of our one–dimensional model given by the system of partial differential equations (4a,4b,10) for different laser and plasma parameters has been performed [33], by means of a central difference method. All parameters were taken within their physically realistic limits. The following parameters were fixed in all runs performed: $n_0 = 0.1n_{cr}$, $L = 100c/\omega_0$, $\kappa T_e = 1\text{keV}$, $\nu_e = 0.001\omega_{pe}$, $\epsilon_1 = 0.01$, $\epsilon_\ell = 0$. The NLFS factor α is estimated to lie, respecting its generics, between $10^{-3}\omega_{pe}$ and $10^{-2}\omega_{pe}$ in physical units, and this value was used in all simulations.

The SRS evolution has expressed most interesting features with variations in the electron quiver velocity v_{osc}, which is an indicator of the pump strength and is proportional to $\sqrt{I\lambda_0^2}$, I being laser intensity value. For a relatively low values of the pump strength, the system attains the steady state, which is demonstrated in Fig. 6 for $v_{osc} = 0.01c$. The SRS reflectivity R is saturated at a rather low level, which is a consequence of the small pump strength used in the simulation. This case corresponds to a fixed point of the system (4a,b,10).

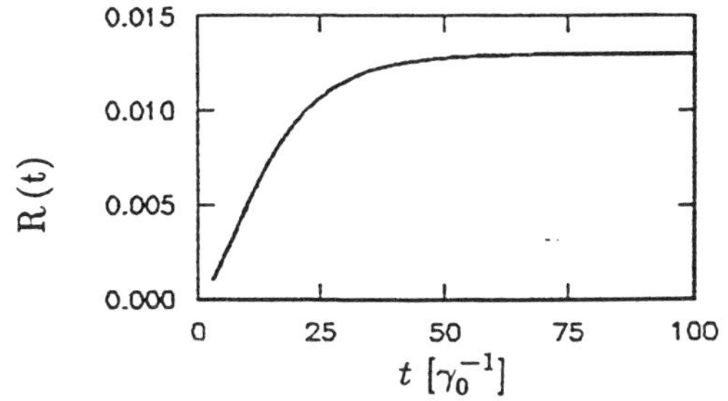

FIGURE 6: Time evolution of the Raman backscattering reflectivity R for the parameters: $n_0 = 0.1n_{cr}$, $L = 100c/\omega_0$, $\kappa T_e = 1keV$, $\nu_e = 0.001\omega_{pe}$, $\epsilon_1 = 0.01$, $\epsilon_\ell = 0$ and $\mathbf{v}_{osc} = \mathbf{0.01c}$.

For larger pump strength values, beyond a certain threshold, we find that the fixed state bifurcates to a periodic one, given in Fig. 7a. A local power spectrum of the reflected energy $|a_1(x = 0, \omega)|$ and a phase portrait (the projection of the phase trajectory onto a complex a_1–plane) (Figs. 7b and 7c, respectively) clearly exhibit a single frequency (limit cycle) coherent dynamics.

Further, we note that a periodic state can make a transition to a more complex regime (Figs. 8 and 9), once the bifurcation threshold for quasiperiodicity is

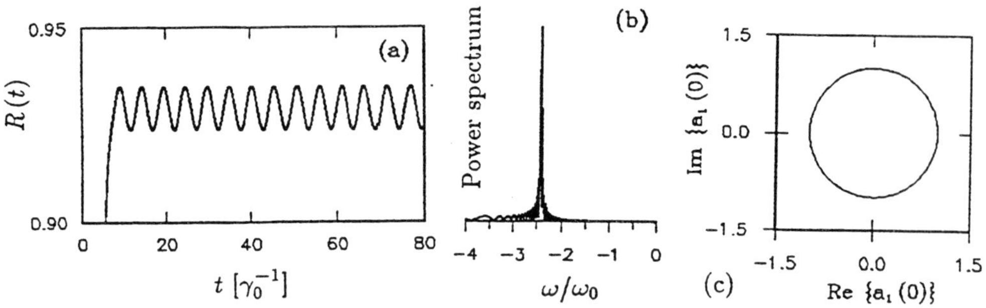

FIGURE 7: SRS reflectivity vs. time (a), power spectrum (b), and phase diagram (c), for the parameters of Fig. 6 and $\mathbf{v_{osc} = 0.02c}$.

approached. On the other hand, besides the fundamental frequency of limit cycle, a second frequency of quasiperiodicity, considerably lower than the original one, begins to make appearance in the spectrum (Figs. 8c and 9c), along with a series of descending higher harmonics of the main frequency modulated by the lower characteristic frequency. Namely, by increasing the pump strength, the reflectivity (Fig. 9a) closely resembles intermittent time series. At that time the phase plot (Fig. 9b) reveals attractors with orbits that do not close, developing features of "stretching and folding" type, characteristic of a low-dimensional chaos [34]. The power spectrum (see Fig. 9b) begins to show the appearance of broad band noise indicative of chaos.

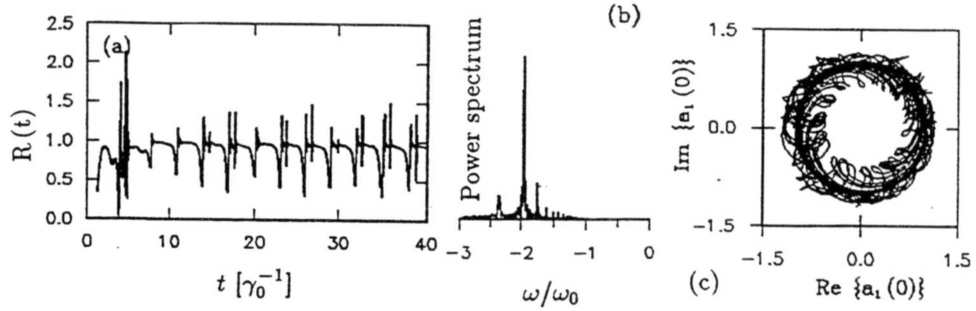

FIGURE 8: SRS reflectivity vs. time (a), power spectrum (b), and phase diagram (c), for the parameters of Fig. 6 and $\mathbf{v_{osc} = 0.0341c}$.

It is noticed that the main frequencies in the power spectrum still remain in the range where $\omega \ll \omega_0$, which justifies the temporal WKB approximation.

Moving deeper into the chaotic regime, distinct intermittent reflectivity patterns are observed in Fig. 10a, with more pronounced spectral broadening and with a chaotic attractor which loses any real structure. Similar behavior of an extended system for the case of the conservative nonlinear Schrödinger equation has been reported recently [35].

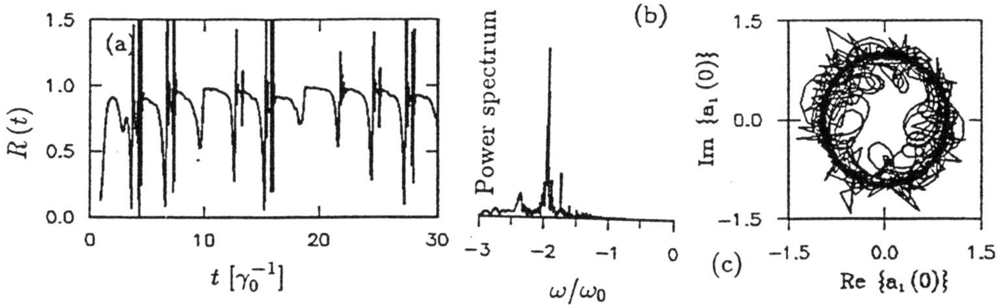

FIGURE 9: SRS reflectivity vs. time (a), power spectrum (b), and phase diagram (c), for the parameters of Fig. 6 and $v_{osc} = 0.035c$.

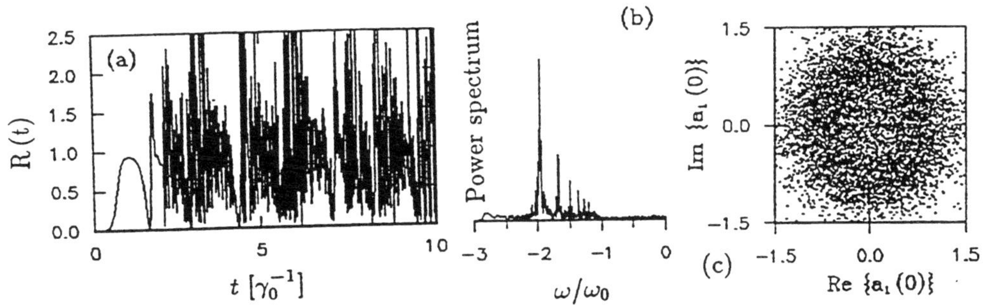

FIGURE 10: SRS reflectivity vs. time (a), power spectrum (b), and phase diagram (c), for the parameters of Fig. 6 and $v_{osc} = 0.05c$.

Finally, for a large pump strength of the order of 10% in v_{osc}/c (Fig. 11), we observe a completely chaotic reflectivity, broad continuous spectra and a phase diagram without any developed structure of a low-dimensional chaos.

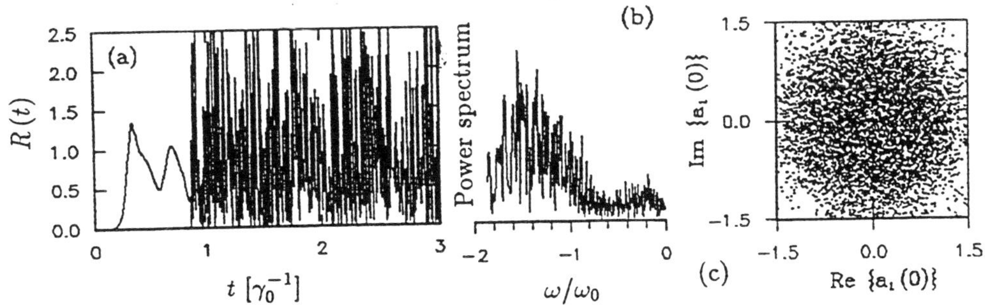

FIGURE 11: SRS reflectivity vs. time (a), power spectrum (b), and phase diagram (c), for the parameters of Fig. 6 and $v_{osc} = 0.1c$.

The simulation time was long enough to allow the transients die away and a nonlinear saturated regime to be realized. The saturation stage is assumed to be attained when the total energies of the three waves ($\int |a_0|^2 \, d\xi$, $\int |a_1|^2 \, d\xi$, $\int |b|^2 \, d\xi$) begin to fluctuate about some average values, and, what is important, all this takes part within a typical laser pulse duration.

To illustrate a transition from a regular state via low–dimensional to a developed spatio–temporal chaos (STC), a few space–time patterns, exhibiting the evolution of the pump and EPW, are given in Figs. 12 and 13. The spatio–temporal plot for an intermittent state (Fig. 12) shows the case in which regularity in space is almost retained in some extent, while, simultaneously, the temporal evolution, in particular that of EPW, can be classified as irreproducible, clearly leading to a low–dimensional chaos. The plots in Fig. 13 reveal the existence of chaos both in space and time, with some coexisting coherent patterns. Namely, the EPW sample shows a pronounced chaotic background; however, with some ridgelike laminar structures drifting to the right (forward direction).

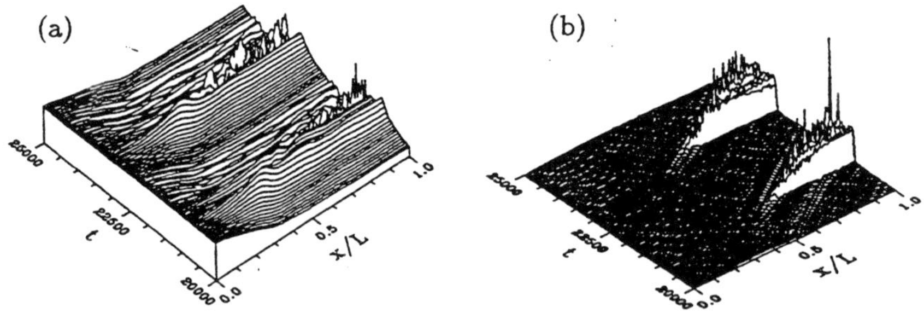

FIGURE 12: Spatio–temporal plots of the amplitudes of pump wave (a) and electron plasma wave (b) amplitudes in the intermittent regime ($v_{osc} = 0.04\,c$.

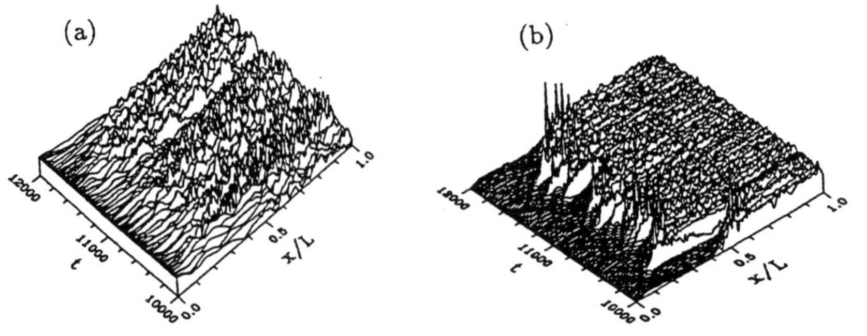

FIGURE 13: Spatio–temporal plots of the amplitudes of pump wave (a) and electron plasma wave (b) amplitudes in the intermittent regime ($v_{osc} = 0.1\,c$.

In conclusion, a general tendency of the described system to transit from regular to chaotic dynamics is revealed, as pointed out in some early works on the subject [30]. All the observed regimes were realized by varying a single control parameter: the pump wave intensity, given through the electron quiver velocity. We note a qualitative agreement of our simple model of SRS with more recent 1D simulations based on much more complex Zakharov–Maxwell models [7–10,35,37], which do not offer clear physical transparency and systematic pattern in behavior.

We are aware that our model suffers from a number of limitations (one–dimensional uniform plasma weakly coupled model, with higher nonlinearities and imperfect matching neglected). Still we believe that these facts do not diminish the general validity of the conclusions of our study. The study of the STC dynamics in more details is in progress, applying the spatio–temporal correlation functions and spectral analyses. Furthermore, the application of a novel technique of bi-orthogonal decomposition [38,39] is underway. Our results are of interest, besides for laser fusion, for particle acceleration and light generation applications as well.

ACKNOWLEDGEMENTS

This work is supported by the Ministry of Science and Technology of Republic of Serbia, under contract No. 0112.

REFERENCES

[1] W.L. Kruer, *Physics of Laser Plasma Interactions*, Addison–Wesley, New York (1988) and references therein

[2] Yu.M. Aliev, V.Yu. Bychenkov, M.S. Jovanović and A.A. Frolov, *J. Plasma Phys.* **48**, 167 (1992)

[3] H.A. Baldis, E.M. Campbell and W.L. Kruer, in *Handbook of Plasma Physics, vol. 3: Physics of Laser Plasma*, eds. A.M. Rubenchik and S. Witkowski, Elsevier Science Publishers B.V. (1991) p. 361

[4] D.W. Forslund, J.M. Kindel and E. Lindman, *Phys. Fluids* **18**, 1002 (1975)

[5] R.P. Drake, E.A. Williams, P.E. Young, K. Estabrook, W.L. Kruer, H. Baldis and T.W. Johnston, *Phys. Rev. Lett.* **60**, 1018 (1988)

[6] H.A. Rose, D.F. DuBois and B. Bezzerides, *Phys. Rev. Lett.* **58**, 2547 (1987)

[7] G. Bonnaud, D. Pesme and R. Pellat, *Phys. Fluids* **B2**, 1618 (1990)

[8] G. Bonnaud, D. Pesme and R. Pellat, *Phys. Fluids* **B4**, 423 (1992)

[9] W. Rozmus, R.P. Sharma, J.C. Samson and W. Tighe, *Phys. Fluids* **30**, 2181 (1987)

[10] T. Kolber, W. Rozmus and V.T. Tikhonchuk, *Phys. Fluids B* **5**, 138 (1993)

[11] R.P. Drake and S.H. Batha, *Phys. Fluids* **B3**, 2936 (1991)

[12] M.S. Jovanović and M.M. Škorić, *Proceedings of the XXI European Conference on Laser Interaction with Matter*, Warsaw (1991)

[13] M.S. Jovanović and M.M. Škorić, *Laser & Part. Beams*, **11**, 227 (1993)

[14] D.J. Kaup, A. Reiman and A. Bers, *Rev. Mod. Phys.* **51**, 915 (1979) and references therein.

[15] R.P. Drake and S.H. Batha, in *Nonlinear and Chaotic Phenomena in Plasmas, Solids and Fluids*, Eds. W. Rozmus and J.A. Tuszynski, World Scientific (1991) p. 345

[16] R.R.E. Salomaa and S.J. Karttunen *Phys. Scripta* **33**, 370 (1986)

[17] F.F. Chen, *Phys. Scripta* **T30**, 14 (1990)

[18] R. Bingham, *Phys. Scripta* **T30**, 24 (1990)

[19] W.L. Kruer, *Phys. Scripta* **T30**, 5 (1990)

[20] T.W. Johnston, P. Bertrand, A. Ghizzo, M. Shoucri, E. Fijalkow and M.R. Feix, *Phys. Fluids* **B4**, 2523 (1992)

[21] P.C. Hohenberg and B.I. Shraiman, *Physica* **D37**, 109 (1993)

[22] M.C. Cross and P.C. Hohenberg, *Rev. Mod. Phys.* **65** (1993) 851.

[23] R.W. Harvey and G. Schmidt, *Phys. Fluids* **18**, 1395 (1975)

[24] S. Hüller, *Phys. Fluids* **B3** 3317 (1991)

[25] R. Blaha, E.W. Laedke, A.M. Rubenchik and K.H. Spatchek, *Europhys. Lett.* **7** 237 (1988)

[26] F.S. Acton, *Numerical Methods That Work*, Harper & Row, New York (1970)

[27] V. Fuchs, *Phys. Fluids* **19**, 1554 (1976)

[28] S. Hüller, P. Mulser, H. Schnabl and A.M. Rubenchik, *Phys. Fluids* **B3**, 3339 (1991)

[29] M. Casanova, G. Laval, R. Pellat and D. Pesme, *Phys. Rev. Lett.* **54**, 2230 (1985)

[30] J.A. Heikkinen and S.J. Karttunen, *Phys. Fluids* **29**, 1291 (1986)

[31] A.J. Lichtenberg and M.A. Lieberman, *"Regular and Stochastic Motion"*, Springer–Verlag, Berlin (1982)

[32] C.C. Chow, Ph.D. Thesis (unpublished), Department of Physics, Massachusets Institute of Technology (1991)

[33] M.M. Škorić and M.S. Jovanović, in *"Laser Interaction and Related Plasma Phenomena"*, Eds. G.H. Miley and H. Hora, Plenum Press, New York, 1994.

[34] P. Bergé, Y. Pomeau and C. Vidal, *"Order within Chaos"*, J. Wiley and Sons, New York (1984)

[35] O. Larroche and D. Pesme, in *Nonlinear and Chaotic Phenomena in Plasmas, Solids and Fluids*, Eds. W. Rozmus and J.A. Tuszynski, World Scientific, (1991) p. 308.

[36] G. Bonnaud, D. Pesme and R. Pellat, *Laser & Part. Beams* **9**, 477 (1991)

[37] B. Bezzerides, D.F. DuBois and H.A. Rose, *Phys. Rev. Lett.* **70**, 2569 (1993)

[38] N. Aubry, R. Guyonnet and R. Lima, *J. Stat. Phys.* **64**, 3 (1991)

[39] M. Rajković, *private communication*

Low Temperature and
General Plasmas
M. Milosavljevic and Z. Petrovic

ISBN 1-56072-399-8

THE IDENTIFICATION OF RESONANTLY ABBSORBED MHD MODES IN STRUCTURED MEDIA

V.K. Brujic-Okretic, University of Surrey, Guildford, Surrey GU2 5SH, UK*

1. INTRODUCTION

In nature, in the Space environment in particular, the atmosphere of stars is often found to be strongly structured. The solar atmosphere which has been thoroughly investigated due to its complexity and major significance, is one example. Structuring occurs because the medium is permeated by a non-uniform magnetic field and is particularly identifiable in the photosphere where magnetic field-lines concentrate into flux tubes. As the atmospheric particles move around with the field-lines ("frozen field" approximation is applicable because of a very high conductivity) the medium itself is structured in the same way as a consequence of strong convective processes. Alongside with structuring a stratification arises due to a very rapid variation in pressure and density. However, the effects of stratification are not so remarkable in either the interior or the higher regions of the Sun's atmosphere - the corona.

An interesting natural phenomenon - the relatively high coronal temperature - has long been investigated by many scientists but no comprehensive explanation seems to have been found so far. Nevertheless, one of the processes that appears to contribute to the coronal heating significantly is the resonant absorption of waves propagating in the region.

We shall concentrate on the investigation of various MHD waves, both bulk and surface, that occur in a theoretical model of such a structured medium and define the criteria for resonant processes to take place. Eventually, we shall identify modes that are most likely to be resonantly absorbed in these circumstances.

Although the medium under consideration is merely a model, the study provides an insight into the physical processes taking part in any configuration of such kind and is, therefore, interesting in its own right.

2. BASIC CONCEPT

The medium under consideration is an inhomogeneous, compressible, perfectly conductive, dissipationless one-component (fully ionized) plasma permeated by a

* The author is on leave of absence from the Faculty of Mechanical Engineering, Unversity of Belgrade, Yugoslavia

unidirectional magnetic field of a variable strength. For sufficiently slow processes plasma can be modelled as a continuous medium (fluid) provided the particle relaxation time (τ) is significantly shorter than the large-scale events variation period (T). Consequently, the mean free path of particles (l) is much smaller than the characteristic dimension of the system (L):

$$\tau < T, \qquad l < L$$

The motion of such a medium is described by the system of ideal MHD equations:

$$\frac{\partial \rho}{\partial t} + \nabla(\rho \vec{v}) = 0$$

$$\rho \frac{\partial}{\partial t}\vec{v} = -\rho(\vec{v}\cdot\nabla)\vec{v} - \nabla\left(p + \frac{B^2}{2\mu}\right) + \frac{1}{\mu}\left(\vec{B}\cdot\nabla\right)\vec{B}$$

$$\frac{\partial \vec{B}}{\partial t} = \nabla \times \left(\vec{v} \times \vec{B}\right) \qquad (2.1)$$

$$\frac{dp}{dt} = c^2 \frac{d\rho}{dt}$$

where $c=(\gamma p/\rho)^{1/2}$ is the sound speed and all other variables are functions of space co-ordinates and time: $\rho(\vec{r},t), p(\vec{r},t), \vec{v}(\vec{r},t)$ and $\vec{B}(\vec{r},t)$.

Let all the equilibrium quantities, i.e. the density ρ_0, pressure p_0, temperature T_0 and the magnetic field strength B_0 vary in only one direction, say x.

Assume the plasma is slightly disturbed externally so that perturbations of the equilibrium state are of the first order. In this case, a linearized system of MHD equations will be sufficient to analyse the plasma behaviour, and the equations then read:

$$\frac{\partial \rho}{\partial t} + \nabla(\rho_0 \vec{v}) = 0$$

$$\rho_0 \frac{\partial \vec{v}}{\partial t} = -\nabla\left(p + \frac{1}{\mu}\vec{B}_0\cdot\vec{b}\right) + \frac{1}{\mu}\left[\left(\vec{B}_0\cdot\nabla\right)\vec{b} + \left(\vec{b}\cdot\vec{B}_0\right)\right]$$

$$\frac{\partial \vec{b}}{\partial t} = \nabla \times \left(\vec{v} \times \vec{B}_0\right) \qquad (2.2)$$

$$p = \frac{c^2}{\gamma}\rho$$

As the medium is homogeneous in *y* and *z* directions the motion ensued in these directions will be oscillatory. Applying a Fourier-Laplace transform to the system of equations (2.2) a standard calculus will lead to reducing the system of partial differential equations to a single ordinary differential equation of the second order:

$$\frac{d}{dx}[\varepsilon(x)\frac{du}{dx}] - \beta(x)u = 0 \tag{2.3}$$

where:

$$\varepsilon(x) = \frac{\rho_0(x)\left[k_z^2(x)v_A^2(x) - \omega^2\right]}{m_0^2(x) + k_y^2} \tag{2.3a}$$

$$\beta(x) = \rho_0(x)\left[k_z^2 v_A^2(x) - \omega^2\right] \tag{2.3b}$$

$$m_0^2(x) = \frac{\left[k_z^2 c^2(x) - \omega^2\right]\left[k_z^2 v_A^2(x) - \omega^2\right]}{\left[c^2(x) + v_A^2(x)\right]\left[k_z^2 c_T^2(x) - \omega^2\right]} \tag{2.3c}$$

$$c_T^2 = \frac{c^2(x)v_A^2(x)}{c^2(x) + v_A^2(x)}\ ; \qquad v_A(x) = \left(\frac{B_0}{\mu\rho_0}\right)^{1/2} \tag{2.3d}$$

$$k^2 = k_y^2 + k_z^2 \tag{2.3e}$$

k_z, k_y are the wave-numbers in z,y directions, respectively; c(x) is the sound speed and v_A is the Alfven speed.

The equation (2.3) is not known to be solvable in a closed form but we shall attempt to solve it in the special case of our interest.

3. SURFACE WAVES

Taking into consideration a typical solar-atmospheric structure presented in the Introduction we can model it in the way illustrated in Figure 1.

The structure consists of three different homogeneous sections: $x \leq 0$ (region I), $0 \leq x \leq D$ (region II), $x \geq D+a$ (region IV) mutually separated by a sharp discontinuity interface at x=0 and by a transitional layer $D \leq x \leq D+a$ (region III) where the continuous parameter-profile is given. The medium is otherwise static, stationary and uniform in *y* and *z* directions.

As the medium is assumed in static equilibrium the total pressure:

$$p_{t0} = p_0(x) + \frac{B_0^2(x)}{2\mu} \tag{3.1}$$

is required to be constant throughout.

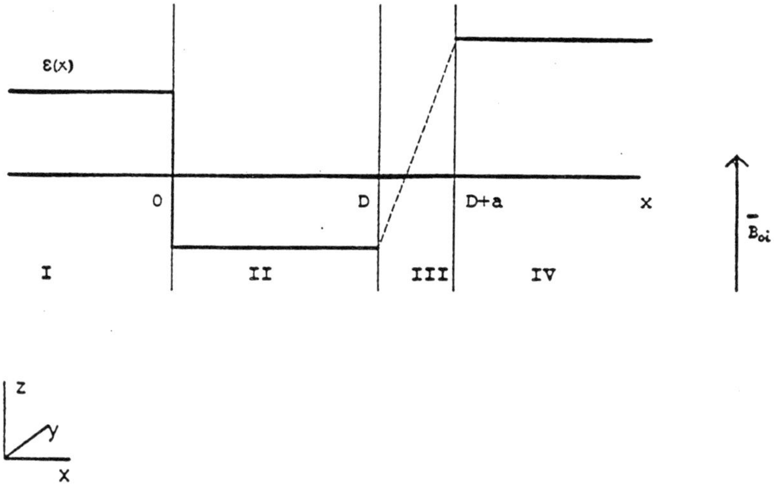

Fig. 1: *Profile of the medium; ε(x) is defined by the relation (2.3a)*

Applied to each of the boundaries this condition imposes constraints on the equilibrium state quantities and can be expressed in terms of the sound (c) and Alfven (v_A) speeds as follows:

$$\rho_{0j}\left(2c_j^2+\gamma v_{Aj}^2\right)=\rho_{0j+1}\left(2c_{j+1}^2+\gamma v_{Aj+1}^2\right) \qquad (3.2)$$

Small amplitude, isentropic perturbations: p, $\rho, \vec{v}(u,v,w)$ and $\vec{b}$ of the equilibrium state obey the differential equation (2.3). In the circumstances described above this equation can be solved in a straightforward manner in each of the homogeneous regions. Its solution in the transitional layer, however, is not so obvious and requires additional approximations. We shall restrict ourselves to the modes with sufficiently long wavelengths compared to the layer thickness: $ka<1$. Then, neglecting the inertia of the region III the total pressure perturbation can be assumed nearly constant across the layer (Hollweg 1987) and the solution to (2.3) reads:

$$u_3(x)=C_3'+C_3^{\bullet}\int_D^x\frac{d\xi}{\varepsilon(\xi,\omega)} \qquad C_3',C_3^{\bullet}=const \qquad (3.3)$$

The complete solution to (2.3) is thus obtained to be:

$$u=\begin{cases}e^{i\kappa_1 x}+R\cdot e^{-i\kappa_1 x}\\ C_2'e^{\kappa_2 x}+C_2^{\bullet}e^{-\kappa_2 x}\\ u_3(x)\\ C_4e^{-\kappa_4(x-D-a)}\end{cases} \qquad (3.4)$$

where $\kappa_i=(\beta_i/\varepsilon_i)^{1/2},\ i=1,2,4$ (3.5)

$R, C_2^{'}, C_2^{''}$ and C_4 are the integration constants determined by the boundary conditions requiring the continuity of both the velocity perturbation x-component (u) and the total pressure perturbation (p_t) across all boundaries. The incident wave amplitude is normalized to unity (Okretic and Cadez 1991).

Let us pause at this point and analyse solutions (3.4). Note that quantities κ_i are all real according to the assumed initial conditions. It is then apparent from the form of the velocity perturbation in the region II that waves decay away from both sides of the boundary x=0 and are, therefore, essentially *surface waves*. In other words, they do propagate along the surface (in *y*, *z* plane) but are evanescent in the direction perpendicular to it (*x*-axes). Their nature is *magnetoacoustic* and, analysing the dispersion relation, two such modes are found to exist: the *slow* mode and the *fast* mode. As the surface waves have been thoroughly investigated in the past (Roberts 1981; Cadez and Okretic1989) we shall leave the discussion on their behaviour here.

4. RESONANT ABSORPTION

Suppose the low frequency MHD bulk wave is incident from the region I of the structure illustrated in Fig. 1. If there was no transitional layer it would have been totally reflected from the boundary at x=0 (along x-axes) and only surface modes would have been generated.

However, in the presence of the layer in which parameters vary smoothly, an anomalous absorption is found to exist due to the continuous spectrum of modes that may be triggered in its interior. In this section, the mechanism of resonant absorption will be explained in more detail.

In fact, what happens is that whenever the frequency of the surface mode (ω_s) generated at the layer's adjacent interface matches the one inside the layer ($_s = \omega(x_r)$; $D \le x_r \le D + a$) an oscillation will be resonantly excited at the location $x = x_r$. As the region III is continuously inhomogeneous there will be no normal modes; instead, some kind of non-linear coupling of such oscillations will take place. The energy of the incident wave will thus be carried away in the resonant process and eventually transported to the plasma occupying this part of the medium.

A measure of this energy exchange is an absorption coefficient, C_a, that can be worked out from the solutions (3.4) and the appropriate boundary conditions.

Proceeding in a standard way we have derived a reflection coefficient, $C_R \equiv |R|^2 = 1 - C_a$, and obtained the following analytical expression for it (Okretic and Cadez 1991):

$$C_R = \frac{m^2 + n^2}{r^2 + s^2} \tag{4.1}$$

where:

$$m = \Delta + (\Delta - 2)e^{-2\kappa_2 D} - D_{12}\delta\left(1 + e^{-2\kappa_2 D}\right)$$
$$n = D_{12}\left[\Delta - (\Delta - 2)e^{-2\kappa_2 D}\right] + \delta\left(1 - e^{-2\kappa_2 D}\right)$$
$$r = \Delta + (\Delta - 2)e^{-2\kappa_2 D} + D_{12}\delta\left(1 + e^{-2\kappa_2 D}\right)$$
$$s = D_{12}\left[\Delta - (\Delta - 2)e^{-2\kappa_2 D}\right] - \delta\left(1 - e^{-2\kappa_2 D}\right)$$

and:

$$= D_{42}\varepsilon_2\kappa_2 \operatorname{Im} J, \quad D_{ij} = \frac{\varepsilon_i \kappa_i}{\varepsilon_j \kappa_j}, \quad i, j = 1,2,4$$

$$J = \int_D^{D+a} \frac{d\xi}{\varepsilon(\xi,\omega)}, \quad \Delta = 1 + D_{42} \tag{4.2}$$

To calculate C_R it is necessary to determine the imaginary part of the integral J. Within the limits of the linear theory it was carried out similarly to what is done in Landau damping derivation (Okretic and Cadez 1991).

The profile of the absorption coefficient contains the information on the fraction of the incident wave energy that is being absorbed in the process.

5. NUMERICAL RESULTS

The relation (4.1) immediately gives that $C_R = 1$ in the limiting case when a=0 (two sharp boundaries), since J=0 and, consequently, δ=0. The incident MHD wave then tunnels through the region II and excites surface modes along the interface x=D. These modes leak back to the region I causing the wave energy incident on the interface x=0 to be totally reflected (Cadez and Okretic1989).

The unperturbed equilibrium state quantities were chosen to allow for the bulk incident MHD wave to propagate through the region I and a surface mode along the layer boundaries to occur. The smooth parameter distribution within the layer is taken to be linear.

In Fig. 2 the absorption coefficient (C_a) variation with the surface wave propagation angle (θ) is shown for various (normalized) frequency values ($\Omega = \left(\omega D / v_{A4}\right)$ keeping the incident angle (α) fixed. (θ is the angle between $\vec{k}\left(k_y, k_z\right)$ and $\vec{B}\left(0,0,B_z\right)$). The distinctive maxima prove the existence of the resonant process, in the first place, and correspond to optimal conditions for the energy transfer. The Fig. 2 also shows that a *total absorption* is reached when the frequency and angles θ and α fall into narrow intervals that correspond to the *fast MHD wave* characteristics for the given configuration.

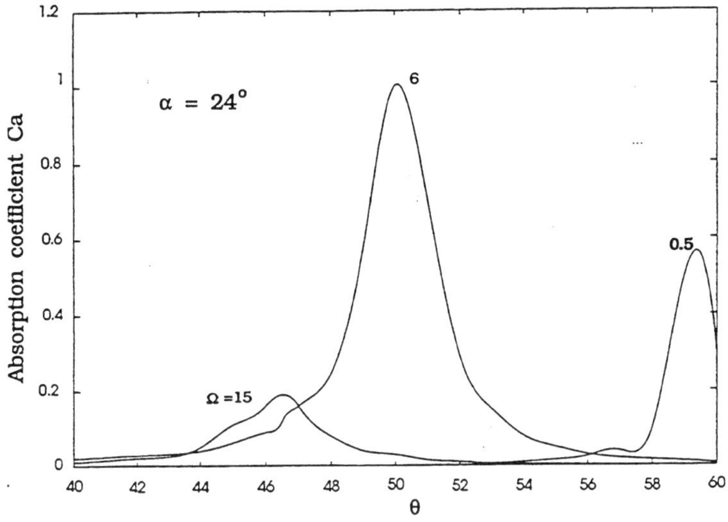

Fig. 2: *The Absorption Coefficient profile* (V.K.Okretic et al. 1993)

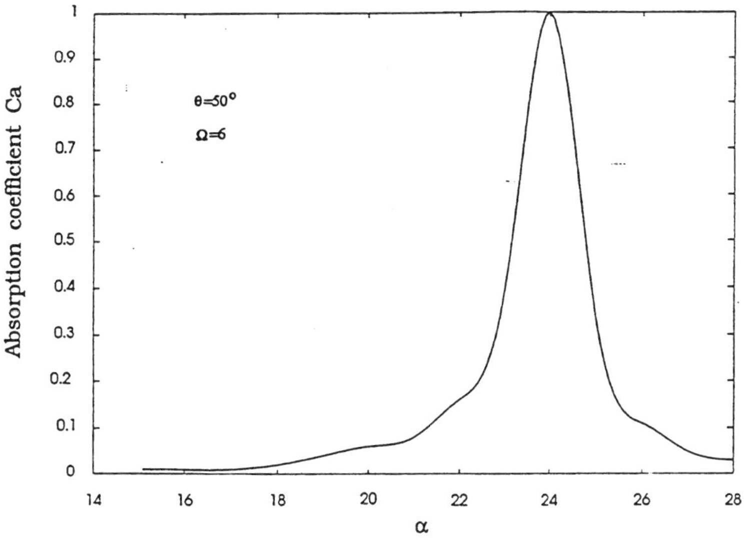

Fig. 3: *The Absorption Coefficient exhibiting the conditions for total absorption* (V.K.Okretic et al. 1993)

This is verified by examining dispersion relations of various MHD modes that can be generated in the structure. The analysis was used to identify modes "allowed" to propagate in the structure. Simultaneously, the criteria for the resonant mode conversion to take place, as well as the mechanism of the wave energy exchange, have been thoroughly investigated.

In Fig. 3 the incident angle α was varied in order to investigate the sensitivity of the absorption process to the direction of the incident wave propagation, when θ and Ω are kept fixed at values corresponding to the total absorption (in the previous case). It is shown that the process is very sensitive to the incident angle value which must fall into a well defined cone so that the resonant absorption may take place.

6. CONCLUSION

A phenomenon of *resonant absorption* of an MHD wave incident on a structured medium that contains both sharp and "smooth" interfaces was investigated. In particular, the emphasis was put on the identification of MHD modes subject to the resonant conversion. Both the absorption coefficient and the dispersion equation have been derived and analysed for that purpose.

It is found that it is the *fast* magnetoacoustic incident wave that is susceptible to the resonant conversion. The process is performed via *surface* modes generated along the interfaces. As the parameters of the medium and the propagation angles have an impact on the modes' phase speeds, the resonant process is proved to be very sensitive to those quantities.

An illustrative numerical experiment has been carried out and a verification of these conclusions are shown on diagrams.

The mechanism of the energy dissipation to the inhomogeneous plasma contained in the layer is associated with non-linear mode-coupling and is being studied at present.

Acknowledgement:

This work forms a part of a PhD thesis that was successfully completed in collaboration with Dr Vladimir Cadez, Institute of Physics, Belgrade, to whom I express my sincere gratitude. I am also grateful to the Faculty of Mechanical Engineering, University of Belgrade, for enabling a study leave during which these investigations have continued. Finally, I wish to thank the University of Surrey, UK, for providing an encouraging and supportive environment and for making the resources available for this project.

References:

[1] **Okretic, V.K. and Cadez, V.M.,** *Solar Magnetic Fields,* Proceedings of the International Conference, Freiburg, Germany, 1993 (Cambridge University Press)

[2] **Okretic, V.K. and Cadez, V.M.,** *Physica Scripta* **43**, 306 (1991)

[3] **Cadez, V.M. and Okretic, V.K.,** *J. Plasma Phys.* **41**, 23 (1989)

[4] **Hollweg, J.V.,** *Ap.J.* **320**, 875 (1987)

[5] **Roberts, B.,** *Solar Phys.* **69**, 27 (1981)

SUBJECT INDEX

I

L

M

N

O

P